Coenraad Jacob Temminck and the Emergence of Systematics (1800–1850)

Emergence of Natural History

VOLUME 4

The titles published in this series are listed at *brill.com/enh*

Coenraad Jacob Temminck and the Emergence of Systematics (1800–1850)

By

Maria Eulàlia Gassó Miracle

BRILL

LEIDEN | BOSTON

Naturalis Biodiversity Center

This publication has been funded in part by Naturalis Biodiversity Center in Leiden, The Netherlands.

Chapters 5 and 6 have been adapted by permission from Springer Nature:
Gassó Miracle, "Significance of Temminck's Work on Biogeography: Early Nineteenth Century Natural History in Leiden, the Netherlands." *Journal of the History of Biology* 41, no. 4 (2008): 677–716. doi:10.1007/s10739-008-9158-8.
Gassó Miracle, "On Whose Authority? Temminck's Debates on Zoological Classification and Nomenclature: 1820–1850." *Journal of the History of Biology* 44, no. 3 (2011): 445–481. doi:10.1007/s10739-010-9265-1.

Cover illustration: *Bec de Calao*. Hand-colored engraving after a drawing by Nicholas Huet. In Coenraad Jacob Temminck and Guillaume Michel Jerôme Meiffren Laugier de Chartrouse, *Nouveau recueil de planches coloriées d'oiseaux*, vol. 2, livraison 64 (1831): plate 521. Image from the Biodiversity Heritage Library. Digitized by the Smithsonian Libraries.

The Library of Congress Cataloging-in-Publication Data is available online at https://catalog.loc.gov
LC record available at https://lccn.loc.gov/2021046572

Typeface for the Latin, Greek, and Cyrillic scripts: "Brill". See and download: brill.com/brill-typeface.

ISSN 2452-3283
ISBN 978-90-04-41917-9 (hardback)
ISBN 978-90-04-44149-1 (e-book)

This book is printed on acid-free paper and produced in a sustainable manner.

Contents

Acknowledgments

The idea for an in-depth study of Temminck and the Rijksmuseum arose, like many other great ideas, from good conversation over a cup of coffee. For this I am forever indebted to Professor Rob P. W. Visser at Utrecht University. The conversation took place many years ago. Subsequently Professor Visser became my PhD supervisor. After his retirement his successors, Frans van Lunteren at the University of Leiden and Bert Theunissen at Utrecht University, guided and supported me with patience, enthusiasm, and expertise as I completed the dissertation that preceded this book. Their help has been invaluable in every way.

My debt is also to my colleagues at the Naturalis Biodiversity Center. Specifically, I would like to thank René Dekker, Jeroen Snijders, Maaike van de Kamp-Romijn, Luc Willemse, Rob de Vos, and Tiny Monquil-Broersen, who made it possible for me to spend numerous hours on researching and writing. I am also indebted to the Naturalis Library and Archives team for granting me swift access to books and manuscripts, and for digitizing part of Temminck's correspondence, especially Caroline Pepermans, Karien Lahaise, Menno Hooft, Sylvia Schwencke, and Marianne van der Wal. Collection managers Esther Dondorp, Pepijn Kamminga, Lisa Lankhaar, and Ronald de Ruiter kindly provided information about the collections under their care, and photographed some of the specimens. I would also like to express my gratitude to the late Chris Smeenk, to John de Vos, Marinus Hoogmoed, Hein van Grouw, and Martien van Oijen; each of them proved to be an inexhaustible source of knowledge.

Many kind and knowledgeable people helped me with my questions, broadened my views, commented on the manuscript, and inspired me, each in his or her own stimulating way. In particular, I would like to thank Justin Jansen, Ad Maas, Robbert Striekwold, Robert-Jan Wille, Bernhard Schär, Esther van Gelder, Hieke Huistra, Martin Weiss, Eric Jorink, Paul Farber, Lissa Roberts, and Paul Smith. I met many of them at events organized by *Gewina*, the Belgian-Dutch Society for History of Science and Universities.

I would like to express my sincere gratitude to Baroness Alberta Maria van Lynden-de Bruïne (1922–2016), the great-great-grandchild of Coenraad J. Temminck, who was so kind as to invite me to her home in Middelburg, where we had the most delightful and interesting conversations about Temminck and his wife, Caatje. The baroness generously granted me access to the family archives.

I am indebted to the archivists and librarians of the Staatsbibliothek zu Berlin, the National Archives of the Netherlands, and the Leiden University

Libraries for providing (digital) access to their holdings. The Biodiversity Heritage Library has also been essential for this research: without this open-access digital library and archive specializing in biodiversity I would not have been able to consult many of the works needed for this book.

I am truly thankful to the editorial team of the Brill book series *Emergence of Natural History* for their support, especially to Andreas Weber, an assistant professor in the Department of Science, Technology and Policy Studies (STePS) at the University of Twente. I also would like to thank Rosanna Woensdregt, the associate editor Biology and History of Science at Brill, for her guidance through the editorial process.

This book could not have been published without the meticulous and insightful copyediting by Anke Timmerman of Type & Forme. She has provided invaluable advice on referencing, style, and content, polishing the manuscript and making it ready for publication. I am very thankful to Dan Connolly of Word for Word Book Services for writing the perfect index for this book.

Finally, I could never have completed this research without the support of my beloved family. My husband, Vincent van der Meij, and my son, Víctor, both looked after and encouraged me with patience and humor throughout the long and convoluted process leading to the finished book.

This publication has been funded in part by Naturalis Biodiversity Center in Leiden, the Netherlands.

Illustrations

Abbreviations

AC	Archieven Curatoren [Trustees Archives], Leiden University
ARA	Algemeen Rijksarchief (ARA), National Archives of the Netherlands, The Hague
BAAS	British Association for the Advancement of Science (now the British Science Association, BSA)
BiZa	Ministerie van Binnenlandse Zaken [Ministry of Internal Affairs]
NHA	Nord-Hollands Archief, Haarlem
NL-HaNA	National Archives of the Netherlands, The Hague
RMNH	('s) Rijksmuseum van Natuurlijke Historie (now Naturalis Biodiversity Center)
SBB	Staatsbibliothek zu Berlin
UBL	Leiden University Libraries
VOC	Verenigde Oost-Indische Compagnie (1600–1799) [Dutch East India Company]
WIC	West-Indische Compagnie (1621–1792) [Dutch West India Company]

Note on Quotations and Translations

In all quoted texts, the orthography and typography are as in the original.

Unless stated otherwise, all translations are by the author. Transcriptions of the original texts (retaining the original spelling) can be found in the Appendix.

Introduction

> Never be surprised by the crumbling of an idol or the disclosure of a skeleton.
>
> LORD ACTON, *Inaugural Lecture on the Study of History*[1]

∴

Proud raptors, flamboyant parrots, small perching birds, and plump pigeons sit patiently on their white-lackered wooden pedestals, looking straight back at the visitor: the historical collection of birds of Naturalis Biodiversity Center is a feast for the eye.[2] Some specimens conserve the brilliant plumage of their youth, others show the passage of time, and all of them carry labels detailing their history. New tags are printed on white, acid-free paper, displaying information in text form and on a matrix code. Old labels are more succinct but infinitely more interesting. On some of them, a former curator has crossed out a species name and scribbled another one beside it – an appalling crime in the museum world, but indicative of the many disagreements on nomenclature that have occurred in the past two centuries. Often, a series of intriguing codes provide specialists with additional clues about the specimen's history: "Cat. Ost. c"; "Ancient Cab."; "Temm. pl. col. 114 f. 1." The first collectors' names have been written in careful, attentive hand: Forsten, Kuhl & Van Hasselt, Büttikofer, S. Müller, Temminck.

Admittedly, we know surprisingly little about these men. Even Coenraad Jacob Temminck, the museum's director for nearly forty years, is not well-known outside the ornithological community. He owned a magnificent natural history collection and was the author of a series of often-cited monographs on

1 Acton, *Lectures in Modern History*, 24.

2 This is the Dutch national research institute for biodiversity and museum of natural history, in Leiden. It was founded in 1820 as 's Rijks Museum van Natuurlijke Historie, and changed its name to 'Rijksmuseum van Natuurlijke Historie' in 1931. Since then, the name changed a further four times. In 1989, the institution was rechristened the 'Nationaal Natuurhistorisch Museum'; in 1998 it became the 'Nationaal Natuurhistorisch Museum Naturalis'; in 2010 the museum merged with the Zoological Museum Amsterdam of the University of Amsterdam and the Dutch National Herbaria of the universities of Leiden, Amsterdam, and Wageningen to form the 'Nederlands Centrum voor Biodiversiteit'; and in 2012, it became the 'Naturalis Biodiversity Center.'

 | DOI:10.1163/9789004441491_002

birds and mammals. At the beginning of the nineteenth century he promoted the foundation of 's Rijks Museum van Natuurlijke Historie and became its first director in 1820. One might think that his contemporaries and biographers knew Temminck well, both as a naturalist and a museum director, giving us an idea of the stature and character of the man. And yet, they have left few portrayals of his work and its significance which are not only cursory but actually contradict each other. Some found Temminck one of the most outstanding ornithologists of his day, while others depicted him as an educated aficionado. Some stated that he played a crucial role in standardizing the methods of zoological nomenclature and classification, but some accused him of plagiarizing the work of others.

It seems that all of them were right, at one point or another. Obviously, Temminck could not occupy all of the above mentioned roles at the same time. He started as an enthusiastic private collector and gradually specialized and progressed in zoological classification. During the first decades of the nineteenth century he secured a place for himself in the arena of natural history. However, by 1850, this field had evolved faster than he could – or was willing to – digest. Temminck's career spanned the fifty-year period during which classification transformed into a natural historical discipline – systematics. It is this link between his reputation and the developments in natural history that makes Temminck's life an interesting study for the historian of science. The integration of Temminck's scientific biography with the development of systematics not only provides a more accurate account of Temminck, but also fills a gap in the history of Dutch natural history which elucidates the history of zoological classification between 1800 and 1850.

1 Portraits of Coenraad Jacob Temminck

Temminck is mostly remembered by zoologists but otherwise nearly forgotten. His name appears often in zoological treatises, both those dating from Temminck's time and those from the following two centuries. Around fifty species and subspecies of animals are named after him, mainly birds and mammals. The list includes a cormorant, a shrew, a woodpecker, several bats, squirrels, mice, the African Ground Pangolin, and the Asian Golden Cat. Some other, very curious creatures also bear Temmincks' name, such as the terrifying Alligator Snapping Turtle, the unsuitably named Kissing Gourami ("kissing" fishes are, in fact, fighting), a longhorn beetle, and a small crustacean parasite of marine fish. Possibly the most famous of all is Temminck's Tragopan (*Tragopan temminckii*), a splendid pheasant from eastern India and central

FIGURE 1 *Ceriornis Temmincki* (Temminck's Tragopan, now *Tragopan temminckii*). Hand-colored lithograph by Joseph Smit. In Elliot, Daniel Giraud. *A Monograph of the "Phasianidae," or Family of the Pheasants*, vol. 1. (1872): plate 24
THE NEW YORK PUBLIC LIBRARY, DIGITAL COLLECTIONS, B14477911

Asia, named and described in 1831 by John Edward Gray, the zoologist and Keeper at the British Museum. Additionally, nearly one hundred of the scientific names coined by Temminck have survived across two hundred years of constant changes in nomenclature and classification.[3]

Historians may, if at all, associate Temminck's name with the Leiden museum and its historical collections. Historians of science have routinely overlooked Temminck, partly because he did not revolutionize natural history or zoological classification, and partly because the available biographies of Temminck, if compared with each other, are disappointing and confusing.

Just how disparate the judgments on Temminck are can be seen from the following examples.[4] In 1826 the ornithologist Charles Lucien Bonaparte listed

3 These concern only names accepted according to the code of zoological nomenclature. GBIF database, accessed 20 April 2020.

4 On Temminck's biographies and reviews see De Beaufort, "Coenraad Jacob Temminck"; Farber, *Emergence of Ornithology*, 104–07; Holthuis, "Three Dutch Authors of von Siebold's

Temminck among the "most interesting and accurate" European authors, along with Linnaeus, Gmelin, Latham, Buffon, and Vieillot.[5] Similarly, the British naturalist William John Swainson found Temminck's early monographs on pigeons and gallinaceous birds "clear and masterly" and thought that Temminck set a good example for other writers.[6] Temminck's avifauna of Europe, the *Manuel d'ornithologie*, cemented his reputation, and most naturalists quickly adopted his system of classification in the 1820s. Swainson even equaled Temminck to Linnaeus, "next to whom, as an ornithologist, he assuredly ranks."[7] This is, possibly, the best compliment a naturalist could ever hope for.

By contrast, thirty years after naming Temminck's Tragopan, John Edward Gray thought that Temminck, an autodidact without academic education, should be "regarded as a patron and amateur rather than a scientific zoologist."[8] Even if Temminck's classification system was outdated by then, what had tarnished Temminck's reputation in this way? Linnaeus' system had long been outdated too, but naturalists still revered his name. Clearly, the measure of eminence and success, and the importance of scientific training had changed since the beginning of the century.

In the Netherlands, too, Temminck was both praised and criticized. His peers apparently could not agree on Temminck's competence, nor could his later biographers, although here the balance is tipped in Temminck's favor. According to one biographer, the cartographer and architect Pieter Harmen Witkamp, Temminck's peers appreciated his works because of the "very accurate descriptions, detailed synonymy and the many descriptions of the way of life and habits of birds, which surpassed everything that had been written previously."[9] Another, Willem Vrolik – the celebrated Dutch anatomist and pathologist – affirmed that Temminck had become a "systematist zoologist of the highest level."[10] Vrolik did not specify why or how Temminck had achieved such scientific prestige. By contrast, Temminck's successor Hermann Schlegel portrayed him as a selfish man, who locked away specimens to prevent his colleagues from working on them and did not hesitate to publish someone else's

Fauna japonica," 689–731; Holthuis, *Rijksmuseum*, 18–25; Van Lynden-de Bruïne, *In vogelvlucht*, 11–21; Raat, "Coenraad Jacob Temminck"; Snellen van Vollenhoven, "Coenraad Jacob Temminck"; Stresemann, *Ornithology*, 110–25; Susanna, "Levensschets van Temminck" and Vrolik, "Levensbericht van Temminck."

5 Bonaparte, *Nomenclature of Wilson's Ornithology*, 1.

6 Swainson, *Natural History and Classification of Birds*, vol. 1, 206.

7 Swainson, *Treatise on the Geography*, 185.

8 Gray, "Revision of the Genera and Species of Viverrine Animals," 505.

9 Witkamp, "Vier-en-twintig voorgangers," 199.

10 Vrolik, "Levensbericht van Coenraad Jacob Temminck," 66.

discoveries as his own.[11] This was his personal perception of Temminck after thirty-three years of collaboration.

About a century later Erwin Stresemann devoted an entire chapter to Temminck in his influential *Die Entwicklung der Ornithologie* and concluded that, in the end, Temminck "remained what he had been in 1820 [...] – a well-informed zoological amateur."[12] The chapter is mostly biographical in nature, full of details about Temminck's social network: his correspondents, his visits to fellow naturalists during his extensive travels through Europe, and his role as the director of the Leiden museum. The information is extracted from his correspondence and Hermann Schlegel's autobiography (as is almost the entire chapter on Schlegel). Stresemann accepted Schlegel's perception of – and aversion to – Temminck and incorporated it into his book. They thereby entered the annals of the history of ornithology. His book, however, offers no insights into Temminck's work on systematics, except for his severe judgment of Temminck's treatment of materials he received from the Dutch colonies: "Temminck did not have that sort of intellect."[13] Stresemann does not mention Temminck in other chapters that might have helped a deeper understanding of Temminck as a naturalist (for instance, chapter 10, *The Effect of Natural Philosophy*).

In addition to his engagement with Temminck in *Die Entwicklung der Ornithologie*, Stresemann transcribed and published Temminck's correspondence with some of his colleagues, in particular his associate, Baron Meiffrein Laugier de la Chartrouse as well as Karl Illiger and his successor Hinrich Lichtenstein, both directors of the Zoological Museum in Berlin.[14] While he does not reflect upon the historical significance of the letters, Stresemann's publications make these primary sources easily accessible and further define the social environment in which Temminck worked and communicated.

In the 1990s Lipke Bijdeley Holthuis, the curator and "the institutional memory" of the Leiden museum, sought to demonstrate that Schlegel's claims were biased.[15] He argued that antipathy and resentment towards Temminck had distorted Schlegel's picture of Temminck. Holthuis' views of Temminck are very different from Stresemann's: "Temminck's significance for science,

11 Schlegel, "Levensschets van Hermann Schlegel," 37–38.

12 Stresemann, *Ornithology*, 147.

13 Stresemann, 147.

14 Stresemann, "Aus C. J. Temmincks Briefen an H. Lichtenstein"; "Briefwechsel von Temminck mit Hoffmannsegg und Illiger"; "Histoire des origines des 'Planches coloriées'"; "Ornithologen-Briefe aus den Jahren 1816 bis 1820."

15 Fransen and Van Oijen, "L. B. Holthuis."

especially for ornithology and mammalogy, is considerable. He published many fundamental works, and he continued to do so until he reached an advanced age."[16] Holthuis repeatedly tried to dismantle Schlegel's (and with it, Stresemann's) portrayal of Temminck, which was "a myth," as he succinctly put it.[17] However, neither Stresemann nor Holthuis discussed the subjects treated in Temminck's publications, his ideas, methods of classification, or how his peers received his work.

It was not until 1982 that Temminck featured in a history of ornithology that also considered the social and intellectual aspects of the practice of classification: Paul Lawrence Farber's *The Emergence of Ornithology as a Scientific Discipline: 1760–1850*. Farber had already explored some factors that influenced the development of ornithology and zoological classification, including taxidermy, the development of collections, and the different 'type' concepts.[18] In *The Emergence of Ornithology* he integrated technical, material, and cultural aspects in the discussion of the conceptual issues that preoccupied naturalists between 1760 and 1850, revealing the complexity of the history of bird classification. On Temminck he wrote: "In his writings, we find evidence not only of one of a number of approaches to systematics, but another feature of this group of naturalists who dominated ornithology in the first decades of the nineteenth century: an increased specialization and rigor."[19]

Indeed, Temminck focused almost exclusively on bird classification, favoring monographs over encyclopedic work, and working either on particular groups of birds (such as pigeons and fowl) or on a specific geographic region (Europe). Farber concludes that Temminck is "one of the key figures in the transition of the study of birds from a sub-branch of natural history to a scientific discipline."[20] But the specifics of how and why Temminck played a vital role in the development of ornithology remain unexplored. Understanding how Temminck contributed to the field of ornithology, and in particular to zoological classification, requires two explorations, one of his life and career, and one of early nineteenth-century systematics.

I started this book hoping that a comprehensive scientific biography of Coenraad Jacob Temminck would provide a starting point for discussing the practices in Dutch zoological classification during his time. However, in the process of studying and analyzing Temminck's career, I have been forced to

16 Holthuis, *Rijksmuseum*, 24.
17 Holthuis, "Three Dutch Authors of von Siebold's *Fauna japonica*," 696.
18 Farber, *Emergence of Ornithology*, "Development of Ornithological Collections"; "Development of Taxidermy"; "Type-Concept in Zoology."
19 Farber, *Emergence of Ornithology*, 85.
20 Farber, 90.

reconsider much wider themes that were central to nineteenth-century natural history: the European colonial expansion, the development of systematics as a discipline, the rise of a meritocracy in natural history, and the status of systematics within natural history and natural philosophy. As a result, what started as an effort to understand the contradictions in the reviews of Temminck's work has led me to open a Pandora's box and examine a series of questions that, until now, have been explored only partially.

2 On Dutch Natural History

A quick survey of the literature on Dutch natural history reveals that the history of biological systematics in the Netherlands has not been examined in detail. Most works dealing with the practice of biological classification in the Netherlands focus on the early modern period.[21] The few existing accounts of the late eighteenth and early nineteenth centuries have focused either on the institutional history of museums and collections or on the biographies of key figures. Lipke Holthuis published the history of the Dutch museum of natural history in 1995, decades after the zoologist Agatha Gijzen had paved the way. In 1938 Gijzen's doctoral thesis presented a history of the museum from its foundation to the beginning of World War I, after Gijzen had spent years plowing through the disorganized archives of the museum and, in effect, finally putting them into order. Her book, entitled *'s Rijks Museum van Natuurlijke Historie 1820–1915*, provides a wealth of information on the collections, their origins, the museum's policies, its first three directors, and its staff.[22] *'s Rijks Museum* discusses the main developments in the museum between 1820 and 1915, focusing on the growth of the collections. As this was also one of Temminck's primary concerns, Gijzen's analysis of archival material has been extremely useful. Unfortunately, Gijzen does not incorporate the scientific and political context of the museum's history and the staff's work in her narrative.

Gijzen's book was followed by several other accounts of the history of the national museum of natural history in Leiden, as well as that of the Dutch national herbarium. The best-known of them is the above mentioned one by Holthuis, but it is certainly not the only one. There are also a handful of institutional accounts and biographies for a few collectors and naturalists, which

21 See, for example, Enenkel and Smith, *Early Modern Zoology*; Van Gelder, *Bloeiende kennis*.
22 Gijzen, "*'s Rijks Museum*."

are overall, unfortunately, insufficient.[23] Additional information about Dutch natural history can be found in a small number of essays on scientific research in the Indonesian archipelago; the earliest dates back to 1879.[24] Nevertheless, all these works (institutional accounts, biographies, and histories of Dutch science in Indonesia) are lamentably silent on Dutch systematics. None of them explore its practices and theories, nor its cultural, political, and international influences. It is hardly surprising that references to Dutch nineteenth-century natural history are practically absent from the literature on the subject.

From Ernst Mayr's classical *The Growth of Biological Thought* from 1982 (which devotes three chapters to classification) to Arthur MacGregor's *Naturalists in the Field* from 2018, Dutch natural history is occasionally mentioned when the country's collections are discussed – and virtually nowhere else. The essays compiled in the comprehensive volume *Cultures of Natural History* (published in 1996), which explore the cultural, institutional, and political context of natural history and its practices, do not include essays on Dutch natural history. Its successor – and equally relevant – *Worlds of Natural History* from 2018 includes two articles concerning natural history in the Low Countries during the early modern period; however, neither of them discusses the Dutch natural historical collections, or the Dutch contributions to the science of classification.[25] Based on references to the Netherlands in works dealing specifically with classification alone – which have proliferated during the last thirty years or so – one might conclude that the Dutch contributions to the field were negligible. Of course, there is no evidence to support this assumption, and as the saying goes, absence of evidence is not evidence of absence. The need for a historiography of Dutch systematics has become ever more

23 For instance, Stresemann devoted a chapter in his book *Ornithology* to the vicissitudes of the *Natuurkundige Commissie* (the Natural Science Committee for the Dutch East Indies), which had been created to explore the natural richness of the Indonesian archipelago. See also Boeseman, "Collectors and Fish Collections of the Rijksmuseum van Natuurlijke Historie"; Dreier, "Voor museum en vaderland"; Fransen, Holthuis, and Adema, "Type-Catalogue of the Decapod Crustacea"; Holthuis, "Three Dutch Authors of von Siebold's *Fauna japonica*"; Klaver, *Inseparable Friends*; Mörzer Bruyns, *Met de Triton en Iris*; Visser, "Het Rijksmuseum"; Yamaguchi, *Von Siebold.*

24 See, for example, Boomgaard, "Making and Unmaking of Tropical Science"; Goss, *Floracrats*; Honig and Verdoorn, *Science and Scientists in the Netherlands Indies*; Ter Keurs, "Agency, Prestige and Politics"; Sirks, "Indisch natuuronderzoek"; Veth, "Overzicht van hetgeen"; Weber, "Hybrid Ambitions," "Collecting Colonial Nature"; Van Wingerden, "Science on the Edge of Empire."

25 Jardine, Secord, and Spary, *Cultures of Natural History*; Curry, Jardine, Secord, and Spary, *Worlds of Natural History*.

pressing because, without it, Dutch natural history would inevitably be routinely excluded from the history of natural history.

Among the growing corpus of literature on natural history and biological classification I would like to highlight two books that have succeeded in capturing and explaining the complexity of systematics in the nineteenth century. One is a re-evaluation of Antoine-Laurent Jussieu's contribution to botany in *The Development of Biological Systematics* by Peter F. Stevens, the other Jim Endersby's depiction of the botanist Joseph Dalton Hooker in his *Imperial Nature*. Stevens deals with many of the theoretical questions I have been pondering, like the development of systematics, or the relation between classification and natural philosophy. Endersby links the material practices of classification to the broader issues of philosophy, empire, and professionalization. Both works focus primarily on botany, and they are also insightful and essential reading for anyone keen to explore the history of zoological classification, an area in which much work remains to be done.

Our understanding of the emergence of zoological systematics (in the Netherlands and virtually everywhere else) remains inadequate due to the fact that the existing literature is largely episodic. Some authors have focused on the history of the classification of certain groups of animals.[26] Others have explored the practice of classification, such as fieldwork and collecting practices, the history of zoological nomenclature, local or national histories, or specific eminent naturalists' contributions.[27] The conceptual history of systematics is also fragmentary. Historians have often focused on specific concepts, such as the role of essentialism in classification and the 'species' and 'type' concepts.[28]

While there is much interest and value in these studies, there is still a need for a more detailed history of the foundations and aims of zoological

---

26 The history of classification of mollusks and insects has been examined, for instance, by Cain, "Development of Systematic Ideas of Variation Illustrated by Malacology"; Lindroth, "Systematics Specializes between Fabricius and Darwin"; and Winsor, "The Development of Linnaean Insect Classification."

27 See, for example, Appel, *The Cuvier-Geoffroy Debate*; Endersby, "Classifying Sciences"; Farber, *Finding Order in Nature*; Gibson, "Careering Naturalists"; Kohler, *All Creatures*; McOuat, "Species, Rules and Meaning," "Cataloguing Power"; Melville, *Towards Stability in the Names of Animals*; Rookmaaker, "Early Endeavours by Hugh Edwin Strickland"; Rudwick, *Georges Cuvier*; Winsor, *Starfish, Jellyfish, and the Order of Life*.

28 Of special interest are Daudin, *Cuvier et Lamarck* and *De Linné à Jussieu*; Farber, "Type-Concept in Zoology"; Hull, "Effect of Essentialism on Taxonomy" (I and II); Kohler, "Reflections on the History of Systematics"; McOuat, "Origins of 'Natural Kinds'"; O'Hara, "Representations of the Natural System"; Winsor, "Creation of the Essentialism Story"; Witteveen, "Naming and Contingency."

classification, an account in which concepts are explored in relation to natural philosophy and embedded in their social and cultural contexts. The philosophical premises of classification and what it represented (species, genera, and families) in pre-Darwinian natural history are still poorly understood. To date the focus has been on anglophone and French natural history and museums.[29] The history of the Leiden museum has yet to be set in its cultural and intellectual context. To understand how and why systematics developed and why its status within natural history has been a matter of discussion for almost two hundred years, we need both a deeper conceptual history and a social history of zoological systematics.

This book takes another step along the trail blazed by authors like Paul Farber, Peter Stevens, Gordon MacOuat, and Jim Endersby – to name but a few – in order to contribute to the background necessary for a better understanding of nineteenth-century systematics. I hope this will be the starting point of a more comprehensive and ever-growing historiography of Dutch natural history.

3 Summary of the Chapters

This book consists of two parts containing seven chapters altogether and a closing chapter with some final conclusions. The first part of this book follows Temminck's path from his early life in his parental home in Amsterdam to his final years as the director of 's Rijks Museum van Natuurlijke Historie in Leiden. Chapter 1 starts, unsurprisingly, with Temminck's childhood. His love of birds started at home, where he was surrounded by stuffed birds and aviaries, and heard the fabulous stories of François Levaillant struggling up the Orange River in South Africa. Later, the inheritance of his father's natural history collection spurred Temminck towards zoological classification through the practice of ordering and cataloging the collection. The earliest collection catalogs became classifications, and classifications turned into monographs that launched Temminck into the international arena of natural history.

Chapters 2 and 3 describe Temminck's professional career. Chapter 2 tracks Temminck's career through a series of posts that, eventually, served him as stepping stones towards the highly regarded and coveted directorship of the brand new Rijks Museum van Natuurlijke Historie. The most important of these posts was that of director *ad interim* of 's Lands Kabinet, Louis

29 See, for example, Limoges, *Development of the Muséum d'Histoire Naturelle*, and Alberti, *Nature and Culture* and "Museum Nature."

Napoleon's collection of natural history, which was one of his pet projects. It allowed Temminck to build a reputation as an excellent museum curator and ornithologist. Chapter 3 explores the circumstances surrounding the establishment of 's Rijks Museum, Temminck's years as its director, and his efforts to put the museum on the international stage. The collections – on the basis of which Temminck constructed his classification – grew exponentially in part thanks to Temminck's influence and creativity, despite the detrimental effect on the museum's finances and its scientific activity of the Dutch political and economic crisis between 1820 and 1850.

Chapter 4 focuses on the scientific output of Temminck and his staff, outlining the relative importance of their work for the naturalists' community. It discusses how Temminck communicated and the platforms he used, which determined his scientific prestige. So did the political and economic situation of 's Rijks Museum. While the museum's director developed his particular vision and goals, the museum drifted away scientifically from the academies, and different branches of natural history slowly became the domain of different institutions. Simultaneously, museum curators and field naturalists specialized and, inevitably, became occupied with separate fields. Each place – the museum, the university, and the field – became a niche for specific disciplines, bringing about conflicting visions on the best approach to studying nature.

The second part of the book takes a wider perspective as it examines the emergence of systematics within the tradition of natural history, with Temminck as a case study for the discussion. By placing Temminck's career in the context of his time, a clearer image emerges: the image his contemporaries held of him, through his publications and their collaborations, quarrels, competition, and ambition. These seemingly mundane issues are intimately linked to broader questions such as the rise and status of pre-Darwinian systematics or the relationship between classification and natural philosophy.

Chapters 5 and 6 discuss Temminck's understanding of essential concepts in zoological classification and his methodology of arranging animals in a natural system. These chapters consider both his writings and his discussions with his peers, and as the most technical chapters in this book, they are necessary for an understanding of the foundation of Temminck's classifications – and that of any other naturalist of the time. Chapter 5 delves into the concepts of type, genus, and species as they emerge from Temminck's law of the geographical distribution of mammals and birds. Questions concerning the fixity and origin of species and the principle of continuity, among others, were all associated with these concepts. Also, Cuvier's influence on Temminck's views becomes evident in this discussion. Chapter 6 follows the arguments held between Temminck and his fellow ornithologists about the definition and delimitation of genera and species, the rules of zoological nomenclature, and their different approaches towards a natural classification system. Above all, this chapter reveals a rather chaotic state of affairs and a lack of understanding among naturalists. It was their determination to reconcile different views and methods that ultimately led to the democratization of natural history: a scientific meritocracy emerged, with Britain in the lead. Approaches varied, discussions stretched out, and, by the middle of the century, Temminck's authority had greatly diminished among his peers.

Finally, Chapter 7 presents a bird's-eye view on the development of the field of systematics within the natural sciences from 1800 to 1850. It examines what 'natural history' meant and the fields included in or excluded from it before discussing the status of systematics, especially in relation to the highly regarded field of comparative anatomy. Here, the instrumentalization of philosophical arguments in systematics was essential, and most naturalists were very aware of this fact. During this period the search for an ultimate natural classification system – an arrangement of animals that would reflect their natural relationships – had become the holy grail for nearly every naturalist, Temminck included.

The Conclusion drawn from these seven chapters focuses on the emergence of systematics as a discipline within natural history. The development and status of systematics during the early decades of the nineteenth century marked out much of the path that systematics took afterward. The consequences are still felt today.

4 A Word of Caution: On Definitions

In the previous passages, I have referred to 'systematics,' 'classification,' 'discipline,' 'naturalist,' and 'natural history.' However, these terms are far from clearly defined, and their meaning has changed over time and from author to author. It is important, therefore, that I provide an explanation for this choice of words and their intended meaning in this book.

Much ambiguity surrounds the terms 'taxonomy' and 'systematics' today, with various definitions available in the specialized literature. For example, Randall Schuch regards both terms as synonyms, while for Michael Simpson, systematics encompasses taxonomy (the description, identification, naming, and arrangement of organisms) and has as its primary goal the reconstruction of the evolutionary history of organisms.[30] Often taxonomy is defined as

30 Schuh and Brower, *Biological Systematics*, 15; Simpson, *Plant Systematics*, 2nd ed., 9–12.

the theory of classification and nomenclature, while systematics includes the arrangement of groups in an orderly, systematic fashion, based on the relationships between groups (currently based on evolutionary kinship).[31] I have avoided using the term 'taxonomy' and have preferred 'systematics' instead, simply because 'taxonomy' was not a commonly used term in Temminck's time. The term "taxonomie" was introduced by Agustine P. de Candolle in 1813, in his *Théorie élémentaire de la botanique*, as a new term for "la Théorie des classifications."[32] The term was sometimes used in textbooks. For example, the Dutch zoologist Jan van der Hoeven used "taxonomia" in his zoological handbook for students.[33] However, overall the term 'taxonomy' is virtually absent from the literature on zoological classification up to 1850.

In general, naturalists used the terms 'system,' 'arrangement,' 'method,' and 'classification,' all of which were also ambiguous. Many zoologists used these terms as Georges Cuvier had defined them in 1816. He made the distinction between a system and a method: "système" applied to artificial classifications (convenient catalogs that did not necessarily reflect actual relations between organisms), while "méthode" indicated a natural classification, arrived at through careful comparison of organisms following the principle of the subordination of characters.[34] The "méthode" thus resulted in apparently natural classifications that reflected the true order of nature, and it was constructed not by division but by synthesis.[35] Although nearly all naturalists were aware of this distinction between natural and artificial classifications, Cuvier's nomenclature was not universally adopted, and some authors even inverted it, causing misunderstandings and confusion.[36] Temminck was one of these authors. He used phrases like "système général d'ornithologie," "système de la nature" and "partie systématique de l'étude de la nature" for nearly all zoological classifications. However, he referred to single-character classifications and the Linnaean arrangement of birds as the "méthodes artificielles" or "classification méthodique," and to Linnaeus' followers, as "méthodistes."[37] By contrast, Temminck used the term "système" to denote any classification, natural

31 Minelli and Contrafatto, *Biological Science Fundamentals and Systematics*, vol. 2, 163; Winsor, *Reading the Shape of Nature*, xii.

32 Candolle, "Introduction," in *Théorie élémentaire de la botanique*, 19.

33 Van der Hoeven, "Inleiding," in *Handboek der dierkunde*, 28.

34 Cuvier, *Tableau élémentaire*, 19–21.

35 See also Mayr, *Growth of Biological Thought*, 190–3; Stevens, *Development of Biological Systematics*, 10–11.

36 For example, Macleay, "Preface," in *Horae entomologicae*, vol. 1, viii–ix; Stevens, 13.

37 See, for example, Temminck, "Introduction de la seconde édition," in *Manuel d'ornithologie*, 2nd ed., vol. 1, vii.

or artificial – the equivalent of 'system' in everyday language, not in Cuvier's sense. In short, we can never be sure of the meaning of these terms if they are taken out of their original context.

Given these ambiguities, which reflect the conceptual confusion of the time, I have adopted the terms 'systematics,' 'system,' and 'classifications,' and I will indicate whether or not the classification was thought to be natural or artificial. Also, the use of 'systematics' rather than 'taxonomy' avoids the implication that nineteenth-century naturalists did not concern themselves with the relationships between the groups of organisms they were describing, naming, and ranking. On the contrary, this was their main concern. Similarly, I will use the terms 'naturalist,' 'zoologist,' 'botanist,' and sometimes 'systematist,' but generally not 'scientist.' However, I translate the Dutch *wetenschap* (which means both 'knowledge' and 'science') as 'science' rather than 'knowledge,' which is such an inclusive concept.

Throughout the book I will also use the terms 'discipline' and 'field,' but I do so in a different sense than Paul Farber's use of 'discipline' in his book *The Emergence of Ornithology as a Scientific Discipline*. Farber summarizes the activities within eighteenth-century natural history as comprising four "different research traditions": nomenclature and systematics, the description of the life histories of individual species following Buffon, comparative anatomy, and physiology. While the last two evolved into separate disciplines, life histories and systematics "ramified into the specialized discipline of ornithology." Farber continues: "Morphologists and physiologists sought general laws regulating form and function, whereas ornithologists sought to classify and describe a particular group of animals."[38] For nineteenth-century ornithology, Farber equates it to "bird systematics" and considered it a scientific discipline.

However, even if ornithology was most closely engaged with systematics at the time, systematics was also a central task in entomology, mammalogy, malacology, and other natural historical fields. Naturalists were dealing with the same issues: the lack of a uniform nomenclature, the search for a natural classification system, and the standardization of methods. Even considering that the main occupation of ornithologists during the period covered by this – and Farber's – book was, indeed, systematics, books on birds also included detailed life histories, breeding information, and other matters generally more appealing to the general public. All of this also falls under the umbrella term of 'ornithology' which included, but was not exclusively occupied with systematics. I will, therefore, treat ornithology as a field and systematics as a discipline. A discipline is defined, for the purposes of this book, as a branch of science

38 Farber, *Emergence of Ornithology*, 125.

having its own set of goals, methods, and techniques, a specific new language adequately describing a subject matter with unique terms, and having a social structure: a community of practitioners, organizations like clubs and societies, platforms for publication like specialized journals, and a place in academic curricula.

Finally, the phrase 'natural history' defies definition. The term was then, as it is now, generally used but seldom defined. Even during the relatively short period of fifty years covered by this book, 'natural history' had an array of meanings, and it still has. Chapter 7 deals with the problem of its meaning in Temminck's time, but I will use, throughout this book, the term 'natural history' in its broadest sense, in which it includes all branches of botany, zoology, and geology.

PART 1

Birds, Cabinets, and Museums

FIGURE 2 *Coenraad Jacob Temminck*. Portrait by Jan Adam Kruseman, 1836. Oil on canvas
NATURALIS BIODIVERSITY CENTER ARCHIVES

CHAPTER 1

From Catalogs to Monographs

"They will not accept you, my appreciated Temminck, because you are not a scholar," said King Louis Napoleon Bonaparte to Coenraad Jacob Temminck. He had just been denied admittance to the Koninklijk Instituut van Wetenschappen, Letteren en Schoone Kunsten (Royal Netherlands Academy of Arts and Sciences).[1] This verdict must have been a very bitter pill to swallow, however royal its source. After all, the Koninklijk Instituut did not apply such criteria to the selection of new members. Good social relations, a high social position, or even family connections sufficed as credentials for some of its members.[2] However, when the Koninklijk Instituut was founded in 1808, Temminck did not qualify as a member for some reason. Politics, personal antipathies, and hidden agendas were probably at play. At the time he was well-known among naturalists. His natural history collection was famous. It comprised a large number of exotic species, skillfully prepared and organized. Visitors included members of the elite of European naturalists, and yet, Temminck apparently did not meet the approval of the institute's directors. But in the following decades he succeeded in securing both an influential – and remunerated – position as director of a national museum and the respect of his fellow naturalists. Between 1805 and 1820, Temminck added to both his social status and his knowledge in order to make sure that "not being a scholar" would never again be an obstacle. In 1836 he finally managed to become a member of the Koninklijk Instituut.

Temminck knew how to increase his status in the changing political landscape of the Netherlands between 1795 and 1810, and he adapted accordingly. He understood that, if he wanted to pursue a career in natural history, he needed to occupy an official position, and he worked relentlessly towards it. He intentionally paved his way towards a professional career in which he could thrive as a non-academic. The first stepping stone was his father's magnificent bird collection.

1 Susanna, "Levensschets van Temminck," 74. Louis Napoleon founded the Koninklijk Instituut van Wetenschappen, Letterkunde en Schoone Kunsten, now the Koninklijke Nederlandse Akademie van Wetenschappen, in May 1808.

2 Van Berkel, *De stem van de wetenschap*, 124–25.

 | DOI:10.1163/9789004441491_003

1 Exotic Birds on Cupboards and Plates

The patrician Temminck family had enjoyed a prominent status in Dutch society for generations. Some of them had been notable members of the Amsterdam and Amersfoort political elite, including, for example, Egbert de Vry Temminck (1700–1785), who was the mayor of Amsterdam between 1749 and 1784 and a director of the Verenigde Oost-Indische Compagnie (VOC) and the West-Indische Compagnie (WIC). Others were wealthy merchants dealing in luxury goods like Spanish wool. Coenraad Jacob Temminck's grandfather, Coenraad Temminck (1687–1762), was the mayor of Amersfoort, and his brother, Matthias Temminck (1734–1814), was a director of the VOC and practiced law in Amsterdam.[3] Through these ties with the VOC and WIC the family had occasionally supported natural historical pursuits as patrons. Egbert de Vry Temminck, who had an interest in botany and owned a private botanical garden in Haarlem,[4] helped Carl Peter Thunberg to take up the position of surgeon onboard the VOC ship *Schoonzigt* so that he was able to conduct botanical research in Southeast Asia in 1771.[5] Thunberg obtained letters of recommendation from De Vry Temminck, which were indispensable for entry to Java and Japan, both areas strictly controlled by the VOC.[6]

Coenraad Jacob Temminck was born on March 31, 1778. His father, Jacob Temminck (1748–1822), was an Amsterdam patrician and the treasurer-general for the VOC, and married to Aleida van Stamhorst (1751–1806), the daughter of the VOC cashier Jacob van Stamhorst. Through his contacts at the VOC and in Amsterdam, Jacob Temminck built one of the most splendid collections of natural history in the Netherlands. It comprised mainly specimens of birds, but also mammals, fossils, fish, several skeletons, and even an aviary with living exotic birds, all situated at the family's home at the Heerengracht in Amsterdam.[7] The cabinet – and the visitors it attracted – played an important role in Coenraad Jabob's early life and, later, in defining his career.

This was all very fashionable and a sign of good taste and wealth. For example, the Amsterdam Commissioner Arnoldus Ameshoff (1749–1819) also owned a huge aviary. His pride and joy were the exotic species of water birds,

3 Elias, *De Vroedschap van Amsterdam*, 785–89, 986. More information on the family can be found in Van Epen, *Nederland's Patriciaat 1868–1930*, vol. 5, 397–401, and Elias, *De vroedschap van Amsterdam, 1578–1795*, vol. 2.

4 Sliggers, "Henry Hope's vermakelijke buitenleven en de Haarlemmerhout," 28.

5 Van Dulm, *'Zonder eigen gewinne en glorie,'* 307; Rookmaaker, *Zoological Exploration*, 148.

6 Thunberg, *Reise durch einen Theil von Europa ...*, 68, 189.

7 Gijzen, "'s Rijks Museum," 32; Smit, Sanders, and Van der Veer, *Hendrik Engel's Alphabetical List*, 271–272.

fowls, pheasants, and pigeons brought to him from all corners of the globe. Coenraad Jacob recalled the extravaganzas hosted at the Ameshoffs' manor house, *Amstelrust*: "I remember attending a dinner as a child at M. Ameshoff's, who, to display the magnificence of his menagerie, served at his table not only *Pauxis, Hoccos* and various species of exotic pheasants, but also Mandarin Ducks from China and Wood Ducks from Louisiana, a feast worthy of the time of Heliogabalus."[8]

The size and quality of Jacob Temminck's collection in Amsterdam attracted the attention of many European naturalists. Many thought it worth traveling to Amsterdam just to see the collection. The naturalist and voyager Georg Forster praised not only the beauty and rarity of the specimens but also the "inimitable perfection of the art of stuffing birds, attained nowhere else."[9] Although none of the visitors nor Coenraad Jacob himself mentioned who was responsible for preparing the specimens, the taxidermic skills needed for such a display were remarkable. Many of these specimens were used as models for the description and illustration of species in ornithology books like Cornelis Nozeman's and Christian Sepp's *Nederlandse vogelen*. The French explorer and naturalist François Levaillant, a friend of the family, also found models for his books among Jacob's birds, and he asked the artists Lebrecht Reinold and Jacques Barraband to visit Amsterdam and make drawings from Temminck's specimens. Barraband produced several plates for Levaillant's *Histoire naturelle des perroquets*, which was published between 1801 and 1805.

The aviary was also remarkable. According to Levaillant, Jacob Temminck had managed not only to keep alive many exotic species of birds, but they also reproduced successfully in captivity.[10]

In 1790 Forster paid a visit to the Temmincks in Amsterdam and introduced Alexander von Humboldt to the Temminck family, as Humboldt himself fondly recalled in a letter to Temminck in 1857.[11] Coenraad Jacob was only twelve years old at the time of this visit. He grew up in a house where visitors, collections, books, and even dinners revolved around ornithology. It is easy to imagine how conversations about exotic animals, strange countries, and undaunted travelers may have ignited the young Coenraad Temminck's fascination with natural history.

---

8 *Pauxi* and *Hocco* are two genera of curassows from the Americas. Temminck, "Histoire naturelle générale des gallinacés," in *Pigeons et gallinacés*, vol. 2, 458, fn. e; see also Van Lynden-de Bruïne, *In vogelvlucht*, 13.

9 Forster, *Ansichten vom Niederrhein*; translated in Stresemann, *Ornithology*, 111.

10 Levaillant, *Second voyage*, vol. 2, 135.

11 Humboldt to Temminck, 29 March 1857, Naturalis Biodiversity Center Archives, Archive C. J. Temminck; Raat, "Humboldt and Temminck," 31.

Very little is known about Coenraad Jacob Temminck's education. As a child he was privately tutored at home; his memories of his French–Swiss tutor were full of disdain. There are no records of him attending school. In 1795 Coenraad Jacob, seventeen years old, started to work as one of the three auctioneers of ships and merchandise for the VOC, a post undoubtedly arranged by his father in order to grant the young Coenraad Jacob entry into the Company.[12] However, that same year the VOC was declared bankrupt, and the democratic Patriots took the lead (mostly members of the bourgeoisie striving to democratize the country and objecting to the rule of the stadtholder, William V Prince of Orange). The Batavian Republic, which had been proclaimed in January 1795 (and was, in fact, a client state of France or 'sister-republic'), became the owner of the VOC in 1796. Three years after its nationalization, the VOC ceased to exist. In April 1798 Temminck received a letter from the president of the Administrative Municipality of Amsterdam, who bluntly informed him that he was dismissed from his post in the VOC as auctioneer, together with many other "useless or unworthy local civil servants and clerks" who were to be replaced "by worthy and skillful patriots."[13] These civil servants apparently had enough influence in the Republic to reverse the discharge by writing a letter of objection. They were reinstated in their posts a month later. Temminck remained in his position until early April 1812, but it was now an honorary post.[14]

Around 1800 Coenraad Jacob was in charge of his father's collection; Jacob Temminck would live for another twenty years. From then on, the Temminck collection grew fast, doubling its number of species in just five years.[15] Coenraad Jacob took up the task of cataloging the cabinet sometime between 1800 and 1804. His manuscript catalog is titled "Catalogue du Cabinet de C. J. Temminck," indicating that the collection had passed from father to son.[16] This catalog lists

---

12 Susanna, "Levensschets van Temminck," 49; Raat, "Coenraad Jacob Temminck," 91; Van Lynden-de Bruïne, *In vogelvlucht*, 13.

13 Administrative Municipality of Amsterdam to Coenraad Jacob Temminck, 6 April 1798, Naturalis Biodiversity Center Archives, Archive C. J. Temminck.

14 Records relating to these events are kept in the Naturalis Biodiversity Center Archives, Archive C. J. Temminck. See also Van Lynden-de Bruïne, *In vogelvlucht*, 14; Holthuis, *Rijksmuseum*, 19.

15 Stresemann, "Analyse," 320–22; Holthuis, *Rijksmuseum*, 19.

16 A handwritten copy of the catalogue, in Coenraad Jacob's distinctive handwriting, is now kept in the Naturalis archives. Justin Jansen dates this manuscript catalog to ca. 1803–1804 in "Unpublished 'Catalogue du Cabinet de C. J. Temminck'." Jansen's dating is mainly based on Temminck's annotations on the margins, especially those that refer to Levaillant's works. However, these annotations may have been added to the manuscript while Temminck was preparing a new and revised version of the catalog, and they do not provide reliable evidence for the dating of the manuscript.

a total of 839 species. The first 450 were already present in Temminck's father's collection, including some of Levaillant's specimens brought from southern Africa in 1784. The manuscript follows Georges Louis Leclerc de Buffon's classification for birds and Jean Baptiste Audebert's for apes and lemurs.

Temminck actively pursued the acquisition of new species by purchase and exchange. Following the decline and final dissolution of the VOC he was no longer able to add specimens to his collection via his father's contacts. Instead, he contacted other collectors and naturalists. His excellent networking skills would prove to be extremely useful. By the time his first publication appeared in 1807, his name was already well-known among collectors. When, in 1802, Temminck received the order of Chévalier de la Légion d'Honneur at the age of twenty-four, this may have been thanks to the reputation he had formed with the help of his cabinet. Consul Napoleon Bonaparte had created the order in the same year. It was – and still is – a recognition for civil and military merits, including excellence in trade or profession, or for contributions to the welfare of the people and the nation. A civilian may obtain this honor via work in the arts, the sciences, or politics. An extensive and scientifically significant natural history cabinet was much more than an exuberant hobby and a source of information for naturalists. It was also considered an expression of cultural refinement. As Temminck's cabinet ranked among the best in Europe, the Légion d'Honneur may have been a recognition of his achievements in putting together the collection.[17]

Coenraad Temminck played a more active role in the production of Cornelis Nozeman and Christian Sepp's *Nederlandsche vogelen* than his father. After Nozeman's death in 1789, before the publication of the second volume, Martinus Houttuyn took over the work. By unfortunate coincidence Houttuyn, in turn, died in 1797, before the third volume was released. Christiaan Sepp and his son, Jan Christiaan, then asked Temminck for help with the fourth volume. In this volume, which was published in 1809, Sepp father and son refer to the cabinet of "the honorable Mr. C. J. Temminck," who had not only allowed them to use specimens from his excellent collection as models for their illustrations, which was so well-known throughout Europe, but had also provided them with "the most accurate ornithological observations." While the fourth volume included many birds from Temminck's collection, almost all the plates of the fifth and last volume were drawn from Temminck's birds.[18]

17 Van Lynden-de Bruïne, *In vogelvlucht*, 16.

18 Nozeman et al., *Nederlandsche vogelen*, vol. 4, 385; vol. 5, ii. The fifth volume followed Temminck's classification of birds as published in his *Manuel d'ornithologie* (1815).

On April 3, 1804 Temminck married Dionysia Catharina Cau (1779–1828), known as Caatje. She came from a wealthy and influential family that was well-connected with the political elite and the VOC. Her father, Iman Meijnard Cau (1755–1791), had married the daughter of a VOC commander and was one of the Forty Eminent Citizens, the *Veertigraad* of Delft, a group with the power to put forward nominations for the role of city mayor and other offices. If the families had arranged the marriage apparently neither Caatje nor Temminck objected. Caatje wrote in her journal on the day of their twentieth wedding anniversary: "I was up early and I remembered with joy the privilege of reliving today, once again, our wedding day."[19] In the privacy of her journals she sometimes referred to her husband as "mijn lieve Temminck" (my dear Temminck).[20] Caatje moved to the Heerengracht to live with her husband and parents-in-law. In 1806 the couple – and the collection – moved from the Herengracht to a smaller house, no. 401, on the same canal. Their new home was filled to the brim with stuffed birds and mammals. Specimens were stalled in the entrance, corridors, and the attic, where visitors could walk around to admire the cabinet twice a week.[21]

Temminck has been portrayed as a quiet, reserved man for whom propriety, order, and precision were of the utmost importance.[22] He preferred work over entertainment, nature over theaters, and hunting over eating. He "led a sober life and was uncommonly neat and exceedingly precise about his clothes and surroundings."[23] Despite his patrician upbringing Temminck disliked having a valet, whom he did not allow to help him dress, not even in advanced age, when illness confined Temminck to his bed – he still received visitors well-dressed then, as if he was ready to leave the house. Temminck could be aloof and had a curious way of suddenly steering a conversation into a different direction. He was not quick to offer his views on a subject unless it mattered to him, but when he did, he did so politely – at least most of the time. When affronted, he did not hesitate to reply accordingly. Intelligent and quick of mind, Temminck could also be condescending, even sarcastic. He received compliments and honors with relative indifference, but then again, he appears to have been masterful in hiding his inner thoughts and feelings, and he seldom revealed any political sentiments.

---

19 Van Lynden-de Bruïne, *In vogelvlucht*, 26.

20 Van Lynden-de Bruïne, 47.

21 Van Lynden-de Bruïne, 14; Witkamp, "Vier-en-twintig voorgangers," 194.

22 Vrolik, "Levensbericht van Coenraad Jacob Temminck," 78. For the following description of Temminck's character see also Susanna, "Levensschets van Temminck"; Snellen van Vollenhoven, "Coenraad Jacob Temminck"; Holthuis, *Rijksmuseum*, 23.

23 Snellen van Vollenhoven, "Coenraad Jacob Temminck," 68.

FIGURE 3 Scarlet Macaw specimen from Temminck's private collection, ca. 1800–1820 (RMNH.AVES.208372)
NATURALIS BIODIVERSITY CENTER. PHOTOGRAPH BY HENK CASPERS

FIGURE 4 Mandrill specimen from Temminck's private collection, ca. 1800–1820 (RMNH.MAM.59782)
NATURALIS BIODIVERSITY CENTER. PHOTOGRAPH BY HENK CASPERS

Regarding his religious beliefs, Temminck was a Mennonite, or *Doopsgezinde*, and thus part of a community whose members mostly followed an austere life and were considered to be trustworthy and honorable.[24] Temminck could be honest to a fault. Even though he seems to have been discerning in his choice of friends, his friends were friends for life, towards whom he acted generously and obligingly. They were always welcome to stay at his home, whether in Amsterdam or at his country house in Lisse, *Wildlust*, even for extended periods. He always lived a frugal life himself, but he lavished his guests with luxuries as if they were attending a wedding.

In the context of his family, Caatje's diaries portrayed Temminck as loving and caring. She accompanied him everywhere, even on long, uncomfortable carriage rides during his tours around Europe, which lasted for months at a time. In return, Temminck attended concerts and soirées with her. The couple remained childless. Caatje died in November 1828, at the age of fifty. Two years later Temminck married Catharina Nepveu (1785–1834), who sadly died four years into their marriage. Temminck's third wife, who was twenty-eight years younger than him, was Agneta Smissaert (1806–1865), the daughter of his friend Marinus Smissaert (1773–1819), the steward at the Royal Palace in Amsterdam. The couple had three sons between 1838 and 1841: Coenraad Jacob, Marinus, and Willem Anne Temminck. To Temminck's disappointment, none of them inherited his "taste for natural history."[25]

2 François Levaillant and Bernhard Meyer

François Levaillant was probably the most influential person in kindling the love of the young Coenraad Jacob Temminck for natural history. Levaillant was a friend of the family and a frequent guest at Jacob Temminck's house. Levaillant and Jacob met around 1780, and Levaillant was instantly impressed with Jacob's collection, which he described as "an assembly of precious objects

24 There is a marginal note in a letter from Temminck to the Ministry of Home Affairs simply reading "Mennonite" (Temminck to Ministry, 27 February 1819, NL-HaNA, BiZa 2.04.01, inv.nr. 4072). In this letter Temminck requested to be excused from taking an oath – a guarantee for his honest use of the funds given to him to buy specimens for the universities. He was permitted to submit a written affirmation instead (Ministry to Temminck, n.d. March 1819, inv.nr. 4072).

25 Van Lynden-de Bruïne, *In vogelvlucht*, 21.

as I have never seen in France. Everything seemed to me extremely rare and exceptionally well preserved."[26]

Jacob Temminck admired Levaillant, so much so that in 1780 he acted as Levaillant's patron in helping him to travel to South Africa on a VOC ship. He wrote letters of introduction for Levaillant and sponsored the voyage. In the following three years Levaillant collected about two thousand specimens of birds as well as mammals, insects, plants, and Khoikhoi ["Hottentot"] artifacts. Several specimens collected during this journey joined Jacob Temminck's collection.[27] During Levaillant's voyages, he and Jacob kept in touch with each other, with Jacob mostly trying to direct Levaillant towards interesting collecting sites, like Madagascar. He arranged a passage on a VOC slave ship for Levaillant which was scheduled to make port at the Cape of Good Hope, where Levaillant would be able to embark. Levaillant happily agreed to this plan at first, but after sensing some antipathy from the slave ship's captain, who appeared to be very reluctant to take him aboard, he decided to leave the trip for another time.[28] In 1795 Levaillant undertook a second expedition to Africa. He published accounts of both his voyages. The books provided vivid descriptions of his travels and were an immediate success; they were translated into several languages. In 1799 Levaillant's first ornithological work, the *Histoire naturelle des oiseaux d'Afrique*, appeared in Paris, and was dedicated to Jacob Temminck – according to Stresemann, Jacob had also paid for it.[29]

Coenraad Jacob Temminck was twelve years old when the earliest of Levaillant's popular books were published. With their vibrant prose about many adventures and an exotic natural world they were the kind of books that excited the imagination. Through his existing friendship with Jacob, Levaillant became friends with Coenraad Jacob as well. After the latter took over the care for the cabinet and started studying birds, Levaillant continued to give gifts in the form of both interesting specimens and valuable information, and he was granted permission to illustrate birds from the Temminck cabinet in his works.[30]

In 1806, as Temminck progressed in his studies, Levaillant recognized his potential: "M. Temminck, although still young, is a remarkable man, different

26 Levaillant, *Voyage de M. Le Vaillant*, 1. See also Stresemann, *Ornithology*, 86, 111; Rookmaaker, *Zoological Exploration*, 259.

27 Levaillant, *Voyage de M. Le Vaillant*, vol. 1, 1–2, 28.

28 Levaillant, *Second voyage*, 361.

29 Stresemann, *Ornithology*, 88; Huigen, *Knowledge and Colonialism*, 137–38. According to Huigen, Levaillant was employed by the VOC as "gunner's mate" during his first voyage (*Knowledge and Colonialism*, 137).

30 Temminck, *Pigeons et gallinacés*, vol. 3, 640.

from all those collectors who, loving only brilliant and beautifully plumaged species, accumulate at great expense collection upon collection, without any taste for scientific study."[31] Initially Levaillant adopted the role of mentor. Temminck wrote: "I owe to Levaillant and his writings my first thoughts and my first steps in natural history; his works and his advice served me as guides in the practical study of this science, and I am pleased to admit that, on his return from Africa, he laid the first foundations of a collection which, passing from father to son, is today the most beautiful ornament of the Museum of the Netherlands."[32] As an expression of gratitude, Temminck named two birds after him, *Indicator levaillantii*, a honeyguide, and a species of snakebird, *Plotus levaillantii* (now *Anhinga melanogaster rufa*). Notably, in what seems to be an absence of filial deference, Temminck made no reference to his father or his father's cabinet in his publications other than in this brief sentence.

Levaillant had yet another, even more valuable gift for Temminck: arsenic soap. Levaillant had learned how to prepare and stuff bird skins from Jean-Baptiste Bécoeur, a French pharmacist and naturalist from Metz and the inventor of the arsenic soap method. This was an excellent procedure for preserving the skins of birds and mammals that protected them from voracious insects without damaging the skins. Although Bécoeur kept his formula secret and never published it, Levaillant had received it directly from him, and Temminck, in turn, learned it from Levaillant.[33] Thanks to this very effective technique, Temminck's specimens aged much more gracefully and stayed in much better shape than those of other collections in this period. Many other specimens disintegrated after being treated with salt and alum. Others lost feathers, rotted from the inside out, or were eaten by vermin.[34] The formula also allowed Temminck (and others after him) to study all the detail of bird plumage, as feathers retained their original color and structure.

Levaillant's influence on Temminck was, however, limited to a passion for birds, their natural history, and taxidermy. He did not provide Temminck with any scientific training. Levaillant, a writer of appealing prose, lacked a methodical or scientific approach to ornithology. He embellished his descriptions with

31 Levaillant, *Oiseaux d'Afrique*, vol. 5, 61.

32 Temminck and Meiffren Laugier de Chartrouse, *Nouveau recueil*, vol. 5, livraison 64, plate 380, text, [1–2]. See also Rookmaaker, *Zoological Exploration*, 259. For the dates of publication of the *livraisons* see Dickinson, "'Nouveau recueil de planches coloriées' of Temminck & Laugier," 7–54.

33 Temminck to Johann Karl Wilhelm Illiger, 4 May 1810, in Stresemann and Thomsen, "Ornithologen-Briefe aus den Jahren 1816 bis 1820," 85. See also Rookmaaker et al., "Ornithological Cabinet of Jean-Baptiste Bécoeur," 152–153.

34 Farber, "Development of Taxidermy," 553.

details he invented. He went so far as to create new species by making artifacts, i.e. specimens made from parts of different species that were glued together. The Swedish zoologist Carl Jakob Sundevall was one of those who accused Levaillant – many years after the fact – not only of making many errors, but also of using plates from others' works to describe new species.[35] When the French ornithologist and dealer in natural history specimens Jules Verreaux visited Temminck in Amsterdam, Temminck showed him eight or nine of Levaillant's artifacts. Although Verreaux encouraged Temminck to write about them, Temminck declined – perhaps because of Levaillant's long-time friendship with the Temmincks – but permitted Verreaux to do so if he wished. Apparently, Verreaux intended to publish about the artifacts in Guerin's *Revue et magasin de zoologie pure et appliquée*. Later Verreaux confessed to his fear of upsetting owners of cabinets (and, presumably, future customers) and refrained from publishing.[36]

Eventually Levaillant's popularity faded, and the last volume of the *Histoire naturelle des oiseaux d'Afrique* did not sell well. His friendship with Temminck also declined with time. In 1809, as Temminck's career began to blossom, Levaillant complained to his friend and associate, the artist Jean-Gabriel Prêtre, about Temminck's attitude:

> Are you still pleased with our fine T., as for me, I am not really happy? I have received a letter from Mr. Gosse from Geneva, who is not very pleased either, and neither is Mr. Boissiere with the exchanges he made with Temminck. It seems he is reprimanding everyone. You undoubtedly know that he is president of an Academy, director of a School of Art, advisor to the King, and Knight of the Union. God's will has it that these honors have changed him a little.[37]

An altogether different dynamic replaced the apprentice/teacher relationship formed during Temminck's youth. Their friendship was strained in part by their different views on classification, and mostly by Temminck's growing self-confidence and, apparently, somewhat pompous attitude towards Levaillant.

35 Sundevall, "Kritisk framställning," 17–18.

36 Edward Wilson to Hugh Strickland, 25 July 1849, in Rookmaaker, *Calendar*, 176.

37 François Levaillant to Jean-Gabriel Prêtre, 15 January 1809, Staatsbibliothek zu Berlin (SBB), Manuscripts Department, Slg. Darmstaedter Afrika 1780: Levaillant, François. It seems that Prêtre did not necessarily agree with Levaillant, as he later worked with Temminck on a book of pigeons and gallinaceous birds. By "directeur d'une École de peinture" Levaillant referred to the Koninklijke Kunst Museum (Royal Art Museum), where Temminck briefly worked as deputy director.

Levaillant detested Linnaeus' binomial system and preferred Buffon's narrative style. Temminck, by contrast, not only adopted Linnaean nomenclature and classification methods, but also dedicated most of his life to studying the classification systems that Levaillant disliked. In his publications Temminck attempted to elevate ornithology to the status of a science and, as Paul Farber puts it, "held the fashionable bird art books of the period in contempt."[38] Temminck also condemned the making of artifacts: it hindered the advance of science and served only to flatter the ego of those who made them.[39] Although Temminck never explicitly criticized Levaillant's works or artifacts he was, in all probability, referring to them. Tension arose between the two men. Temminck relied on Levaillant for information on the birds he worked on, and at times he could be impatient waiting for it. In June 1812, for instance, Temminck sent a list of five questions for Levaillant. Judging from Levaillant's answer a month later, Temminck had been quite insistent. After giving Temminck the requested information, Levaillant complained that, while Temminck demanded prompt answers, he himself had been waiting for months for money that was owed to him. He wrote to Temminck: "you have to admit that you are a devil of a man."[40]

When Levaillant died in 1824, impoverished and without friends, no obituaries were published.[41] Temminck wrote a short notice, hidden in a footnote, lamenting the harshness with which Levaillant's work had been judged. He seemed disappointed by the ornithologists' indifference towards Levaillant's passing:

> Perhaps it is because of too frank expressions about writings of this kind [i.e. Levailant's], or because of severe and often overly bitter criticism, that this learned ornithologist has been reduced in his old age to the barest minimum, without any kind of encouragement and without being able to pick the smallest fruit of his labors, while his compatriots receive rewards, expressions of thanks and titles, always awarded for real merits and to famous men of whom France is proud.[42]

38 Farber, *Emergence of Ornithology*, 90.

39 Temminck, "Introduction," in *Manuel d'ornithologie*, 2nd ed., vol. 1, xxi.

40 Levaillant to Temminck, 27 July 1812, Naturalis Biodiversity Center Archives, Archives C. J. Temminck.

41 Rookmaaker, *Zoological Exploration*, 181.

42 Temminck and Meiffren Laugier de Chartrouse, *Nouveau recueil*, vol. 3, livraison 4, plate 23, text, [4].

FIGURE 5 *Plotus LeVaillantii.* Hand-colored engraving after a drawing by Nicholas Huet. In Coenraad Jacob Temminck and Guillaume Michel Jerôme Meiffren Laugier de Chartrouse, *Nouveau recueil de planches coloriées d'oiseaux*, vol. 5, livraison 64 (1825): plate 380

IMAGE FROM THE BIODIVERSITY HERITAGE LIBRARY. DIGITIZED BY THE SMITHSONIAN LIBRARIES

The second key figure in Temminck's formation was Bernhard Meyer, a physician, dentist, apothecary, and naturalist from Offenbach, Germany. He was best known for his contribution to the botanical work *Oekonomisch-technische Flora der Wetterau* (1799), which he compiled with Gottfried Gärtner and Johannes Scherbius, as well as for his superb collection of birds, nests, and eggs. Temminck and Meyer probably met through their collections, but Meyer may have also been a friend of Jacob Temminck. In 1804, Coenraad Jacob Temminck and his wife Caatje stayed in Offenbach for six months as Bernhard Meyer's guests. During this visit Temminck also met the physician and naturalist Johann Philipp Achilles Leisler and, most likely, the ornithologist Johann Wolf. These meetings would later inspire Temminck to develop a line of research of his own.

At the time of Temminck's visit to Germany, Meyer and Wolf were working on a manual of the birds of Germany, the *Naturgeschichte der Vögel Deutschlands in getreuen Abbildungen und Beschreibungen*, which would be published in 1805 (the introduction is dated May 1804). This work was very different from Levaillant's writings with their elaborate descriptions and literary style. Meyer and Wolf set a list of rules of how to describe and depict the birds "with fidelity." First-hand observations and accurate descriptions and measurements were essential; Temminck had provided the authors with additional information on certain species.[43] The *Naturgeschichte der Vögel Deutschlands* included a systematic index, and each species description consisted of a list of references to the literature, a morphological description, and remarks on distribution, feeding habits, reproduction, predators, varieties, usefulness to man, and other comments. Overall it seems that in Offenbach, Meyer introduced Temminck to the methodology of naming and classifying species, i.e. the scientific and practical skills he had lacked.[44] A few years later Temminck drew his inspiration from Meyer's and other German works to write a treatise on European birds, his *Manuel d'ornithologie*, first published in Amsterdam in 1815. Temminck noted in the introduction that he had based his work on Meyer's classification of the *Naturgeschichte* while using additional works as a guide.[45]

---

43 Wolf and Meyer, *Naturgeschichte der Vögel Deutschlands*, 33, 38.

44 Vrolik, "Levensbericht van Coenraad Jacob Temminck," 67; Susanna, "Levensschets van Temminck," 50.

45 Temminck, "Avant-propos," in *Manuel d'ornithologie*, ix–x.

3 From Listing to Classifying

After his return from Offenbach, Temminck started updating and finishing the collection catalog he had started around 1800, soon after taking charge of his father's collection.[46] That first catalog of the collection provides a glimpse into the young Temminck's perception of ornithology. In an opening section entitled "About the class of birds in general" he described the problems he had faced while working on bird classification from his collection. While some ornithologists based their descriptions on coloration, which created confusion in species in which juveniles have different patterns than the adults, others based their classifications on habits and nutrition, which yielded equally unsatisfactory results. At this point, despite the fact that improvements were being made on Buffon's system, Temminck did not consider the achievement of a complete natural history of birds possible in the foreseeable future. Buffon had suggested that naturalists should study birds methodically – one country at a time.[47] Temminck agreed. In his view, the road ahead was long and complicated, and natural historians needed to combine classification with observations from the field describing each bird's life history – observations like those provided by Levaillant and Audebert.[48] This manuscript suggests that he yearned for accuracy, comprehensiveness, and structure in the systematic arrangement of birds.

Standardization preoccupied Temminck. After 1804 he applied himself to the manuscript catalog with the specific purpose of publishing it. He was encouraged by the many requests he received from interested naturalists who were keen to receive a description of his cabinet.[49] He set upon the task of updating his first manuscript catalog, putting what he had learned in Offenbach into practice. Temminck revised and re-arranged the groups of species and genera, abandoning Buffon's families, and decided to follow the arrangements by John Latham, Karl Illiger, Louis Vieillot, François Levaillant, Charles Sonnini, Jean Baptiste Audebert, and George Edwards instead.[50] The species, each with its unique catalog number, were listed by their French and Latin names. The manuscript catalog included 839 species, but the new inventory closed with number 1072. With each species, Temminck listed references to its existing

46 "Catalogue cabinet Temminck," n.d., Naturalis Biodiversity Center Archives, C.J. Temminck, NL-LdnNAT_Temminck_47. The manuscript catalog is the only surviving list of the Temminck collection as it was at the turn of the century.

47 Buffon, "Plan de l'ouvrage," in *Histoire naturelle des oiseaux*, vol. 1, xxxiv–xxxv.

48 Temminck, "Catalogue cabinet Temminck," 2.

49 Temminck, "Préface," in *Catalogue systématique*, iii.

50 Temminck, *Catalogue systématique*, v–vi.

descriptions and illustrations, as well as other publications. By the end of 1806 he had finished the inventory. The catalog appeared in Amsterdam in 1807 under the lengthy title *Catalogue systématique du cabinet d'ornithologie et de la collection de quadrumanes avec une courte description des oiseaux non-décrits suivi: d'une nôte d'oiseaux doubles et de quelques autres objets d'histoire naturelle offerts en échange.*[51]

This new, improved *Catalogue* was not merely a list of the specimens in Temminck's collection but included descriptions of a few species that were new for him. However, at that time, his primary objective was rather mundane: publicity. He wished to let amateurs, collectors of natural history, and European museum directors know, firstly, which specimens he had in his cabinet and, secondly, which ones were available for exchange.[52] He sought to augment his collection not with more specimens but with new species. For this purpose he offered his duplicates, which he listed in an appendix, in exchange for species he did not yet possess. This plan was certainly successful. In 1810 Temminck explained to the German zoologist Johann Karl Wilhelm Illiger how the *Catalogue* had helped him build a network of collectors and naturalists for specimen exchange. Publishing a catalog of his collections was, it seemed, yielding fruit. Unfortunately, Temminck lost two shipments of birds, one from Africa and the other from Sulawesi and Java, as the American ships that carried them were wrecked at sea.[53]

Any species new to Temminck were described in a special section at the end of the book, titled "Caractères physiques des oiseaux non-décrits" (Physical characters of undescribed birds). Although he was following Linnaean principles in the catalog listings, Temminck described the new species without giving them a Latin binomial; with few exceptions, these new species received only a French name. Why exactly he decided to do this is unclear. Perhaps Temminck lacked confidence at the time, or perhaps he planned to describe these species later. Whatever the reason, this proved to be an unfortunate decision: the French ornithologist Louis Pierre Vieillot later gave the same species Latin names without acknowledging Temminck, effectively appropriating his work.

One example is a cuckoo specimen from Java, number 380 in Temminck's collection. Temminck could not find a reference to this species in the existing literature, and therefore decided to describe it as a new species, naming it *Coucou roux à bec rouge*. And that was the end of the matter, at least for a

51 Jansen, "Unpublished 'Catalogue du Cabinet de C. J. Temminck'"; Stresemann, "Analyse."

52 Temminck, *Catalogue systématique*, iii.

53 Temminck to Illiger, 4 May 1810, in Stresemann, "Briefwechsel von Temminck mit Hoffmannsegg und Illiger," 225.

while. Ten years later, Vieillot described the same bird as a new species and gave it a Latin binomial: *Cuculus melanogaster*. The descriptive characters he used were the same as Temminck's, as was the country of origin of the species, Java. However, the locality was erroneous, as this particular species lives in Central and South America – a detail that led Temminck to miss previous descriptions of the cuckoo, and that also escaped Vieillot.[54] In fact, this cuckoo had been known to European naturalists for almost half a century as the Cayenne Cuckow (and it is now known as *Piaya melanogaster*). It had been described before by Mathurin-Jacques Brisson as *Cuculus cayanensis*, by Buffon as the *Coucou de Cayenne*, by Johann Friedrich Gmelin as *Cuculus cayanus*, and by Latham as the *Cayenne Cuckow*.[55] It is likely that the erroneous locality for specimen no. 380 had misled Temminck, an error that Vieillot simply duplicated. Temminck was quick to respond in 1817, sparing neither examples nor expletives, in a small publication with the long and telling title *Observations sur la classification méthodique des oiseaux: et remarques sur l'Analyse d'une nouvelle ornithologie élémentaire par L. P. Vieillot, auteur de divers ouvrages d'ornithologie, et un des collaborateurs du Nouveau dictionnaire d'histoire naturelle*. Vieillot's actions had started a life-long hostility.

The cuckoo was, unfortunately, not an isolated case. Without a Latin binomial, Temminck's newly described species were not accepted as valid discoveries, and other ornithologists who followed the Linnaean rules of nomenclature soon claimed authorship for these new species. Some of them may have been unaware of Temminck's catalog while others proceeded covertly without a reference to Temminck. This was a blow that would haunt Temminck for a long time.

The manuscript catalog and the notebooks that Temminck wrote while preparing the final version of the *Catalogue systématique* are testimonies of the evolution of his thought. The process apparently started with the numbering of the collection objects and culminated in systematics. The main force for Temminck's pursuit of knowledge was his cabinet. Its arrangement, cataloging

54 Hartlaub, "Systematic Index," vol. 2, 1–7; Temminck, *Catalogue systématique*, 58, 208; Vieillot, "Le Coucou à ventre noir," in *Nouveau dictionnaire*, 236–37. Van den Hoek Ostende, Dekker, and Keijl (in "Type-specimens of Birds, Non-Passerines," 147) have confirmed that Vieillot's description was based on Temminck's description, and therefore, on Temminck's specimen no. 380 (now RMNH.AVES.88261).

55 Brisson, *Ornithologie*, vol. 4, 122, pl. VIII; Buffon, *Histoire naturelle des oiseaux*, vol. 12, 81; Buffon, Daubenton, and Martinet, *Planches enluminées d'histoire naturelle*, vol. 3, pl. 211; Latham, *General Synopsis of Birds*, vol. 1, 542; this species was also described by Gmelin in the 13th edition of Linnaeus, *Systema naturae*, vol. 1, 417.

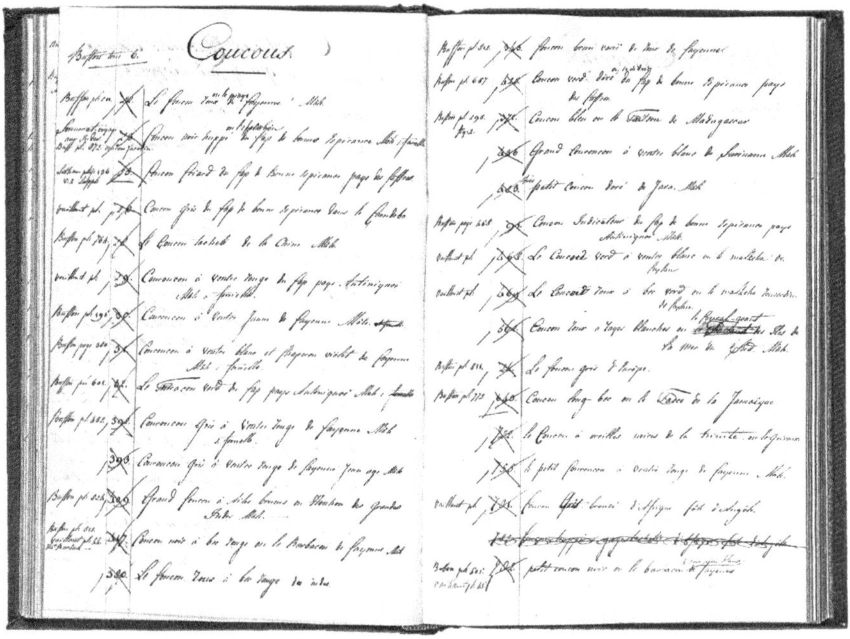

FIGURE 6 *Coucous*, in "Catalogue du cabinet de C. J. Temminck," Coenraad Jacob Temminck's manuscript catalog of his collection, ca. 1800–1804. He was unable to identify no. 380 (*Le Coucou Roux à bec rouge*) and no. 393 (*Coucou Gris à ventre rouge*)
NATURALIS BIODIVERSITY CENTER ARCHIVES, NL-LdnNAT_TEMMINCK_47_050

the specimens, grouping males, females, and juveniles by species, and placing related species in the same group fueled his interest in systematics. By focusing on classification instead of traveling like Levaillant, Temminck's thoughts were necessarily directed towards the acts of naming, ordering, and arranging. Describing new species was a natural consequence of ordering them, as he encountered species not mentioned in other works. After this point, Temminck's interest moved from collecting to publishing.

Browsing through the existing literature, Temminck concluded that one of his favorite groups of birds required renewed attention. In 1804, influenced by what he had learned from Meyer and Leisler, Temminck started working on a monograph on the gallinaceous birds (game birds, fowl, and poultry). He announced it already in the *Catalogue systématique*: "its title will be *Histoire générale et complette* [sic] *des Gallinacés* […] and in it, the Amateurs will find about eighty new species, most of them never mentioned before by any other author."[56] By 1806 the Swiss-French natural history painter Jean-Gabriel Prêtre, who was a friend of Levaillant, had prepared some of the plates for the latter's work.[57] Prêtre had also worked for Vieillot, illustrating his *Histoire naturelle des plus beaux oiseaux chanteurs de la zone torride* (1805) and *Histoire naturelle des oiseaux de l'Amérique septentrionale* (1807). It was probably Levaillant who introduced the artist to Temminck. Prêtre's credentials as an artist were more than satisfactory: apart from being Levaillant's friend and collaborator, he was also one of the natural history illustrators officially employed at the Muséum National d'Histoire Naturelle in Paris.

In preparation for the monograph, Temminck, accompanied by his wife, Caatje, traveled to Paris to meet Prêtre. They arrived on June 29, 1807, and stayed for a month.[58] Between visits to the Panthéon, the "Muséum Napoleon," and the opera, Temminck met two of his father's friends, Levaillant and Jean-Baptiste Leschenault de la Tour. The latter provided Temminck with bird skins and notes from his voyage to Java for his planned book.[59] Levaillant thought that Temminck was the person best suited to taking on a monograph on fowl, so he left the treatment of several beautiful species yet to be described to Temminck; he was someone, Levaillant thought, who would spare no effort in the work.[60] Temminck had, indeed, collected all specimens he could find for his cabinet and had traveled across Europe to visit as many collections as possible. His careful comparison of numerous specimens before publishing was, in Levaillant's view, what marked Temminck as the right person to write a monograph on fowl.[61] Indeed, Temminck's attention to minute detail would turn out to be his greatest strength.

Temminck and Caatje also visited the Muséum National d'Histoire Naturelle, where Temminck was able to discuss the work on the plates with Prêtre. Another person resident in Paris at that time was Mlle Pauline de Courcelles, a

---

56 Temminck, "Preface," in *Catalogue systématique*, vi. Temminck probably used the term *amateurs* here in the sense of "enthusiasts," and not in the modern sense of dilettantes or non-professionals.

57 Stresemann, *Ornithology*, 116.

58 Van Lynden-de Bruïne (1922–2016), the great-great grandchild of Coenraad Jacob Temminck, transcribed the diaries of Temminck's first wife Caatje and published them in 2001 under the title *In vogelvlucht door Europa: de reisjournalen van Dionysia Catharina Temminck-Cau, 1807–1824*. Caatje wrote her captivating diaries between 1807 and 1824 during the couple's travels through Europe. The original diaries are in the Temminck Family Archives, A. M. van Lynden-de Bruïne private collection.

59 Temminck, "Introduction," in *Pigeons et gallinacés*, vol. 1, 12. Leschenault de la Tour's birds were listed in Temminck's *Catalogue systématique*. See Stresemann, "Analyse."

60 Levaillant, *Oiseaux d'Afrique*, vol. 6, 118.

61 Levaillant, 118–19.

gifted aquarellist and a disciple of Jacques Barraband, who had illustrated several of Levaillant's books. Temminck and de Courcelles met, and the meeting changed Temminck's plans. They decided to collaborate.

Temminck diverted from his original plan for a monograph on gallinaceous birds to include pigeons. Confident in his new partner, Temminck embarked on a tour through Switzerland and Italy before returning to the Netherlands, leaving the supervision of the engravings and the printing of the part dedicated to pigeons to de Courcelles. The first volume was dedicated to King Louis Napoleon, and it included a *rapport* by Georges Cuvier and the Comte de Lacépède dated June 15, 1807. It is a letter of commendation, in which these two famous naturalists testify to the artistic and technical quality of the plates by de Courcelles. The earliest installments appeared in 1808, but they were not exactly what Temminck had expected, and the partnership broke up amidst allegations of theft and piracy.

Between 1808 and 1811 fifteen installments on pigeons appeared. While the Temmincks traveled, Pauline de Courcelles altered the installments of the *Histoire des pigeons* in such a way that she appeared to be the sole author. De Courcelles – now married and known by the name of Mme Knip – had removed the index, Temminck's dedication to King Louis Napoleon, and by the ninth installment about forty pages of text. She also gave the work a new title, *Les Pigeons par Madame Knip, née Pauline de Courcelles, le texte par C. J. Themminck*. The fact that Temminck's name was misspelled on the title only added insult to injury. Knip arranged for Temminck to receive a few copies with the original text and unaltered title, ensuring his ignorance of the changes she made while presenting her version to Queen Marie Louise. As a result, two different versions of the book appeared: one authored by Knip, and the original as it had been conceived by Temminck with the complete text and his index, the *Histoire naturelle générale des pigeons, par C. J. Temminck*. Of this second version only twelve copies were printed.[62]

Temminck discovered Knip's deception when he traveled to Paris to prepare the next two volumes on gallinaceous birds sometime in 1811 or 1812. He was, understandably, infuriated. He tried to expose the deception and the artist – unsuccessfully. After all, Knip enjoyed the patronage of the queen:

62 The twelve copies from the original version included eight sets that de Courcelles herself sent to Temminck, plus four other copies. Temminck's own copy is kept at Naturalis Biodiversity Center. See Dickinson et al., "Coenraad Jacob Temminck versus Pauline Knip," and Temminck's own version of the episode in Temminck, *Pigeons et gallinacés*, vol. 3, 640–43.

> All means I employed to appeal against such an arbitrary act were without effect, and my voice could not rise against an intrigue supported by powerful protectors; the journalists refrained from placing my claims in their papers; even my response to the article by the new author in these newspapers was banned.[63]

At this point the collaboration ended. Temminck adopted a defensive strategy, writing to his friends and contacts and informing them of what had happened. If anyone wanted to purchase the *Pigeons* "en grand format," he could provide them with the original fifteen installments – at an exorbitant price of forty francs each – without the arbitrary changes of the other edition. Temminck admitted defeat and, reluctantly, to his naiveté: "the annoyance I have earned as a reward for my disinterestedness towards the Publishers must be placed among the familiar events of the day."[64] Her deception earned Mme Knip the patronage of the Impératrice-Reine Marie Louise and the title of *premier peintre* – much in the same way as Temminck's dedication of the same work to Louis Napoleon had helped him win his monarch's favors.

Temminck picked up his original idea of a monograph on fowl, combined it with what he had already written on pigeons, and published the *Histoire*

FIGURE 7 Coenraad Jacob Temminck and Pauline Knip-de Courcelles, *Histoire naturelle générale des Pigeons* (1808). Temminck's personal copy, one of the twelve copies made of the book without the alterations made by Pauline Knip-de Courcelles
NATURALIS BIODIVERSITY CENTER, RBR D00216. PHOTOGRAPH BY LIZA LANKHAAR

63 Temminck, *Pigeons et gallinacés*, vol. 3, 643.
64 Temminck to Johann Centurius von Hoffmannsegg, 29 July 1812, in Stresemann, "Briefwechsel von Temminck mit Hoffmannsegg und Illiger," 258.

naturelle générale des pigeons et des gallinacés, which appeared between 1813 and 1815. It was a much more modest work than he had originally intended, published in three volumes (the first presenting the original text on pigeons) and illustrated with merely eleven engraved plates showing anatomical features. Temminck did not include drawings of specimens, possibly because he wished to publish the work as quickly as possible. Quite unexpectedly, his conflict with Madame Knip worked to Temminck's advantage.

Since Temminck published the *Histoire naturelle générale* in octavo, without colored plates and with just a few engravings of anatomical details, the three volumes were immediately much more accessible to many naturalists than the luxurious and costly folio edition. Karl Illiger, by then director of the Zoological Museum in Berlin, tried to mitigate Temminck's frustration with the rather humble revised edition of the book: "You must have been so annoyed [with what happened] that you had it reprinted. Anyway, I congratulate ourselves, poor ornithologists, to whom a sumptuous publication is out of bounds, that as a result, your undoubtedly valuable remarks on an interesting class [of birds] will now be accessible."[65] Indeed, the new monograph was received with great enthusiasm among naturalists and secured Temminck some prestige among ornithologists.

4 Temminck's Earliest Monographs

From 1813 onward, beginning with the publication of the *Histoire naturelle générale des pigeons et gallinacés*, Temminck's name became as familiar as Linnaeus', Cuvier's, and Brisson's to ornithologists.[66] His peers agreed in their praise of Temminck's accurate definitions of genera, his precise treatment of the species, and his unraveling of the synonymy as the most significant merits of his work. The English naturalist William Swainson, for example, wrote in 1836: "M. Temminck's Monographs of the *Gallinaceous* Birds are clear and masterly, and should be taken as a pattern for all such dissertations. The description of the species [...] are scientific and accurate, while, in the *Synopsis*, each is technically characterized by a short specific diagnosis in Latin."[67]

Temminck's monograph raised the standards of the time. The introductory sections were full of pious dedications and idealistic notions, encouraging

65 Illiger to Temminck, 14 September 1812, in Stresemann, "Briefwechsel von Temminck mit Hoffmannsegg und Illiger," 260.
66 Farber, *Emergence of Ornithology*, 85–86.
67 Swainson, *Natural History and Classification of Birds*, vol. 1, 205–06.

his fellow-men to pursue enlightenment through the observation of nature. Nature, according to Temminck, brings man closer to God and keeps the young away from unspecified but tantalizing "séductions."[68] Temminck also included a number of pragmatic remarks about nature in general and pigeons and fowl in particular. Pheasants, peacocks, quails, roosters – they were as ornamental as they were delicious. He here followed the tradition of earlier works on natural history. Moral and aesthetic comments would remain a fundamental justification for writing natural history books until the second half of the nineteenth century.[69] Judging from Temminck's previous work and from his general disinclination to apply himself to philosophical quests, however, the introduction appears to have been primarily an attempt to justify his particular interest in describing and classifying birds. Towards the end of the general introduction, after almost twenty pages of discourse, Temminck explicitly stated that the prevailing disorder in these groups of birds needed to be revised, the genera needed to be reconsidered, and the nomenclature disentangled. He did so with such rigor and accuracy, far progressed from the *Catalogue systématique* and spurred by the lessons he had learned with Vieillot's plagiarism and Knip's theft, that his reputation greatly increased, from that of a cabinet owner to that of a skilled ornithologist.

The second volume of this first monograph, the *Histoire naturelle des gallinacés*, was published in 1815, almost simultaneously with Temminck's following work, entitled *Manuel d'ornithologie, ou tableau systématique des oiseaux qui se trouvent en Europe; précédé d'une analyse du système général d'ornithologie, et suivi d'une table alphabétique des espèces*. Inspired by ornithologists like Latham, Illiger, Meyer, and Wolf, Temminck had worked on this handbook of European birds for years. The *Manuel d'ornithologie* was the first comprehensive treatise on the avifauna of Europe to be published. Here Temminck included all known species as well as descriptions of new species, revised the synonymy, and provided extensive references. For each species, the relevant entry was based exclusively on information gleaned from actual specimens, all seen by him while traveling around Europe between 1807 and 1814.[70]

Temminck finished the *Manuel* towards the end of 1814.[71] In its introduction he complained that although many excellent works were devoted to ornithology, none covered all European birds. During this period a mania for exotic animals had led to the publication of numerous treatises on new foreign species,

68 Temminck, "Discourse préliminaire," in *Pigeons et gallinacés*, vol. 1, 9.
69 Farber, *Emergence of Ornithology*, 137.
70 Temminck, "Avant-propos," in *Manuel d'ornithologie*, viii–xi
71 Temminck, "Dédication," in *Manuel d'ornithologie*.

to the detriment of new knowledge on the European fauna. Levaillant also complained – somewhat ironically, as he worked almost exclusively on tropical birds – about how French naturalists had ignored their local avifauna and preferred exotic ornithology.[72] Temminck pointed out that European species seemed to be forgotten by naturalists:

> One goes to search in the regions of the torrid zone or to the ice of the poles to add to the numerous species already known, by means of which one augments the nomenclature catalogues without any goals useful to science. They are sterile acquisitions, that amateur curiosity collectors may esteem, but which will be, for a long time, foreign to the domain of science.[73]

Realizing that a complete treatise on European birds was long overdue, Temminck decided to produce one, as it would be of interest to amateurs and collectors alike. He provided "not only concise and precise descriptions of each species, but also of each variety, both sexual and age variations, as well as those that are accidental."[74] By the time Temminck wrote the *Avant-propos*, he had abandoned the philosophical remarks of 1813. There were no religious dedications, no attempts to convince the reader of any moral gain from the study of nature. The description of the European birds was, in itself, sufficiently useful; it did not demand any practical or spiritual justification. In the *Avant-propos* Temminck substituted all philosophical remarks with explanations of his views on nomenclature, the need to examine seasonal and age-related variation in birds, and the need to review existing genera, followed by an *Index systématique* summarizing his classification system. Systematics was, it seems, coming of age.

Both monographs were enthusiastically received in continental Europe and Britain and became ornithological reference works within a few years. Reviewers praised them for the accurate and exhaustive synonyms list, as well as Temminck's ability to delimit both genera and species with consistent use of characters. According to William Swainson, "The best and *neatest* generic definitions are those of Temminck's *Manuel*: the most verbose and over-labored, those of the *Species Avium* [Wagler's]. Even those of Illiger are too tedious."[75]

72 Stresemann, *Ornithology*, 296.

73 Temminck, "Introduction," in *Manuel d'ornithologie*, 2nd ed., vol. 1, i–ii; translated in Farber, *Emergence of Ornithology*, 85.

74 Temminck, "Avant-propos," in *Manuel d'ornithologie*, viii.

75 Swainson, *Natural History and Classification of Birds*, vol. 1, 246 (emphasis original).

The *Histoire naturelle des pigeons et gallinacés* and the *Manuel d'ornithologie* were routinely cited as references in practically every ornithological treatise published in the following decades.[76]

Interestingly, in the *Manuel*'s "Avant-propos" Temminck distinguished between those "studying ornithology" and the "amateurs."[77] This remark reflects the changes taking place in natural history during the first two decades of the nineteenth century. Luxurious, fashionable books were still very much in vogue while monographs and more specialist works started to appear. Like Meyer's, Temminck's monographs were born out of the necessity of arranging and ordering a group of birds (pigeons and fowl), or in the case of the *Manuel d'ornithologie*, to complete existing regional avifaunas. They were, in fact, a continuation of the traditional encyclopedic works, but Temminck also recognized the impossibility of including all birds in one publication:

> How can anyone dare to claim that a single man can embrace the study of the whole creation? To describe with precision and accuracy a single class of the animal kingdom, the life of the most active man would barely be enough. The best means that naturalists can choose to disseminate the results of large numbers of facts seems to be by publishing *Monographs*.[78]

As ornithological books and species names multiplied rapidly, systematic arrangements differed from each other, and the nomenclature became more and more complex to follow. The need arose for works that provided a good overview of existing names, compared published systems, carefully described species with no room for misinterpretation, and were illustrated with accurate depictions of species based on real specimens. Up to this point Temminck's work had helped fuel the development of classification and nomenclature by standardizing methods and synonymy. While he was leaving Levaillant's methods and views behind, the goal remained mostly the same. At this stage of his career Temminck hoped to achieve an exhaustive and synthetic knowledge of birds. This was, however, not his original goal but rather the by-product of his practical work on his collection – enlarging and arranging it – and his ambition to obtain and describe all existing species.

---

76 Examples of ornithological works based on Temminck's classifications are Gould, *Birds of Europe*; Jardine, *Natural History of Game-Birds*; Latham, *General History of Birds*; Naumann and Naumann, *Johann Andreas Naumann's Naturgeschichte der Vögel Deutschlands*; and Swainson, *Zoological Illustrations*.

77 Temminck, "Avant-propos," in *Manuel d'ornithologie*, viii.

78 Temminck, "Discours préliminaire," in *Monographies de mammalogie*, vol. 1, viii–ix.

The popularity and status that Temminck enjoyed after publishing the *Histoire naturelle des pigeons et gallinacés* and the *Manuel d'ornithologie* would later help him achieve another, more prosaic goal: a paid career in natural history. As the practice and methods of classification evolved into a more rigorous, accurate discipline, its practitioners sought to be recognized as authorities and professionals. In Temminck's time, however, what constituted a profession and who was qualified to practice it was far from clear-cut. His career illustrates the state of affairs at the time.

CHAPTER 2

From Collector to Director

After Temminck saw his first two books published, and more importantly, received acclaim, he was determined to pursue natural history as a profession. Although his social prestige provided the opportunity of occupying a comfortable post as a civil servant, Temminck soon abandoned that path to follow his vocation. However, for a patrician naturalist without academic qualifications a professorship at the universities was not feasible. Putting his background and skills as an author and curator to work, he sought contacts in learned societies and national museums, probably hoping to open up an appointment, and preferably one that was both prestigious and remunerated. It is possible that his family fortune diminished after the dissolution of the VOC, or perhaps Temminck understood the prestige attached to a professional affiliation. Conveniently, he held influence in social and political circles, mainly due to his family connections. Temminck's plan succeeded. His experience with his natural history cabinet, in combination with his rising status in zoological circles and his relentless character, helped him to create a position he would occupy for nearly forty years. His path proved to be an obstacle race.

1 Appointments and Politics

Temminck dedicated the successful *Manuel d'ornithologie* to the directors and members of the Hollandsche Maatschappij der Wetenschappen (Holland Society of Sciences) in Haarlem, as a "tribute to the association of dedicated scholars and protectors, whose work has the noble purpose of enlightening, protecting and encouraging the sciences with the light of their spirit."[1] Temminck knew the Society of Sciences well. Ten years before the publication of the *Manuel* he had been selected to be one of the directors of the Hollandsche Maatschappij, on June 22, 1805.[2] It was an honor, and he signed his earliest publications, as "Directeur de l'Académie Royale des Sciences et des Arts de Haarlem."

1 Temminck, "Dédication," in *Manuel d'ornithologie*.

2 Koninglijke Maatschappy der Wetenschappen te Haarlem, "Directeuren van de Koninglijke Maatschappy," xxxi. Temminck remained a Director at least until 1813. By 1820, the second edition of the *Manuel* did not include Temminck's title as Director.

 | DOI:10.1163/9789004441491_004

The board of directors was a group of influential men interested in science and was in charge of the management and the finances of the society. The directors held important posts in the government and city administrations, the VOC and the WIC, and formed a collective of patrons of science – which raises the question of why Temminck's father, Jacob Temminck, was never one of them.[3] The scientific work was carried out by 'ordinary' members who did not contribute financially to the society and had no influence on its strategy.[4] Members from other countries were also welcome to join the society, which had an international outlook from its foundation. The society's meetings provided a platform for scientific discussion and the exchange of news. Scholars joined discussions during their meetings in Haarlem and had the society's journal, *Natuurkundige Verhandelingen*, at their disposal for the publication of meeting minutes and scientific reports.

Temminck owed his role as a Society director to his social position and his generous donation to the society, although the fact that he was known as a competent manager of natural history specimens may also have been a contributing factor. The society owned a cabinet of natural history intended for study by the members. At the time of Temminck's appointment the director of the cabinet was Martinus van Marum, one of the most influential men of science in the Netherlands. A physician, chemist, inventor, and natural historian, Van Marum was also the secretary of the society and the director of Teylers Museum.[5]

The Haarlem Society was well-established and had a good reputation in the Netherlands as a center of knowledge in general, and of science in particular, and was ranked among the most active and sought-after learned societies in the country.[6] In 1806 King Louis Napoleon had tried to merge the Hollandse Maatschappij der Wetenschappen – then under Martinus van Marum's direction – with all other Dutch scientific societies to create a single organization, a Royal Institute for Science, following the example of the Institut de France. Louis Napoleon planned to dismiss the directors of the Haarlem Society from their duties, but they resolutely rejected the king's project. In 1808 Napoleon founded an entirely new society, the Koninklijke Instituut van Wetenschappen,

3 Van Berkel, *De stem van de wetenschap*, vol. 1, 34.

4 This structure is still in use today, with three hundred directors and more than three hundred and fifty members. See, for instance, Bierens de Haan, *De Hollandsche Maatschappij*.

5 Martinus van Marum published a catalogue of the cabinet in 1806. It included 130 mounted mammals, 525 birds, 100 reptiles and amphibians, 50 snakes, almost 600 mollusks, 220 kinds of corals and sponges, and 70 kinds of crustaceans and worms. See Bierens de Haan, *Geschiedenis van een verdwenen Haarlemsch museum*.

6 Bierens de Haan, *De Hollandsche Maatschappij*; Weiss, "The Masses and the Muses."

Letteren en Schoone Kunsten.[7] It was a forum for exchange and networking for and by academics and scholars, and it looked like there was no place for Temminck.[8]

After his coronation Louis Napoleon surrounded himself with important men and aimed to win the approval of Dutch society. Among other things he established the Orde der Unie in 1806, which was a chivalric order recognizing those who had proved their value for the nation and the king through scientific, artistic, academic, or economic achievements.[9] Somehow Temminck attracted the king's attention: on August 25, 1808, Temminck was appointed *Chambellan* of the king. This was an honorary title granted to nobility and civilians of outstanding achievements, and Temminck had dedicated his *Histoire naturelle générale des pigeons* to the king, the "Protecteur des Arts, par l'estime et le respect le plus profound" (Protector of the Arts, with the deepest esteem and respect).[10]

Just two days later Temminck was knighted and named *Ridder van de Orde der Unie* (Knight of the Order of the Union), and a week after that he was appointed the interim director for Louis Napoleon's Koninklijk Kunst Museum, a position he held until its new director, Cornelis Apostool (who at that time traveled abroad) was ready to occupy the post.[11] Among his other duties Temminck was to compile a catalog of the museum's collection. He remained in the Kunst Museum until January 1809, when Apostool became the formal director.[12] Within only ten days Temminck had received two honorary titles as the king's advisor and as a knight of the Order of the Union, and immediately afterward the interim directorship of the Museum. It is possible that the titles were means to facilitate – or maybe, to justify – Temminck's appointment at the museum.

---

7 Now the Koninklijke Nederlandse Akademie van Wetenschappen. Sliggers, "Van individu tot instituut," 9.

8 See, for example, the role the Institute played in the career of Caspar G. Reinwardt, a professor of chemistry and natural history at the Athenaeum Illustre (Amsterdam) and a close friend of Van Marum's, in Weber, "Hybrid Ambitions," chap. 3.

9 Koninklijke Orde der Unie, "Beschrijving van de plegtigheden," in *Gedenkschriften van de Koninklijke Orde der Unie*, 174.

10 Temminck, "Dédication," in *Pigeons*.

11 Van Lynden-de Bruïne, *In vogelvlucht*, 16. Communication on the Royal Decree from Louis Napoleon to Coenraad J. Temminck, September 4, 1808, Naturalis Biodiversity Center Archives, Archive C. J. Temminck.

12 Apostool was appointed director of the Koninklijk Museum in Amsterdam on August 25, 1808 but did not effectively take the position until 1809. Moes and Van Biema mistakenly reported that Egbert Temminck, *Inspecteur der Nationale Gebouwen*, was appointed as interim director (*De Nationale Konste-Gallery*, 120).

Louis Napoleon's choice of Temminck as deputy director may have been suggested either by Everhardus Temminck (probably Coenraad's cousin) or by Marinus Smissaert, Temminck's future father-in-law. Everhardus Temminck (1758–1837) was the inspector-general of national domains and national buildings, and a supervisor at the Dutch Nationale Konst-Galerij, one of the collections that Louis Napoleon incorporated into the new Koninklijk Kunst Museum.[13] Marinus Smissaert was also a Knight of the Order of the Union, and since June 1808 he had been a steward at the Royal Palace in Amsterdam. He was in charge of supervising the move of the Louis Napoleon's royal art collections for the planned Koninklijk Museum, from Huis ten Bosch in The Hague to the Royal Palace on Dam Square in Amsterdam.[14] Smissaert also worked as a hunting officer for Louis Napoleon in Amstelland between 1807 and 1808. Since Temminck was also a hunting enthusiast, this may have been a common ground for them to meet.[15] Smissaert and his wife, Maria Feitama, *Dame du palais* to queen Hortense, are known to have been close to the Temmincks. Caatje described Maria Feitama as a "true friend."[16] Whether Temminck and Smissaert knew each other before 1808 or whether the friendship blossomed only after Temminck's appointments, is unclear. In either case, Temminck could make good use of these titles to promote his career in natural history. For a start the royal appointments served to secure him an appointment as the manager of a national collection.

1808 was a momentous year for Temminck. On September 25 he joined the Société Impériale des Naturalistes de Moscou as an ordinary member, and on November 30, he became a member of the Wetteranische Gesellschaft der gesammte Naturkunde in Hanau. These were just the initial two of a long list of learned societies that Temminck would join. Between 1808 and 1850 he became a member of no fewer than thirty-eight different societies and academies, all dedicated, at least in part, to the study of natural history. They were located anywhere from Washington to Saint Petersburg and from Cape Town to Batavia (now Jakarta), including a considerable number of European societies.[17] As for many other amateur naturalists, networking was essential for Temminck.

---

13 Moes and Van Biema, *De Nationale Konste-Gallery*, 120.

14 Moes and Van Biema, 114.

15 "Portret van Marinus Adriaan Perpetuus Smissaert (1773–1819)," R. K. D. Nederland Instituut voor Kunstgeschiedenis, last modified January, 2015, https://rkd.nl/nl/explore/images/151688.

16 Van Lynden-de Bruïne, *In vogelvlucht*, 21; Molhuysen and Blok, "Smissaert," 1329.

17 The original documents are in the Naturalis Biodiversity Center Archives, Archive C. J. Temminck.

Temminck retained the position as the king's advisor until July 1810, when the Louis Napoleon's rule ended and Napoleonic France incorporated the Dutch territories. Despite the volatile situation in Europe, Temminck and Caatje traveled to Germany and Switzerland between May and October 1811. There are no records of any official engagements for Temminck until February 1812, when he obtained a job with the French government as "entreposeur principal des Tabacs pour le Département des Bouches de la Meuse" (main warehouse-keeper of tobacco for the department of Bouches-de-la-Meuse) which, in Holthuis' words, "considering the little amount of tobacco that reached the country, cannot have been very time consuming."[18]

Despite his engagements with the French government, Temminck was a proud Dutchman. After the fall of the French emperor in 1813 he wrote to Baron Cornelis R. T. Kraijenhof, a military commander in Amsterdam (and Temminck's neighbor) and volunteered to serve in the cavalry, as a captain, to defend the nation's independence. While waiting for Kraijenhoff's answer, Temminck contacted his friend Anton Reinhard Falck, a statesman under King Louis Napoleon and advisor to Prince Willem I.[19] Both Anton's and Coenraad's fathers worked for the VOC, and the sons had been friends since childhood. In his letter to Falck, Temminck expressed his nationalist sentiments. He had served under Louis Napoleon and accepted honors from the French government "out of financial necessity, but this has not stopped me from being a Dutchman."[20] It is doubtful whether Temminck had a need for a remunerated job. In any case, with Falck's recommendation, Temminck joined the regiment of Lieutenant-Colonel Willem François Boreel.

After the return of the Prince of Orange from England, Temminck resumed his work in his cabinet, but not for long. His studies were again interrupted by the unstable political situation in Europe. Napoleon escaped from Elba, and Temminck volunteered once more for military service, now for the War of the Seventh Coalition. In 1815 he organized a cavalry regiment, which was on active service in Paris.[21] Temminck also paid for the regiment, which makes his earlier claim of having found himself in financial straits rather questionable.

---

18 Holthuis, "Three Dutch Authors of von Siebold's *Fauna japonica*," 693. Bouches-de-la-Meuse (now in the present-day Dutch province of South Holland) was one of the nine departments formed when the Kingdom of Holland was annexed by the First French Empire in 1808.

19 Susanna, "Levensschets van Temminck," 56.

20 Temminck to Anton Reinhard Falck, February 1813, Naturalis Biodiversity Center Archives, Archive C. J. Temminck; Holthuis, *Rijksmuseum*, 19.

21 Holthuis, "Three Dutch Authors of von Siebold's *Fauna japonica*," 694.

Peace was finally established, and the sovereign prince proclaimed himself King Willem I of the Netherlands in March 1815.

2 The Direction of 's Lands Kabinet

Towards the end of 1815 Temminck, together with Martinus van Marum, was appointed interim director of 's Lands Kabinet of Natuurlijke Historie, formerly Louis Napoleon's natural history collection in Amsterdam, or the Cabinet du Roi. The collection consisted mainly of minerals, shells, and insects, while vertebrates were only a tiny fraction of the whole.[22] Since 1808 's Lands Kabinet had been under the direction of Caspar G. C. Reinwardt, who became a professor of natural history, chemistry, and botany at the Athenaeum Illustre in Amsterdam in the summer of 1810. He was a member both of the Koninklijk Instituut and of the Haarlem Society, and a good friend of Van Marum's.[23] Reinwardt himself had recommended Temminck and Van Marum for their appointments. Temminck and Reinwardt had met each other a few years earlier, while Temminck worked on the *Histoire naturelle des pigeons et gallinacés* and the *Manuel d'ornithologie*. Temminck had consulted Reinwardt on details of bird anatomy.

In December 1814 Reinwardt received a letter from Anton R. Falck – by then the state secretary for the provisional Dutch government – asking him to accompany the commissioner-general, Cornelis Theodorus van Elout, to the newly formed Dutch East Indies. After the restoration of the Dutch colonial empire by the Treaty of London of August 13, 1814, the king and his government had turned their attention to the 'East Indies,' intending to restore colonial administration. The government turned to Reinwardt to investigate the most pressing issues in the colonies in the fields of science, agriculture, and industry.[24] Falck offered Reinwardt the position of advisor to the colonial government, with the sweeping title of "Director for Agriculture, Arts, and Sciences in the Island of Java and Neighboring Islands." Reinwardt accepted, and in October 1815, before leaving for the colonies, he wrote to the minister of home affairs recommending that Van Marum and Temminck – both men of experience with collections – replace him during his absence. Reinwardt was to remain the director of 's Lands Kabinet. He presented Temminck as the

22 For more information about the Cabinet du Roi see Smit, Sanders, and Van der Veer, *Hendrik Engel's Alphabetical List*, 164; Sliggers, "Van individu tot instituut," 9–10.

23 Weber, "Hybrid Ambitions," 98–101.

24 Weber, 112.

Director of the Hollandsche Maatschappij, the author of several works on natural history, and the owner of an excellent collection of birds and mammals.[25]

Temminck's and Van Marum's tasks consisted of curating the collection, expanding the collection, and supervising the renovation of the Trippenhuis, the building where the collection was to be moved from the Royal Palace in March 1816.[26] They would be assisted by Anthony Johannes d'Ailly, who worked with the insect collection, and Joachim Sandra on the mineralogy collection. Reindert Draak assisted with the vertebrates and learned taxidermy from Temminck along the way.[27] Reinwardt had recommended all these men to Falck. Van Marum took care of the finances and the insects, while Temminck was responsible for the rest of the collection.[28] This division of domains, however, could not prevent an enduring disagreement between the two directors. If Van Marum was too proud to share the command, so was Temminck.

Only four months into their appointment as deputy directors of 's Lands Kabinet van Natuurlijke Historie, Van Marum and Temminck were already quarreling. To start with, they disagreed about who was in charge. Van Marum stated that Reinwardt had explicitly told him he was to be his successor; Temminck declared that he knew nothing about such agreement. The discussion was not, as one might expect, about gaining control, but, curiously, about avoiding responsibility. Van Marum had not wanted to take up the position – apparently, he had accepted only to please Reinwardt – and he wished to leave the management of the collections to Temminck. But Temminck was not interested, either. He had no desire to be in charge of the minerals, fossils, mollusks, and other invertebrates. These, he thought, should be in Van Marum's care, with Temminck only in charge of the vertebrates. While Van Marum thought this disagreement was very unpleasant, Temminck remained calm and tried

---

25 Reinwardt to the minister of home affairs, 4 October 1815, NL-HaNA, BiZa 1813–1848, 2.04.01, inv.nr. 4032A; Koolmees, "Caspar Georg Reinwardt," 46.

26 Van Marum to Reinwardt, 21 June 1816, NHA, Archive Martinus van Marum 529-20b; Van Heiningen, *Correspondence of C. G. C. Reinwardt*, 143. Holthuis (*Rijksmuseum*, 14–15) erroneously states that the move to the Trippenhuis took place in 1814.

27 Anthony Johannes d'Ailly (1793–1851) was a chemist from Amsterdam and the owner of a significant insect collection. Reindert (or Rijndert) Willebrordus Draak (1786–1866) worked as a taxidermist for the Hollandsche Maatschappij in Haarlem between 1818 to 1834, and built his own private collection (Sliggers, "Het naturaliënkabinet van de Hollandsche Maatschappij," 117). In 1838 Draak became the director of the Amsterdam zoo Natura Artis Magistra, and remained in that role until he was dismissed in 1840 after a series of conflicts with the board of Artis. See, for example, Rijksbureau voor Kunsthistorische Documentatie, "Directeuren van Artis: Reindert Draak"; Nieuwendijk, *Zoo was Artis*.

28 Temminck to the minister of public education, national industry, and colonies, 23 March 1819, NL-HaNA, BiZa 1813–1848, 2.04.01, inv.nr. 4073; Koolmees, "Caspar Georg Reinwardt," 47.

FIGURE 8 *Martinus van Marum, director of the Teylers Museum.* Portrait by Charles Howard Hodges, ca. 1826. Oil on canvas
TEYLERS MUSEUM, DIGITALE COLLECTIE, KS 1999 004

to settle the issue.[29] Temminck and Van Marum might have felt some antipathy towards each other previously – they had met before at the Hollandsche Maatschappij – but during the few years of working together their animosity

29 Van Marum to Temminck, 7 February 1816, NHA, Archive Martinus van Marum 529-20b; Temminck to Van Marum, 14 February 1816, NHA, Archive Martinus van Marum 529-20b; Van Heiningen, *Correspondence of C. G. C. Reinwardt*, 31–32.

towards each other grew even more. The main problem was that they had very different ideas of the implications of their appointments.

It all started, it seems, with humidity. The Trippenhuis housed 's Lands Kabinet and the Koninklijke Kunst Museum, where Temminck had worked in 1808 as the temporary cover for Cornelis Apostool, as well as the headquarters of the Koninklijk Instituut van Wetenschappen. Apostool had populated all available space, including Reinwardt's former residence, with the collections of the Kunst Museum. The residence was reserved for the collection of weights and measures of the 'First Class' (*Eerste Klasse* or the Science Department) of the Koninklijke Instituut. He had filled Reinwardt's rooms with objects from the art collection and refused to clear them.[30] Temminck complained that the insect collection, which had been removed from other rooms by Apostool, had been tucked into a space so damp that, without heating, "one could soon pick up mushrooms from the barrels."[31] Temminck found this unacceptable. Somewhat sarcastically he wrote to Reinwardt: "one need not have the slightest fear of an exhibition equal to those of the Museum of Paris, Vienna or London, and something like that should someday be, I trust, part of your plan."[32] In this situation Van Marum and Temminck adopted opposite positions. Temminck wanted to address the problem and impede Apostool's expansion at the Trippenhuis. Van Marum preferred a less direct approach. He wrote to Reinwardt explaining that he had decided not to allow anyone to visit the collection in order to avoid bad publicity and that he would rather leave the expansion of the collection to Reinwardt.[33]

While Van Marum wanted to interfere as little as possible, Temminck had considerably more ambitious plans for 's Lands Kabinet. Van Marum claimed that Reinwardt had left instructions to preserve the cabinet as it was until his return, a claim that disconcerted and even offended Temminck. He did not know anything about such instructions. Temminck was convinced that his role should be much more than that of a "simple subordinate supervisor or custodian," and that he had much more to offer: looking after Reinwardt's interests and promoting scientific research, no less. For Temminck, holding the fort for a few years until Reinwardt's return would not do. He wrote to Reinwardt explaining that he would not have accepted the position under different circumstances: being a mere guardian of the collection would be "a waste of his

30 Van Berkel, *De stem van de wetenschap*, vol. 1, 129.

31 Temminck to Reinwardt, 25 November 1816, NL-HaNA, BiZa 1813–1848, 2.04.01, inv.nr. 4032B; Koolmees, "Caspar Georg Reinwardt," 48.

32 Temminck to Reinwardt, 25 November 1816.

33 Van Marum to Reinwardt, 15 April 1817, NHA, Archive Martinus van Marum 529-20b.

FIGURE 9 *Gevelaanzicht van het Trippenhuis*, Amsterdam (Façade of the Trippenhuis). Drawing by Reinier Vinkeles, 1803
COLLECTION ATLAS SPLITGERBER. STADSARCHIEF AMSTERDAM, NO. 010001000919

time." Temminck complained that it was because of Van Marum that he had to abandon some of his plans, as the latter could not accept Temminck's decisions "because of his old-fashioned view of things."[34]

Apart from an antipathy towards each other, the two directors had very little in common. Van Marum was sixty-seven years old at the time, the secretary of the Hollandsche Maatschappij of Haarlem and the keeper of its natural history

34 Temminck to the minister of public education, national industry, and colonies, 23 March 1819, NL-HaNA, BiZa 1813–1848, 2.04.01, inv.nr. 4073.

cabinet, as well as the director of Teylers Museum. He had a brilliant career behind him, was a celebrated man of science in the Netherlands, and spent much of his time in Haarlem. Temminck, by contrast, was thirty-nine years old and had no academic background. He did have the financial resources to work and publish independently. He had built an outstanding reputation for himself, both as a keeper of collections and as an ornithologist, but since the direction of 's Lands Kabinet was a temporary appointment he needed to make the best of his current position if he ever wanted to be offered an official post. In short, Temminck had his own agenda.

In the following years Van Marum's and Temminck's different views of their roles at 's Lands Kabinet increased the tension. They disagreed on the acquisition of new collections, the scientific value of the objects, their curation, on who was to give orders to whom, and on the best use of every inch of available space. They disagreed on almost every decision that had to be taken at the museum.[35] They both wrote flurries of letters of complaint to both Reinwardt and Falck who, with patience and diplomacy, tried to keep the peace. Van Marum accused Temminck, among other things, of plotting to take over Reinwardt's position so that he would become the director of 's Lands Kabinet. Temminck emphatically denied this in a letter he wrote to the commissioner-general towards the end of 1816.[36]

At one point both Van Marum and Reinwardt complained that Temminck was only looking after his affairs and that he was overly occupied with his cabinet and his publications.[37] They were right. Towards 1816 Temminck seemed to abandon his responsibilities as deputy director and was mostly concerned with his own enterprises. This is Temminck's description of how he spent his time in October 1816, one year after his and Van Marum's appointment:

> I am the factotum of my numerous collections, sometimes general and sometimes soldier. I am the one who mounts all the birds, fish, and quadrupeds; in search of one or the other I vary my many trips, sometimes by the seashore, or beside our lakes, or near our swamps; all the many animal skins that I use for exchange or as presents for my friends are prepared by my own hands; in the winter I concentrate on mounting quadrupeds and exotic birds, which I receive dried and sometimes badly prepared; then I work at classifying and preserving my collections, and the evening

35 Temminck to the minister of public education, national industry, and colonies, 3 March 1819, NL-HaNA, BiZa 1813–1848, 2.04.01, inv.nr. 4073; Koolmees, "Caspar Georg Reinwardt," 46–51.

36 Temminck to the commissioner-general of education, arts, and sciences, 4 December 1816, NL-HaNA, BiZa 1813–1848, 2.04.01, inv.nr. 4032B.

37 Van Heiningen, *Correspondence of C. G. C. Reinwardt*, 184.

> is devoted either to my literary work or to my social duties. [...] I still have not succeeded in getting a good assistant, because in our country there is absolutely no taste for the study of natural history. The only idol worshipped by my compatriots is gold; it is considered almost ridiculous for anyone to concern himself with science, except simply as a hobby and a relaxation from office work.[38]

Notably, Temminck makes no mention of his tasks at 's Lands Kabinet here, which seem to have been subsumed into his plans and collection. In 1819 he even employed Reindert Draak, the keeper at 's Lands Kabinet, as a doorman at his house in the Heerengracht. Once a week Draak received visitors to Temminck's private collection, as "poor Reindert is now married and his wife expecting, and he needs more than three guilders a week."[39] Reinwardt was not pleased with Temminck's charitable initiative, which he thought took up too much of Draak's time.[40]

Temminck had become curiously uninterested in 's Lands Kabinet. Although it would have been in Temminck's interest to excel in the direction of 's Lands Kabinet, apparently the quarrels with Van Marum were too much to bear. Towards the end of 1816 Temminck noted that he had seen Van Marum only twice in six months and that to avoid further escalation, Temminck had decided to "leave the field of honor."[41] Perhaps years of frustration with Van Marum had driven him to seek refuge in his projects. Resourceful as ever, Temminck managed to take a long break and leave the country, probably with permission of the commissioner-general of public education, Ocker Repelaer van Driel.

3 Collections for the Universities

From September 1817 until August 1818 Temminck traveled with Caatje to Germany, Switzerland, Austria, and Italy. Van Marum was not pleased, but it

---

38 Temminck to Johann F. Naumann, 14 October 1816, in Stresemann and Thomsen, "Ornithologen-Briefe aus den Jahren 1816 bis 1820," 106. Translated in Stresemann, *Ornithology*, 118–19.

39 Temminck to Reinwardt, 25 November 1816, NL-HaNA, BiZa 1813–1848, 2.04.01, inv.nr. 4032B.

40 Reinwardt to Van Marum, 29 October 1815, NHA, Archive Martinus van Marum 529-20b; Van Heiningen, *Correspondence of C. G. C. Reinwardt*, 121.

41 Temminck to Reinwardt, 25 November 1816, NL-HaNA, BiZa 1813–1848, 2.04.01, inv.nr. 4032B.

seems that he and Temminck had agreed on this leave.[42] Besides, at the beginning of 1817 Temminck had received a new assignment from the government: he was to enlarge the zoological collections of the 'southern universities' (the Universities of Liège, Ghent, and Leuven, now in Belgium).[43] In order to do this, Temminck needed to acquire specimens abroad. However, the main goal of this long tour was one closer to Temminck's heart: he was working on a second edition of his *Manuel d'ornithologie*, and he traveled to gather as much information as possible.[44] This leave also allowed Temminck to put some distance between himself and Van Marum.

Temminck was asked to build three similar collections for the southern universities by purchasing and labeling specimens for the relevant professors, Franz Joseph Adelmann in Leuven, Henri-Maurice Gaede in Liège, and François-Pierre Cassel in Ghent. Temminck was in charge of the budget and accountable to the Ministry regarding every invoice and transaction. For this purpose he made use of his extensive network and journeyed through Europe to see nearly every natural history cabinet there was to visit. He actively accumulated specimens by asking for duplicates at museums, like the Muséum in Paris, as well as purchasing them at auction in London, where the Bullock's collection was sold.[45]

After his return from the tour of 1817 to 1818, Temminck sent two chests with natural history specimens for Professor Adelmann to the University of Leuven. When the specimens arrived in Leuven, the professor found they had "beaucoup de défauts," missing eyes and feathers. Temminck argued, laconically, that the university required "didactic material," and that was what he had provided. In reaction to Adelmann's complaint Temminck suggested that the minister send the specimens to Liège and refund a third of the money to Leuven. Falck, now minister of education, arts, and sciences, sided with Temminck and dismissed Adelmann's protests, arguing that it was not unusual for feathers to

42 Reinwardt to Van Marum, 12 December 1818, NHA, Archive Martinus van Marum 529-20b; Van Heiningen, *Correspondence of C. G. C. Reinwardt*, 279–80. Reinwardt here referred to a letter dated March 10, 1818 that he had received from Temminck.

43 According to Poot, Vanpaemel, and Waelkens (*Een walvis in de stad*, 30), Temminck was charged with enlarging the collections of these universities in 1818 (Anton R. Falck became minister of education on March 19, 1818). However, Temminck mentions this appointment in a letter to Johann F. Naumann: 5 March 1817, in Stresemann and Thomsen, "Ornithologen-Briefe aus den Jahren 1816 bis 1820," 111.

44 Temminck to Naumann, 27 November 1817, in Stresemann and Thomsen, "Ornithologen-Briefe," 119.

45 Temminck to the Ministry of Home Affairs, 27 February 1819, NL-HaNA, BiZa 1813–1848, 2.04.01, inv.nr. 4072.

be lost during transport. The specimens stayed in Leuven, and Adelmann, to his exasperation, never received any other specimens from Temminck.[46]

Temminck adopted a similar approach for the University of Liège. He wanted to send birds from the Achtienhoven collection, which was, according to Temminck, of inferior quality. It included many misidentified specimens, which only increased confusion about the species nomenclature and therefore had no place in 's Lands Kabinet.[47] Unsurprisingly, this resulted in yet another fight with Van Marum, who intended to keep the Achtienhoven collection in Amsterdam. Temminck consulted the Achtienhoven family regarding their wishes, and they could not recall ever having seen an instruction that the collection should be sent to Amsterdam.[48] Since Temminck declared that he was in charge of the vertebrates, the minister again permitted him to deal with this task as he saw fit, and that ended the dispute.[49]

Temminck considered assembling the university collections as a burden and a waste of his time: "In charge of organizing the cabinets for the Universities of Ghent, Liège, and Leuven, I was forced to spend a considerable time on this task, which was why I could not work for myself; now that the organization of this part comes to an end, I eagerly resume the thread of my previous relations...."[50] Once more, official duties seem to be less important to Temminck than his projects, which gives us an impression of his sense of entitlement, a consequence of his social status and upbringing. It even seems that Temminck resented university professors: "I do not have any students at my service as the distinguished professors do; all at once Generalissimo and soldier of my collection, I have to do everything myself."[51] Temminck's lack of interest reveals something other than a rather phlegmatic attitude. It reflects his understanding of the function of natural history collections at the universities and has its roots in the fragmentation of natural history into different subjects.

46 Poot, Vanpaemel, and Waelkens, *Een walvis in de stad*, 30.

47 Temminck to the Ministry of Home Affairs, 27 February 1819, NL-HaNA, BiZa 1813–1848, 2.04.01, inv.nr. 4072; Koolmees, "Caspar Georg Reinwardt," 48. J. B. J. Achtienhoven (1751–1801) was an art dealer from Amsterdam and the owner of a natural history cabinet of birds, mammals, insects, and shells (Smit, Sanders, and Van der Meer, *Hendrik Engel's Alphabetical List*, 1).

48 Temminck to the minister of public education, national industry, and colonies, 23 March 1819, NL-HaNA, BiZa 1813–1848, 2.04.01, inv.nr. 4073.

49 Ministry of Education to Temminck, 7 March 1819, NL-HaNA, BiZa 1813–1848, 2.04.01, inv.nr. 4072.

50 Temminck to Naumann, 5 March 1817, in Stresemann and Thomsen, "Ornithologen-Briefe aus den Jahren 1816 bis 1820," 111.

51 Temminck to Lichtenstein, 18 September 1818, in Stresemann, "Aus C. J. Temmincks Briefen an H. Lichtenstein," 255.

The universities were focusing mainly on comparative anatomy. They were compelled to do so. Since 1815, after the implementation of a reform of the Dutch education system, each faculty of mathematics and natural sciences was required by law to include in its curriculum the "natural history of animals and minerals, which should include, regarding zoology, *anatomia comparata*."[52] Article 193 of the reform also dictated that universities should keep educational collections. Therefore, Temminck argued, professors were not in the least interested in creating comprehensive collections – i.e. collections covering the entire spectrum of nature – but collected specimens for the sole purpose of illustrating lessons.[53] Therefore, specimens of inferior quality would do perfectly well. After all, researchers and students did not just look at the collections: they also necessarily handled specimens and preparations as part of their direct examination.[54]

Then again, systematics required perfectly preserved specimens for the identification of useful external 'characters' (morphological features) in order to classify them, while comparative anatomy called for intact specimens, not stuffed ones that had had their internal organs removed. Consequently, Temminck advised the creation of "double collections": one for comparative anatomy, one for systematics. Accordingly, he requested that Reinwardt send him two specimens of each fish species from Java: one preserved in spirit for "anatomy," and one mounted for "reference," prepared to preserve the external characters.[55] In short: different disciplines required different collections. Natural history was slowly developing from a broad discipline into a cluster of different subjects, each with its own practices and methods, some of them object-based. Each subject was practiced in a different setting and by different specialists. Systematics was mostly practiced outside the universities, where comparative anatomy had relegated it to a secondary position.

The University of Groningen was the only Dutch university where systematics received ample attention, mainly due to the efforts of Theodorus van Swinderen. He became professor of natural history at the University of Groningen in 1814, a post he occupied until his death at the age of sixty-six. A man of many interests, Van Swinderen was also a doctor of linguistics, law, mathematics, and physics. He worked actively for the improvement of public

52 "Besluit regelende het hooger onderwijs," in *Staatsblad der Verëenigde Nederlanden*, no. 14, art. 63, 41.

53 Temminck to Reinwardt, 25 November 1816, NL-HaNA, BiZa 1813–1848, 2.04.01, inv.nr. 4032B.

54 Huistra, *Afterlife of the Leiden Anatomical Collections*, chap. 1.

55 Temminck to Reinwardt, 25 November 1816, NL-HaNA, BiZa 1813–1848, 2.04.01, inv.nr. 4032B.

education and applied himself among many other things to the preservation of various historical monuments in Groningen.[56] Thanks to him zoological systematics was an important part of the natural history curriculum, and he also created a university museum of natural history to display the systematic order in nature. The academic collection was Van Swinderen's pet project and soon grew into a considerable museum, open to students and the public once a week: on Saturdays for students, on Wednesdays for the general public.[57] For years he earnestly collected, bought, and exchanged material, with the help of two of his students, the Dutchman Johan Conrad van Hasselt and a talented young naturalist from Hanau, Heinrich Kuhl. By the early 1820s the museum was renowned throughout Europe.

Kuhl arranged the collection according to the most recent insights in classification, including those of Temminck and Illiger, and both students' careers were furthered by their involvement in Van Swinderen's museum.[58] For Van Swinderen it was of the utmost importance that the arrangement of the objects of the museum reflected classification: "I always show the Museum to visitors in such a way that they see the collection in this Systematic arrangement."[59]

It was thanks to Van Swinderen that Temminck finally received an academic title. In 1819, after his successful publication of the *Manuel d'ornithologie*, Van Swinderen recommended both Kuhl and Temminck for the title of *Matheseos Magister, Philosophiae Naturalis Doctor honoris causa*.[60] The Faculty of Mathematics and Natural Science and the University Senate accepted the recommendation, and both received their titles on August 6, 1819.[61] In honor of Van Swinderen and their friendship, Temminck named an African rodent, the Greater Cane Rat, after him in 1827, the *Thryonomys swinderianus*. Kuhl named, perhaps a bit more aptly, a Lovebird native of South Africa after his teacher and mentor, *Psittacus swinderianus* (now known as *Agapornis swinderianus*).

A doctor's title *honoris causa* from the University of Groningen and not from Leiden reflects the different interests of the university professors. Van Swinderen was enthusiastic about systematics and Temminck's work while Sebald Justinus Brugmans, at the University of Leiden, favored comparative anatomy over systematics. Brugmans was the professor of botany, chemistry,

56 Botke, *Theodorus van Swinderen*.

57 Van Swinderen, *Index rerum naturalium*, vi, xvii.

58 Klaver, *Inseparable Friends*, 13.

59 Van Swinderen, *Index rerum naturalium*, xi; see also Botke, *Theodorus van Swinderen*, 65, 67, 72.

60 Klaver, *Inseparable Friends*, 15; documents in Naturalis Biodiversity Center Archives, Archive C. J. Temminck.

61 Klaver, 16.

FIGURE 10 *Psittacus Swinderianus* Kuhl. Hand-colored engraving after A. Prevost. In Heinrich Kuhl, *Conspectus Psittacorum* (1820): plate 2
IMAGE FROM THE BIODIVERSITY HERITAGE LIBRARY. DIGITIZED BY THE SMITHSONIAN LIBRARIES

medicine, and natural history, and also the head of the anatomical collection of the university.[62] His lessons focused on applied sciences, mostly compared anatomy and applied botany. Van Swinderen was one of the few university professors teaching systematics at a university. He wrote: "I have set as my main goal the competence of the young students, and therefore I have spared no effort to gather as many genera as I could in the short period of six years (because before then, there was nothing here), to give the students a Systematic overview of the whole science, which is, in my view, the first goal of Academic studies."[63] Interestingly, he did not add anatomical or pathological samples to this academic cabinet while these were so essential to the University of Leiden and in Brugmans' collection.

Temminck's ties with the University of Groningen remained strong after 1819, as were his controversies with the rest of the Dutch universities. In the meantime, and mostly thanks to Reinwardt's efforts, Van Marum and Temminck remained grudgingly in their posts at 's Lands Kabinet.

4 Building Up a Network

If there was anything Temminck could not do without it was an extensive network of naturalists and collectors. He needed them to obtain information, books, specimens, and access to private and public cabinets. His contacts in the Netherlands included, among others, Joan Calkoen from Amsterdam, who was a real estate broker and the owner of a zoological cabinet containing mostly insects; the Sepp family, who were collectors and publishers; Anthony J. d'Ailly, the chemist who assisted Van Marum with the insect collection in 's Lands Kabinet; Theodorus van Swinderen, the professor of natural history in Groningen; Arnoldus Ameshoff from Amsterdam, the collector who had dazzled young Temminck with his culinary extravaganzas; Johan Raye van Breukelerwaerd, a cabinet owner and Temminck's neighbor; and the Van Lenneps, a family of merchants and politicians with a taste for natural history.[64] Few of them, however, attempted to describe new species or to publish systematic works as Temminck had done. For fellow systematists he needed to go beyond the Netherlands. His correspondents were like-minded naturalists from Germany, Switzerland, Italy, and France.

---

62 See, for example, De Jonge, "Sebald Justinus Brugmans"and *Het vergeten fenomeen*.

63 Van Swinderen, *Index rerum naturalium*, xiii.

64 For more information on Dutch cabinets of natural history see Smit, Sanders, and van der Veer, *Hendrik Engel's Alphabetical List*; Visser, "Het Rijksmuseum."

Following his initial advertising campaign for the exchange of duplicates in his *Catalogue systématique*, Temminck soon corresponded and exchanged specimens with dozens of European naturalists. He built his network through the simple but effective means of correspondence and travel. The latter provided the best results. Visiting naturalists and collectors at their place of residence meant that he could examine their collections in person. A few became his friends, some became his correspondents for many years, and all kept him well-informed at one level or another. Also, very conveniently, he could directly take care of the transport of exchanged specimens during his travels.

Temminck aspired to build a complete collection comprising nearly all known species of birds, mammals, and fish. It was a constantly ongoing enterprise, especially since large numbers of new species poured into Europe from the colonies. He exchanged specimens with nearly every other collector he knew. Traveling allowed him to study specimens first-hand in any private or public cabinets to which he could gain entry. Temminck took great pride in the fact that his research was only based on direct observation of original objects.[65] He needed to examine the species previously described by others, especially if he wanted to succeed in disentangling the nomenclature of the time.

At least once a year Temminck stayed in Paris, where two of his younger sisters lived (Aletta Cornelia and Elisabeth Jacoba, who were known among the Temminck family by their cheerful nicknames 'Ceetje' and 'Cootje').[66] Temminck usually arranged meetings with his publishers in Paris as well as his illustrator, Jean-Gabriel Prêtre. He also visited his friends Levaillant and Leschenault de la Tour, studied the magnificent collections at the Muséum d'Histoire Naturelle, and met professors Georges Cuvier and Étienne Geoffroy Saint-Hilaire. On several occasions Temminck and his wife were invited to dinner at Cuvier's house or, as Caatje often put it, "at Mrs. Cuvier's."[67]

Besides his trips to Paris Temminck also undertook longer tours through Europe for months at a time, mostly through France, Germany, Switzerland, and northern Italy. On other occasions his trips were shorter, but he always planned visits to interesting cabinets, private or public, and never missed an opportunity to meet old friends or make new acquaintances, building a

65 Temminck, "Introduction," in *Manuel d'ornithologie*, 2 ed., vol. 1, xxvii–xxviii.

66 Elisabeth Jacoba (1783–1849) and Aletta Cornelia (1784–1848) Temminck, both unmarried. See Van Lynden-de Bruïne, *In vogelvlucht*, 12, 19.

67 Curiously, no correspondence between Temminck and Cuvier or Geoffroy Saint-Hilaire seems to have survived, although it is clear from the report included in the *Histoire naturelle des pigeons* (1813), as well as from Temminck's writings and Caatje's diaries, that they met several times, in the Muséum and at Cuvier's home. See Van Lynden-de Bruïne, *In vogelvlucht*, 220, 223, 261; Outram, *Letters of Georges Cuvier*.

network of naturalists and collectors as he went along. These meetings shaped his career.

During his first trip to Offenbach, Temminck met naturalists including Bernhard Meyer, Johann Philipp Achilles Leisler, and Johann Wolf, who taught him bird systematics and inspired him to write his monographs. Later encounters were equally decisive for Temminck's career, like meeting Madame Knip or the young and talented Heinrich Kuhl and Heinrich Boie in Germany. None of this, of course, would have been possible without independent funding. Temminck traveled for long periods, even an entire year, regardless of whether or not he was employed. His social position and wealth allowed him to dedicate a substantial amount of time to his personal projects. In 1804 he stayed with Bernard Meyer in Offenbach for six months while being employed as a civil servant. As the director of 's Lands Kabinet, he undertook an eleven-month tour through Europe. Although Temminck had a keen interest in ornithology and considerable taxidermic skills, his professional life would have been, in all probability, very different if he had not been born into privilege.

The tours also had downsides. Traveling by road was at times dangerous and always slow. Temminck boasted about having suffered no fewer than eighteen accidents with his carriage without ever sustaining an injury.[68] He had a carriage specially adapted for the transport of delicate natural history specimens – some still fresh, others reeking of decay – in which he spent days of uncomfortable journeys, most of the time accompanied by his wife.[69] Temminck was always concerned about the transport of mounted specimens and the damage they might suffer, especially if they were his own: "I'm a little reluctant, frankly speaking, to risk the integrity my unique birds on a long journey during which they might be totally spoiled, or lose their freshness and purity."[70]

Occasionally inns posed serious threats both to the travelers' health and their morale. Beds could be infested with fleas and bedbugs, so that they had to find creative solutions to this problem. In Lausanne Caatje made an improvised bed out of two chairs which, according to her, worked quite well following three exhausting days in the carriage. In Turin the couple slept the first three nights of their stay on a table. In addition to nightly discomfort, they often had

68 Susanna, "Levensschets van Temminck," 75, n. 9.

69 Susanna, 74.

70 Temminck to Naumann, 14 October 1816, in Stresemann and Thomsen, "Ornithologen-Briefe aus den Jahren 1816 bis 1820," 106.

to get out of their carriage to continue on foot.[71] The roads were precarious and dangerous, but nothing seemed to affect Mrs. Temminck's cheerful disposition.

During the longer tours Temminck and Caatje combined pleasure and work. They planned excursions (they even once climbed the Mont Blanc), visited theaters, and went on sightseeing tours. They paid their respects to Conrad Gessner's tomb in Zürich and gasped at the sight of the Turin shroud. They fit these activities efficiently between visits to naturalists, collectors, museums, and universities. As a result Temminck knew practically everyone in Europe: taxidermists, dealers, cabinet owners, professors, and amateur naturalists.

At the University of Turin Temminck met Franco Andrea Bonelli, the professor and curator of the university's zoological collections. Temminck also made sure to stop in Bern to meet the amateur entomologist and ornithologist Karl Friedrich August Meisner. He was the author of a systematic catalog of the birds of Switzerland present in the Burgerbibliothek in Bern. This institution's collection also included minerals, models of mountains, clothes from Haïti, and some Roman artifacts, which Caatje found "very curious."[72] Meisner's collection catalog was quite similar in format and contents to Temminck's.[73]

In Hanau they paid a visit to Johann Leisler and his collection of birds, eggs, and minerals, "which is very satisfactory."[74] In Frankfurt Temminck bought Karl Illiger's *Prodromus systematis mammalium et avium*, which proved to be "of great service" for his classification of fowl.[75] Illiger had arranged to send to Temminck some specimens from the collection of the German naturalist Johann Centurius Hoffmann, Graf von Hoffmannsegg, who had also been the founder of the Zoological Museum of Berlin in 1809. However, the dispatch had been delayed for months. Illiger thanked Temminck for his patience and apologized: "a collection like yours, remarkable in the number of species it contains and by the beauty of its specimens, makes it extremely difficult to find anything worthy of being added to it."[76] Illiger finally found specimens meeting Temminck's high standards amongst those collected by Friedrich Wilhelm Sieber in Brazil.[77] Unfortunately, Illiger and Temminck did not have

---

71 Van Lynden-de Bruïne, *In vogelvlucht*, 68, 89.

72 Van Lynden-de Bruïne, 93.

73 Meisner, *Systematisches Verzeichniss der Vögel*.

74 Van Lynden-de Bruïne, 124.

75 Temminck to Hoffmannsegg, 29 July 1812, in Stresemann, "Briefwechsel von Temminck mit Hoffmannsegg und Illiger," 259.

76 Illiger to Temminck, 14 September 1812, in Stresemann, 259.

77 After this, Sieber and Temminck stayed in contact. Sieber also collected some specimens for 's Rijks Museum (Sieber to Rijks Museum van Natuurlijke Historie, 5 March 1825, Naturalis Biodiversity Center Archives, Correspondence RMNH).

the opportunity to build a lasting friendship, as Illiger died in 1813, when he was only thirty-eight years old.

Amongst Temminck's long-term correspondents was the German physician Philipp Jakob Cretzschmar, the founder and director of the Senckenberg Natural History Society. They corresponded for fifteen years. In 1816 Temminck also started corresponding with Johann F. Naumann, who was a talented German engraver, a great ornithologist, and a skilled taxidermist. They would not meet in person until two years later, when Temminck visited Naumann in Ziebigk. Naumann had prepared excellent engravings for his father, Johann Andreas, to illustrate his *Naturgeschichte der Land- und Wasser-Vögel des nördlichen Deutschlands*.

Apparently, the illustrations Temminck had received from Paris were inferior to Naumann's, so with a hint of envy, Temminck tried to involve Naumann in the second edition of the *Manuel d'ornithologie*. He suggested a partnership: Temminck would send Naumann any specimens he wished to have for his projects, and in return Naumann would sell the roughly two hundred engravings he already had made for his father's book to Temminck, "at the lowest price possible."[78] Temminck reasoned that they would both profit from the deal, as they were selling the *Naturgeschichte* and the *Manuel* in different countries. Naumann sold his book only in German-speaking countries, and it was not available in France, Italy, the Netherlands, or England, where Temminck's *Manuel* was distributed. Similarly, Temminck's work was not available in Germany or Switzerland. By combining their knowledge and Naumann's engravings, they could make sure to reach a broader audience.[79] Without a trace of modesty, Temminck told Naumann that his name would be on the "most perfect work that exists on the natural history of European birds" – that is, the second edition of Temminck's *Manuel*.[80] Temminck was not merely exaggerating in order to convince Naumann; he genuinely believed that his book was the most exact, complete, and rigorous work available at the time.

Unfortunately, Naumann did not accept the proposition. He did not wish to risk damaging the copper plates by using them over and over again. However, Naumann could make new engravings if Temminck sent him his specimens. Temminck could not agree to this proposal because he did not want to put the safety of his rarest and most beautiful specimens at risk. Neither Temminck nor Naumann would endanger their most precious possessions, whether

78 Temminck to Naumann, 23 July 1816, in Stresemann and Thomsen, "Ornithologen-Briefe aus den Jahren 1816 bis 1820," 101.

79 Temminck to Naumann, 23 July 1816.

80 Temminck to Naumann, 23 July 1816.

plates or birds. Eventually Temminck employed Jean-Charles Werner for making the necessary plates; he was a natural historical artist employed at the Paris Muséum.[81] In the following years Temminck and Naumann remained to be friends, corresponded with each other, and exchanged information, specimens, plates, and books.[82]

Between October 1817 and August 1818 the Temmincks took their longest European tour, which was necessary for the preparation of the revised *Manuel* and for leaving the Trippenhuis and Van Marum behind. When he was asked to collect specimens for the universities of the southern Netherlands, Temminck quickly seized the opportunity. With the help of the commissioner-general of education, Ocker Repelaer van Driel, he arranged to travel to Germany, Switzerland, and Italy, where he already had the necessary contacts. Caatje, of course, accompanied him. They started by going south, to Antwerp and Brussels, where Temminck met Repelaer van Driel and Jan-Willem Janssens, the former governor of the Cape of Good Hope and the Dutch East Indies and the provisional commissary-general of war until 1815, as well as a natural history enthusiast. After all, political contacts were as important as scientific ones.

Their stay in Germany was the most profitable, scientifically speaking. Prince Maximilian Wied-Neuwied had returned from Brazil a year earlier with many interesting specimens. The first visit to the prince led to a long-term friendship and a lively exchange of specimens. The Temmincks were the Meyers' guests every time they passed through Offenbach, and they visited Frankfurt a couple of times. Temminck also had the opportunity to meet another of Meyer's guests, a young German zoologist and curator of the zoological collection at Heidelberg University, Heinrich Boie. In the previous year Boie, a former law student who proved to be an excellent zoologist, had published a review of twenty-nine pages of Temminck's *Manuel* in the *Heidelbergische Jahrbücher der Literatur*, which included some criticism. The review was well written and even better argued. Temminck not only accepted the criticism but praised Boie for it, the two men started corresponding, and they then met at the Meyers in August 1818.[83] In Heidelberg Temminck also visited the German anatomist and physiologist Friedrich Tiedemann, a professor of anatomy, zoology, and physiology at Heidelberg University. Tiedemann had a splendid collection of anatomical and zoological specimens which he later donated to the university.

During this fruitful trip Temminck also met the physician, explorer, and naturalist Martin Hinrich Carl Lichtenstein, with whom he enjoyed a long-lasting

81 Temminck to Naumann, 28 August 1816, in Stresemann and Thomsen, 102–05.

82 Temminck to Naumann, 18 May 1817, in Stresemann and Thomsen, 114.

83 Boie, "Manuel d'ornithologie."

friendship afterward; they corresponded for nearly forty years.[84] Lichtenstein had been a professor of zoology at the University of Berlin since 1811 and was appointed the director of the Berlin Zoological Museum in 1813, following Illiger's death. According to Temminck, Lichtenstein occupied "a distinguished position in the sphere of science under the triple title of scientist, naturalist, and traveler."[85]

In Berlin, by happy coincidence, Temminck also had the opportunity to meet Heinrich Kuhl (Van Swinderen's student) who was touring the natural history museums in Germany.[86] Kuhl, who was a very young and gifted naturalist, had studied natural history at the Hohe Landesschule of Hanau and afterward enrolled at the University of Groningen in 1816. Temminck took a particular interest in Kuhl and invited him in 1819 to join him for two months in his country retreat *Wildlust* in Lisse.[87] Temminck thought very highly of him: "The young Kuhl of Hanau, recently employed in one of our universities, will accompany me and assist me this year; he is a young man of whom I have a high opinion, and I want to arrange to send him someday to make trips to the little-known areas of our colonies in the Indies."[88] Indeed, Kuhl dreamed of a trip to the tropics, and few years after their first encounter Temminck recommended Kuhl for a position in the Natuurkundige Commissie voor Nederlands-Indië, a commission which had been founded in May 1820 for the exploration of the natural resources of the Dutch East Indies.

It was also during this tour in 1817 to 1818 that Temminck finally visited Naumann in Ziegbik, following two years of correspondence. At that time he also met Blasius Merrem, a professor of political economy and botany at the University of Marburg and a well-known polymath: Merrem was also an ornithologist, a herpetologist, and a mathematician. In Switzerland Temminck and Caatje carried on with their visits. They called on Friedrich Meisner in Bern, who once more showed them the museum.[89] They spent the winter in Lausanne, where Temminck's father now lived with his three sisters Ceetje, Cootje, and Suze. In the new year they moved on to Italy to visit Franco Andrea Bonelli again, who was still a professor at the University of Turin. Thanks to Bonelli's efforts the university museum had improved much since their last visit. This time Caatje was shocked by the sight of a preserved human body,

84 Stresemann, "Aus C. J. Temmincks Briefen an H. Lichtenstein."
85 Temminck and Meiffren Laugier de Chartrouse, *Nouveau recueil*, vol. 5, livraison 60, plate 361, text, [1].
86 Klaver, *Inseparable Friends*, 13.
87 Temminck to Naumann, 19 April 1819, in Stresemann and Thomsen, "Ornithologen-Briefe aus den Jahren 1816 bis 1820," 129.
88 Temminck to Naumann, 5 March 1817, in Stresemann and Thomsen, 111.
89 Van Lynden-de Bruïne, *In vogelvlucht*, 145.

the body of the former faithful servant of the prince of Borghese: "The piece was well made, but I thought it was a gruesome sight, because although I have seen mummies and other embalmed bodies, never before have I encountered a human stuffed and exhibited in the same manner as birds."[90] After almost a year of traveling the couple returned to the Netherlands, and Temminck resumed both his preparations for the *Manuel d'ornithologie* and, very reluctantly, his post at 's Lands Kabinet.

Bonelli and Temminck continued their friendship, and in 1824, as a token of his appreciation, Temminck named an eagle after Bonelli, *Falco Bonelli*, which has been known ever since as Bonelli's eagle. Temminck had seen specimens of this eagle in the forests outside Paris and in the north of France as well as a specimen from Bonelli's collection, a drawing of which was later published in Temminck's *Les oiseaux d'Europe* (a volume with illustrations by Jean-Charles Werner to complement the second edition of the *Manuel d'ornithologie*).[91] The portrait Bonelli had painted of himself shows him proudly with that particular specimen on his right and is still held by the Museo Regionale di Scienze Naturali in Turin.

Now Louis Pierre Vieillot, the French ornithologist who had appropriated Temminck's new species from the *Catalogue systématique*, protested to the ornithological community that he had described the same eagle just two years previously: first in a lecture to the Société Linnéenne de Paris on August 22, 1822, and later in the published description in the *Tableau encycopédique et méthodique des trois règnes de la nature*.[92] A correction and the text of Vieillot's letters of reclamation, addressed to the French naturalist René-Primevère Lesson, were published in the *Bulletin des Sciences Naturelles* in 1828. Vieillot was now the recognized discoverer and authority. The bird retained its vernacular epithet of 'Bonelli's eagle,' and Vieillot's Latin binomial, *Aquila fasciata*, was accepted after some time – but not by Temminck.[93]

In his description Temminck had failed to mention that Vieillot had been the first to recognize the new species. According to Vieillot, however, Temminck was perfectly aware of that fact:

> [the specimen] had been reported to me by Mr. Dupont the elder, who received it from Mr. Bonelli, asking me whether I considered it to be a new species, and it was only after my decision that it was sent to Mr. Temminck. Do not believe that I attach great importance to being the

90 Van Lynden-de Bruïne, 163.
91 Temminck and Werner, *Les oiseaux d'Europe décrits par C.J. Temminck* , vol. 1, plate [24].
92 Vieillot, "L'Aigle à queue barrée," in *Tableau encyclopédique et méthodique*, vol. 3, 1192.
93 Lesson, "Réclamation de M. Vieillot"; Aimassi, "Original Description of Bonelli's Eagle."

FIGURE 11 *F. Andrea Bonelli*. Portrait by Giovanni Battista Biscarra, 1830. Oil on canvas UNIVERSITY OF TURIN, DEPARTMENT OF ANIMAL BIOLOGY. PHOTOGRAPH BY AND COURTESY OF LUCA GHIRALDI

first to discover a new species; but I must avoid being taken for an author who appropriates the facts of others by changing the names, a means employed very often by certain savants.[94]

94 Vieillot to Lesson, June 28, 1828, in Lesson, "Réclamation de M. Vieillot," 144.

FIGURE 12 *Aigle Bonelli.* Hand-colored engraving after a drawing by Jean-Charles Werner based on Bonelli's specimen depicted on his portrait. In Coenraad Jacob Temminck and Jean-Charles Werner. *Les oiseaux d'Europe décrits par C. J. Temminck* (1848): plate [24]
IMAGE FROM THE BIODIVERSITY HERITAGE LIBRARY. DIGITIZED BY THE NATURAL HISTORY MUSEUM LIBRARY, LONDON

It was a case of the pot calling the kettle black; however, Vieillot was, indeed, the first to describe this species.

There was yet a third party that had also described the species. In a far from elegant maneuver Temminck wrote in 1838: "A miniature lithograph of a specimen younger than the one of plate 288 has been published recently in a new work on the birds of prey in Europe, under the name of Intermediate Eagle; the author undoubtedly did not know that we had mentioned this bird under the name indicated above."[95] Indeed, Pierre Boitard had described the same specimen that Vieillot and Temminck had seen as *Aquila intermedia* in his *Histoire naturelle des oiseaux de proie d'Europe* in 1824.[96] However, Temminck failed to mention in the *Nouveau Recueil* the author and the title of this new work on European raptors. It appears that Temminck, annoyed by Boitard's decision to name the bird, had decided not to include a proper reference to Boitard's *Aquila intermedia*. While Temminck was unaware of Vieillot's description, he obviously knew Boitard's, but Temminck still claimed authorship. Almost ten years after the reinstatement of Vieillot's binomial Temminck persisted in calling the bird "*Falco bonelli* Temm.," ignoring both Vieillot's and Boitard's descriptions as well as Lesson's rectification.[97] Needless to say, this kind of behavior did not win the sympathy of his peers.

Temminck traveled specifically to grow his network, but sometimes his perpetual quest for new and interesting species determined the route. He had made sure to meet Prince Maximilian Wied-Neuwied right after his return from Brazil in 1817, and along the same lines, he visited the former soldier and explorer Alberto Ferrero della Marmora and his Sardinian birds in Genova. Temminck was also a regular visitor of the Naturhistorisches Museum in Vienna where he studied, amongst other things, the specimens collected by the naturalist and explorer Johann Natterer, who had just arrived from Spain.[98] On other occasions Temminck made short trips for specific purposes. In 1819 he went to the London auction of the famous William Bullock collection of more than three-thousand birds together with Kuhl and Lichtenstein. The auction took place over the course of almost two months, from April 29 until

95 Temminck and Meiffren Laugier de Chartrouse, *Nouveau recueil*, vol. 1, livraison 49, plate 288, text, [1], fn. 1. For the dates of publication of each livraison, see Dickinson, "'Nouveau recueil de planches coloriées' of Temminck & Laugier."

96 Boitard, *Oiseaux de proie d'Europe*.

97 Temminck, *Manuel d'ornithologie*, 2 ed., vol. 3, 19; Temminck and Werner, *Oiseaux d'Europe*, suppl. 1st vol., "Aigle Bonelli," plate [24].

98 Stresemann, *Ornithology*, 119.

June 11.[99] In total he acquired 536 birds, keeping the most interesting ones for himself, and sending the few remaining others to the southern universities.[100]

At the auction another fortuitous meeting took place. Temminck was introduced to Guillaume Michel Jérôme Meiffren Laugier, baron de Chartrouse, a politician, collector, and naturalist from Arles. Kuhl and the baron had met a few months previously in Paris during Kuhl's visit to the collections of the Muséum.[101] Meiffren Laugier had a considerable collection of birds in his hotel in Paris and had been drawn to the auction in the hope of enlarging it. He gracefully described himself as "a useless worshiper of science" and was also the head of an enticing project for which he tried to enroll both Lichtenstein and Temminck.[102] He wished to publish a series of colored engravings of birds as a continuation of Buffon's *Planches enluminées d'histoire naturelle* with additional, new species. He needed the expertise of recognized ornithologists for its preparation, and he spoke to Temminck and Lichtenstein about the project in London. Sometime between that summer and the following March Temminck and the baron reached an agreement. By spring 1820 Temminck had prepared an index with the species to be included in the work. For some reason Lichtenstein decided not to participate in the enterprise.

Temminck and the baron signed the official contract on April 24, 1820. This contract stipulated that the work would consist of monthly *livraisons*, or fascicules, of six plates each, which would be sold at nine francs per *livraison*. Both men would choose the species to be included. Temminck was responsible for the scientific part, while Meiffren Laugier took over the management and finances, and the notarial and editorial matters. Costs and profits were to be equally divided between them. The artists chosen for the preparation of the plates were both artists at the Múseum in Paris: Jean-Gabriel Prêtre, Levaillant's friend and engraver of his and Temminck's previous works, and Nicolas Huet the Younger, who was of artistically notable descent as the son of Jean-Baptiste Huet and the grandson of the Nicolas Huet the Elder.[103] In total the *Nouveau recueil de planches coloriées* comprised six hundred plates which were published in one hundred and two *livraisons* between 1820 and 1839.[104]

99 Steinheimer, "Lichtenstein and his Ornithological Purchases," 90.

100 Steinheimer, 90, 93; Stresemann, *Ornithology*, 121.

101 Stresemann, "Origines des 'Planches coloriées'," 34.

102 Meiffren Laugier de Chartrouse to Lichtenstein, 26 December 1819, in Stresemann, "Origines des 'Planches coloriées'," 35.

103 *Contract 'Planches coloriées' Temminck, C.J. en Meiffren Laugier, G.M.J. Baron de Chartrouse*, Naturalis Biodiversity Center Archives, Archive C.J. Temminck, NL-LdnNAT_Temminck_38.

104 Dickinson, "'Nouveau recueil de planches coloriées' of Temminck & Laugier."

Temminck often stayed in Paris during the preparation of this work, sometimes for long periods. Here he enjoyed the freedom of working in the Muséum with the permission of Cuvier and the other "Messieurs du Jardin du Roi."[105] He took on the task of relabeling some specimens and changing the labels according to his views: "all of Illiger's names have been adopted, and all of Vieillot's new names crossed out."[106] It seems that Temminck's zealous attitude and arrogance made him a difficult partner to work with. In April 1821 Meiffren Laugier complained to Lichtenstein: "Temminck is not the same [man]."[107] Apparently, while Temminck accepted no criticism on his part of the work, the baron had to put up with Temminck's fault-finding on the administrative side which was, as they had agreed, his domain. Their partnership survived nonetheless and the work was published as scheduled, with minor changes.

Overall, travel enabled Temminck to build a network of acquaintances, make good friends, and start collaborations. Also, together with the direction of 's Lands Kabinet, it allowed an old idea to grow to colossal proportions.

5 The Concept of a National Museum

Prior to his appointment at 's Lands Kabinet Temminck had dreamt of a museum that would serve to put the Netherlands, which was still in the process of establishing its independence and sovereignty, on the European map. Initially Temminck's concept for a national museum of natural history seemed to be fueled by patriotic feelings and personal ambition, but the present course of natural history played a decisive role in Temminck's ideas.

On April 18, 1814 Temminck wrote a very brief letter to the State Secretary Anton R. Falck in which he proposed the possibility of creating a national museum. Temminck appealed to the general public interest and the "regeneration of the nation" (Napoleon was in exile on Elba at the time and the French period in the Netherlands was coming to an end). With the desire to be useful to his country, he wrote, he offered his private cabinet to form part of a national collection so that it might be of more general use to the country. His timing was deliberate. After the re-establishment of independence, the Netherlands became a centralist monarchy. It was safe to assume that public institutions

105 Temminck to Lichtenstein, 20 May 1820, in Stresemann, "Origines des 'Planches coloriées'," 39. The Jardin du Roi had been renamed as Jardin des Plantes twenty-seven years earlier, in June 10, 1793.

106 Temminck to Lichtenstein, 20 May 1820, in Stresemann, 39.

107 Meiffren Laugier de Chartrouse to Lichtenstein, 18 April 1819, in Stresemann, 46.

would be reinstated and that the government would soon create new ones. In his letter to Falck Temminck suggested that he was the best person to direct such a new museum: "Should our Sovereign, supported by favorable times, plan on establishing a Museum dedicated to the natural sciences, he might be inclined to entrust me with its direction."[108]

Falck responded to the proposal by pointing out that there was already a national collection in existence, namely the Cabinet du Roi under the direction of Reinwardt (which became 's Lands Kabinet within the following year). In response to this, Temminck recognized Reinwardt's qualifications with the elegant admission that they "suffice to induce me to abandon my project."[109] He confessed he had been under the incorrect impression that the Cabinet du Roi belonged to the city of Amsterdam. Temminck notified Reinwardt of his plan and Falck's response, to avoid giving the impression that he was dispensing career advice to Reinwardt, and put the matter to rest – for the time being.

Temminck's proposal of a national museum was, of course, not an original idea. National history museums were appearing all over Europe during the early decades of the nineteenth century. A few years before Temminck's suggestion to Falck, Count Hoffmannsegg had recommended the establishment of a collection of natural history to the German State Councilor Wilhelm von Humboldt and proposed Karl Illiger as its curator. Hoffmannsegg's collection would serve as a founding collection to help the realization of the project. The king accepted, and in March 1810 Illiger became the first director of the Zoological Museum of Berlin, also known as the Museum für Naturkunde or the Humboldt-Museum. This was part of the University of Berlin, and Illiger was also appointed professor of natural history.[110] Temminck had been in touch with both Hoffmannsegg and Illiger since 1810. The foundation of the Berlin Museum possibly gave Temminck the idea for a Dutch national museum and of himself as its director. Similarly, in 1819 Baron Gustaf Paykull, an entomologist, ornithologist, and marshal of the court, convinced the king of Sweden to found a national natural history museum in Stockholm, the Naturhistoriska Riksmuseet, via the donation of his private zoological collection. It does not seem too far-fetched to assume that this also informed Temminck's vision.

108 Temminck to Falck, 18 April 1814, Naturalis Biodiversity Center Archives, Archive C.J. Temminck.

109 Temminck's reply to Falck's answer is written at the bottom of his draft letter to Falck, 18 April 1814, without a date.

110 Stresemann, *Ornithology*, 103.

Falck rejected Temminck's initial suggestion simply because he considered 's Lands Kabinet under Reinwardt's direction the national collection of natural history. While Temminck acknowledged Reinwardt's position, he did not share Falck's vision of 's Lands Kabinet. Having worked with the collection – and argued with Van Marum – for a number of years, and after being commissioned to build collections for the southern universities, Temminck saw so many shortcomings in the organization of the Dutch academic cabinets that his belief in the need of a central museum grew stronger. Between 1814 and 1816 his arguments shifted from patriotic sentiment to a deeper understanding of the role of a national museum for the natural sciences and particularly systematics.

By 1816 Temminck had formed a clear idea of what such a national museum would look like and described it in a letter to Reinwardt. Temminck saw how university professors were competing with each other for specimens intended to illustrate their lessons in anatomy and pathology. He even went so far as to declare that some professors were taking advantage of their position by using specimens for "ostentation."[111] What universities should do instead, he proposed, was build collections with examples of genera of European animals, either fresh or preserved in methylated spirit, to be used for dissections and comparative anatomy. However, Temminck believed that as a complement for the didactic collections there should be a national museum, which would function as a reference collection:

> Also, a National Museum could be founded, which, with its duplicate specimens and as a result of the care of its directors, could serve to enlarge the University Cabinets. It should be built in such a manner as to serve as a source of information for those who desire to increase their knowledge in the different subjects of Natural Science, by means of comparison.[112]

As the focus of the academic cabinets lay mostly in comparative anatomy and physiology, and they were well-stocked with preparations of organs and pathological specimens, these collections were unsuitable for systematics. Temminck envisaged a new kind of institution that would serve as a reference collection for universities, independent researchers, and the general public. Temminck referred to the Muséum in Paris and the Egyptian Hall in London as

111 Temminck to Reinwardt, 25 November 1816, NL-HaNA, BiZa 1813–1848, 2.04.01, inv.nr. 4032B. Here Temminck was referring to the reform of the Dutch education system of August 1815 ("Besluit regelende het hooger onderwijs").

112 Temminck to Reinwardt, 25 November 1816.

good examples.[113] This time, however, he did not put himself forward as a candidate for the directorship of such a museum, or at least not explicitly. He said that there would be "directors" – plural – in charge of this national museum.[114] Temminck may have had the Paris Muséum in mind, where twelve professors of equal rank worked on different subjects.

Temminck sent this description of the ideal national museum to Reinwardt, and he also made sure to send a copy of this particular letter – all sixteen pages of it – to Repelaer van Driel, the commissioner-general of education, arts, and science. Temminck's main reason for sending this second copy was, he wrote, that he wanted to refute the rumors that were circulating about his wish to replace Reinwardt. Temminck hoped that his letter to Reinwardt would prove to the commissioner that Van Marum's accusations of Temminck and his supposed plan to dethrone Reinwardt as director of 's Lands Kabinet were false.[115]

That particular letter for Reinwardt included, besides his ideas for a new national museum, details on his ichthyological research, accounts of his quarrels with Van Marum and Apostool, instructions for the collection of fish and marine invertebrates, and some news about the collections for the southern universities. He also counseled Reinwardt about approaches to the study of zoology, and advised that he might focus on comparative anatomy. This lengthy missive, unusually chaotic for Temminck, was sent to Batavia as proof of Temminck's commitment and loyalty to Reinwardt. Rather conveniently, it also helped convey his ideal plan for a national museum to Repelaer van Driel. This time, too, just as in 1814, Temminck's proposal fell on deaf ears.

There was yet another advocate for a national museum for natural history: professor Sebald Brugmans from the University of Leiden, who was in charge of the university's natural history cabinet. In March 1817 Brugmans wrote to Falck suggesting the creation of a central academic museum that would be open to the public. By reorganizing and expanding the existing collection at Leiden University they could create, without too much expense, a museum "in accordance with the value of our collection, the splendor of our University and with the current state of Science."[116] Brugmans' expansion plans were well received, and although there was no intention yet to transform the Leiden collection

113 The Egyptian Hall was commissioned by William Bullock as a museum for his private collection, and it was erected in Piccadilly, London, in 1812.

114 Temminck to Reinwardt, 25 November 1816, NL-HaNA, BiZa 1813–1848, 2.04.01, inv.nr. 4032B.

115 Temminck to the commissioner-general of education, arts, and science, 4 December 1816, NL-HaNA, BiZa 1813–1848, 2.04.01, inv.nr. 4032B.

116 Brugmans to Falck, 31 March 1817, UBL-AC2, 70; Dreier, "Voor museum en vaderland," 19–20.

FIGURE 13 *Sebald Justinus Brugmans, hoogleraar Botanie, Natuurlijke Historie, Medicijnen en Chemie te Leiden*. Portrait by Charles Howard Hodges, 1815. Oil on canvas
LEIDEN UNIVERSITY LIBRARIES, DIGITAL COLLECTIONS, ICONES LEIDENSES 205

into a national museum, the government paid the generous amount of 57,390 guilders – a small fortune – for the expansion of the university cabinet.[117]

The natural history collections of the University of Leiden had been previously enlarged with part of the Stadhouders Kabinet (the Cabinet of the Stadholder) when Brugmans brought with him from Paris in 1815. The cabinet belonged to William V, the Prince of Orange and last stadholder of the Dutch Republic, who had gathered it during the second half of the eighteenth century. The French confiscated the cabinet and took it to Paris, to the Muséum d'Histoire Naturelle. After Napoleon's downfall the Stadholder's son, now King Willem I, asked Brugmans to travel to France and retrieve his father's cabinet. Although the mission was not without its difficulties – Cuvier and Lamarck were opposed to it, and Alexander von Humboldt became the mediator between the different parties – Brugmans succeeded in taking back at least part of the original material.[118] The king ordered that the cabinet was to be given to the University of Leiden.[119] The fact that Willem I did not use his father's cabinet to enlarge the collection of 's Lands Kabinet may be an indication of the king's sympathies and gratitude towards Brugmans for his efforts in Paris.

Somehow the idea of a central national museum started to take root. The king was, after all, determined to raise the prestige of his new country by boosting its cultural institutions. During the summer of 1819 Falck – now minister of education, industry, and colonies – was considering a future for the collection at the University of Leiden and asked Temminck's advice on the matter. Following a few initial conversations a distressed Temminck wrote to Falck on July 17 opposing Brugmans' ambitious plans of turning the Leiden University Cabinet into a national museum.

His arguments revolved around three main points. First, Temminck referred to the French proverb "Qui trop embrasse mal étreint" (He who embraces too much grasps imperfectly).[120] The university cabinet was acceptable for didactic purposes, but converting it into a national museum would require whoever was in charge to master all the different fields of natural history, to give lectures, manage the collections, master taxidermic and curatorial skills, and fulfill all official duties. This was nothing short of a herculean task. Second, giving more prominence to the University of Leiden would only exacerbate the rivalry with other universities, as all were required by law to have their

117 Trustees Policy note, 23 May 1818, UBL-AC2, 71 II; Dreier, 20.

118 Pieters, "Naturaliënkabinet van Stadhouder Willem V," 39–40; Raat, "Humboldt and Temminck," 21.

119 Copy of the Royal Decree of 3 November 1815, UBL-AC2, 69 I.

120 Temminck to Falck, 17 July 1819, NL-HaNA, Collectie 048 Falck, 2.21.006.48, inv.nr. 85; Colenbrander, *Gedenkschriften van Anton Reinhardt Falck*, 413.

own collections. The competition between the professors and the lack of skills amongst them prompted Temminck to call these academic collections "towers of Babel." Third, and perhaps more important, Temminck argued that a national museum – or, as he put it, a "national monument" – should be centralized and independent of university management. Only then could such a museum be of service to all universities and to "the learned practitioner of diverse fields."[121] The city chosen for the location of the museum, Brussels or Amsterdam, should reflect the national character of the institution. In other words, if the academic museum of the University of Leiden should become the new national museum, it would be to the detriment of the other universities, of practitioners of natural history, and of the quality of the collection itself.

Temminck did not explicitly propose anyone as a director for his ideal national museum in that particular letter, not even himself. He suspected that Brugmans intended to merge the Leiden cabinet with 's Lands Kabinet, which would put Reinwardt's position in peril. By stating that "Reinwardt was in no way to be employed as a subordinate or substitute after his return" from Java, Temminck implied that Reinwardt should remain to be the director.[122] Concerning Brugmans, Temminck criticized his skills as a collection manager, arguing that the Leiden cabinet was poorly curated and that Brugmans did not have the necessary knowledge or experience to keep it in good shape. Although Temminck did not put himself forward for the position, he clearly stated that he had plans for himself and his collection. He added, resolutely, that "if a national museum is not founded, or if its establishment seems altogether defective, no one can blame me if I decide to pursue my passion for expanding my knowledge in another country."[123] He simply threatened to leave the Netherlands, collection and all.

Temminck was wary of further discussions and of the competition with both Brugmans and Van Marum who, he thought, would stop at nothing to get him out of the way. He was surprisingly outspoken in this particular letter to Falck. Perhaps Temminck sensed that, as he lacked both an academic education and a good relationship with Brugmans and Van Marum, there would be no place for him in a national museum based at the University of Leiden. He was worried about the future of his private cabinet: he would not let it become a part of Brugmans' museum, even if it meant that he had to move to another country.

Neither Brugmans nor Temminck considered 's Lands Kabinet suitable as a national museum as it was too small and incomplete and contained mostly

121 Temminck to Falck, 17 July 1819.
122 Temminck to Falck, 17 July 1819.
123 Temminck to Falck, 17 July 1819.

invertebrates and minerals. Even after the purchase of Reinwardt's private collection – which consisted mainly of anatomical samples and skeletons, but also of insects, minerals, and a few other specimens – it would remain incomplete and lack the grand character that both Brugmans and Temminck deemed necessary for a national institution.[124] Besides, the Trippenhuis, where the Kabinet was housed, had no room for expansion: it also housed the constantly growing art collection of Apolstool's Kunst Museum and the Koninklijk Instituut's scientific instrument collections. Therefore, it seemed most logical to merge 's Lands Kabinet with other existing collections. Brugmans thought that this should be the Leiden University Cabinet, which had already been enlarged with the old Stadholder's cabinet and was, at the time, one of the most prominent academic cabinets in the country.

Temminck objected to 's Lands Kabinet transformation into an academic collection: for him a national museum should be part of an independent institution under ministerial administration. The role reserved for Reinwardt, who was still the director of 's Lands Kabinet and a professor at the Athenaeum Illustre in Amsterdam, was another matter for debate. In any case, Temminck felt obliged, as the interim director of 's Lands Kabinet and as Reinwardt's friend, to oppose Brugmans' plan to unite 's Lands Kabinet with the academic cabinet in Leiden strongly.[125]

However, fate intervened. The day after Temminck finished his letter to Falck Brugmans felt unwell after a long day of working in his garden. He passed away unexpectedly just four days later, on July 22, 1819. He was fifty-six years old. His death was attributed to colic, and according to Brugmans' colleagues, the result of his drinking too much beer after the consumption of fruit.[126] Conjectures about a successor immediately followed. Reinwardt seemed to be the most likely candidate for the position. Although Reinwardt was still in Java, Van Marum wasted no time and informed him of Brugmans' death and told him in the same paragraph that a letter was on its way to Java inviting him to accept the professorship in Leiden.[127]

---

124 Temminck and Van Marum were commissioned to produce a catalogue of and a valuation for Reinwardt's collection when he sold it to the government because, in his view, a director of a national collection should not also own one privately; Gijzen, "'s Rijks Museum," 31.

125 Temminck to Falck, 17 July 1819, NL-HaNA, Collectie 048 Falck, 2.21.006.48, inv.nr. 85; Colenbrander, *Gedenkschriften van Anton Reinhardt Falck*, 413.

126 Van Marum to Reinwardt, 15 August 1819, NHA, Archive Martinus van Marum 529-20b. Briefwisseling, 5–29, 1776–1837. Nord-Hollands Archief, Haarlem; Van Heiningen, *Correspondence of C. G. C. Reinwardt*, 325.

127 Van Marum to Reinwardt, 15 August 1819.

For Reinwardt this invitation was not only unexpected, but also unwanted. His plans were altogether different. He wished to stay in Java for a longer period to finish his writing and publish the results of his research. He was perfectly satisfied with his current position in Amsterdam and considered a professorship in Leiden a daunting prospect.[128] A few months and several letters from an insistent Van Marum later, he nevertheless accepted the position in Leiden and became a professor at the university as well as the director of its Botanical Garden.

With Reinwardt's new appointments the proposals for a national museum changed. By the end of 1819 the need for a national museum was no longer questioned, but it seems that Temminck's objections to Brugmans persisted in Falck's mind. He had kept the letter sent to him by Temminck in July 1819, which Falck had noted should be taken "cum grano salis," but it had been "the seed encapsulating the now fulfilled ideal for the great museum of natural history."[129] Falck's plans included the merging of the cabinets from the University of Leiden and 's Lands Kabinet as well as Temminck's private collection, which Temminck himself had offered for this purpose before. Although Falck admitted that a divorce of the museum directorship from the university professorship would be the best option, he wished to grant a final say on the management to the University Board of Trustees (*Curatoren*). The board would act as supervisors to the museum's director as well as intermediaries between the Ministry and the director of the museum. This was not a satisfactory arrangement for Temminck, who predicted that there would be problems whenever a disagreement regarding budgetary questions arose, not to mention, heaven forbid, the risk of letting unsupervised professors use the collection for teaching.[130]

At this point both Reinwardt and Temminck were candidates for the directorship of the museum. Unsurprisingly Temminck – although he thought more highly of Reinwardt than he did of Brugmans – still argued that the burden of the management of the museum was incompatible with the demands of a full professorship. He objected to Falck: "It seems to me that a Director of the Botanical Gardens, charged with the professorship of chemistry, botany, mineralogy, comparative anatomy, and all subjects of zoology, has an infinite amount of work to do."[131] Besides, Temminck insisted, an independent director

---

128 Reinwardt to Van Marum, 3 June 1820, NHA, Archive Martinus van Marum 529-20b; see also Weber, "Hybrid Ambitions," 166–67.

129 Colenbrander, *Gedenkschriften van Anton Reinhardt Falck*, 411.

130 Gijzen, "'s Rijks Museum," 6–7.

131 Temminck to Falck, 29 October 1819, NL-HaNA, ARA, 2.14.03, inv.nr. 5; Gijzen, 8–9.

would prevent competition with professors of other universities. Now putting himself forward as the best candidate, Temminck added that twenty years of experience had taught him that a vast network of collectors and museums was much more useful than funds when it came to expanding a collection. And he had spent years building up such a network. On top of this, Temminck assured Falck, he could never part with his collection. It was the basis of his research and the fruit of many years of work and worries. In other words, if Temminck was unsuccessful in securing the position, he would not donate his collection to the national museum.

Six months later, after reaching an agreement with Temminck, Falck sent a report including his final proposal to the king: his private collection ("the richest of Europe") would be part of the new museum provided that he would receive a modest yearly salary of two thousand guilders as director of the museum plus a life annuity of another three thousand guilders per year for him – or his widow.[132] Falck's arguments in favor of this proposition were exactly the same as Temminck's. First, neither the Leiden University collection, nor 's Lands Kabinet, nor the merger of both could match in importance other European museums like those in Paris and Vienna. Second, a university professor could not take on the amount of work that the direction of the museum required. Third and final, a new museum would serve as a central repository for the universities' didactic material, thus preventing rivalry between the universities. Since King Willem I was very keen to match France, Austria, and every other country in Europe in all possible domains, Falck insisted on the attractive nature of Temminck's proposition and its benefits for the country, its king, its commercial interests, and its colonies. For financial and practical reasons Falck suggested Leiden as the place where the new museum should be established.[133] Five days later 's Rijks Museum van Natuurlijke Historie was founded by royal decree with Temminck as its director.

132 Falck to King Willem I, 4 August 1820, NL-HaNA, ARA, 2.14.03, inv.nr. 5; Gijzen, 10. For comparison, Reinwardt received 24,000 guilders per year while working in Java, which, according to Andreas Weber, "far exceeded his salary as professor at the Athenaeum Illustre in Amsterdam" (Weber, "Hybrid Ambitions," 122).

133 Falck to King Willem I, 4 August 1820, NL-HaNA, ARA, 2.14.03, inv.nr. 5.

CHAPTER 3

National Museum, National Expeditions

The interim directorship of 's Lands Kabinet and his growing prestige as naturalist gave Temminck confidence to pursue an old idea he had concocted at the beginning of 1814: the country was in dire need of a new national museum of natural history. His intimate knowledge of university collections, the limited extent and quality of those of 's Lands Kabinet, and his visits to museums abroad had confirmed his impressions. In the end it took Temminck six years and a number of arguments to see his vision realized. He put all his influence in government affairs to work. After careful consideration of Minister Falck's report of August 4, 1820, King Willem I issued Royal Decree no. 75 on August 9, ordaining the foundation of 's Rijks Museum van Natuurlijke Historie. The museum collection was to consist of Temminck's cabinet, 's Lands Kabinet, and that of the University of Leiden, and it was to be of service to academic education. The royal decree met almost all of Temminck's conditions – all but two. The new museum was placed under the responsibility of the Leiden University trustees, and it was intended for the use of professors and students, two measures explicitly disapproved of by Temminck. Nevertheless, Temminck was as resourceful as he was ambitious, and he built upon his dream until the new museum ranked amongst the best in Europe.

1 The Birth of 's Rijks Museum van Natuurlijke Historie

The royal decree by which 's Rijks Museum came into being consisted of only five brief articles. Three of them specified the financial arrangement with Temminck, one determined the collections to be merged for the founding collections, and one ordained that the museum should send its duplicates to the universities.[1] In all the decree was about five hundred words long – one-third of Falck's report. It was concise and clearly insufficient, as it left many delicate issues open to interpretation. The trustees, for example, were alarmed by the possibility of the dispersal of Brugmans' collection. When the University of Leiden acquired Brugmans' collection it was stipulated that it should not, under no circumstances, be given either in part or in its entirety

1 Royal Decree no. 75, 9 August 1820, NL-HaNA, ARA, 2.14.03; Gijzen, "'s Rijks Museum," 12–14.

 | DOI:10.1163/9789004441491_005

to another university.[2] Even more confusing was the fact that the decree did not clearly define the shared responsibilities of the trustees and Temminck. In December 1820, after four months of exhausting negotiations, Falck issued a ministerial decree clarifying the relationship between the new director and the Board of Trustees of the Leiden University, as well as the composition and function of the collections. With this decree Falck found a way to please – or at least, appease – both Temminck and the trustees.

Notably, this second decree divorced the anatomical collections from the rest of the natural history specimens. 's Rijks Museum collections were to consist of zoological specimens (excluding human remains), skeletons, fossils, and minerals, but "no pathological nor physiological preparations of animal organs should be part of it."[3] These anatomical and pathological preparations were to be donated to the Cabinet of Physiology at the University of Leiden, which was in the care of Gerard Sandifort, a former student of Brugmans and professor of anatomy and medicine in Leiden. Other objects from Brugmans' collection, like the skeletons and prepared animals, were placed in 's Rijks Museum. The director was not allowed to pass them on to any other institutions.[4]

Falck gave particular directions to the professors in *anatomia comparata*: they had free access to the museum objects, but they could not remove them from the museum. Any specimens needed for their lessons should be retrieved from storage and placed back only by museum staff. Those professors who wished to study specimens for an extended period of time could borrow them. Similarly, students of natural history were granted access to the objects in situ.[5] There were, in short, actual restrictions for the use of the collection for teaching comparative anatomy, but these did not apply to research. In this way Temminck could rest assured that the professors would not use the museum collections for comparative anatomy lessons. These arrangements reflected the universities' preference for teaching and researching comparative anatomy above systematics, which was considered the domain of *aficionado* naturalists and collectors of prepared specimens – a preference that would have far-reaching consequences for the development of systematics to the present day.

---

2 C. M. van Dam to Leiden University, 14 October 1819, UBL-AC2, 70; Gijzen, 34.

3 "Extract uit het Register der Handelingen en Resolutien van den Minister, voor het Publieke Onderwijs, de Nationale Nijverheid en de Koloniën, No. 30 Den 31 December 1820," in Gijzen, 17.

4 Temminck, "Uit het kabinet van wijlen professor Brugmans," a catalogue of the specimens transferred from the Brugmans collection to 's Rijks Museum, Naturalis Biodiversity Archives, Archive C. J. Temminck, NAT_ARC_TEM_02460-64.

5 "Extract uit het Register, No. 30 Den 31 December 1820," in Gijzen, 17; Huistra, *Afterlife of the Leiden Anatomical Collections*, 63.

With the director's role, Temminck found himself in a position he did not enjoy. The director was in charge of the finances, the curation of specimens, and redistribution of duplicates, but the final responsibility for the museum lay with the trustees. Temminck had to answer to them and consult them about any significant change in management. Temminck had to report to the trustees on matters of staffing, funding, and acquisitions. The Ministry of Education was also to receive a copy of these reports.[6] As a man in his forties, and a proud man, who had always worked autonomously, Temminck found this situation hard to accept. His relationship with the university would prove challenging in the following years.

Reactions to the new museum varied with the different parties' interests. The Koninklijk Instituut van Wetenschappen, for instance, wrote a letter to the king to remind him that 's Lands Kabinet had been created for the Institute's First Class members who had an interest in natural sciences and mathematics. With the move of the collection from Amsterdam to Leiden, the Koninklijk Instituut feared losing access to the natural history specimens.[7] As their objections remained unanswered, Gerardus Vrolik, the secretary and a member of the First Class as well as a former pupil of Brugmans, even tried to stop the imminent transfer of 's Lands Kabinet from the Trippenhuis to Leiden.[8]

The University of Leiden, by contrast, was content. Its collection remained in Leiden to become a national museum, and under the supervision of their trustees, very much like Brugmans had envisioned.[9] However, after Brugmans' death, there was yet another Leiden contender for the directorial position: Paulus Godfried van Hoorn. He had been Brugmans' student and, later, his friend; the two men lived in the same house. He had completed a doctorate in medicine and became the assistant director of the University Cabinet, where he specialized in geology and mineralogy.[10] Since he was Brugmans' assistant, he had expected to succeed him in the University Cabinet. When this cabinet

---

6 Gijzen, "'s Rijks Museum," 17–20.

7 Secretary First Class of the Koninklijk Nederlands Instituut to the minister of public education, national industry, and colonies, 15 September 1820, NL-HaNA, BiZa 1813–1848, 2.04.01, inv.nr. 4100; Koolmees, "Caspar Georg Reinwardt," 50. The First Class was concerned with the natural sciences in the broadest interpretation, including mathematics, chemistry, astronomy, zoology, botany, anatomy, physiology, geology, medicine, and pharmacy.

8 Van Marum to Reinwardt, 17 September 1820, NHA, Archive Martinus van Marum 529-20b; Van Heiningen, *Correspondence of C. G. C. Reinwardt*, 444.

9 Curatoren van Hogeschool Leiden to the minister of public education, national industry, and colonies, 2 September 1820, NL-HaNA, BiZa 1813–1848, 2.04.01, inv.nr. 4100; Koolmees, "Caspar Georg Reinwardt," 50.

10 Van Hoorn, "Levensberigten," 4.

was merged with the Temminck and Amsterdam collections, Van Hoorn was offered work as an assistant director under Temminck, the same role he had occupied under Brugmans. This hurt Van Hoorn's pride. He claimed the position as director of the Mineralogy Department and stated that "my Doctorate, and the position I have otherwise enjoyed in Society, prevent me from accepting a subordinate position."[11] Van Hoorn's doctorate was, in his view, sufficient proof of his abilities to grant him a higher status in an official institution – Temminck's honorary doctorate from Groningen University was apparently not enough in Van Hoorn's eyes to justify Temminck's position as the museum director. Van Hoorn threatened to resign. After consulting with the minister and the board of trustees, he reluctantly accepted the position of curator of the mineralogy department.[12] Unfortunately, this turned out to be an unsatisfactory arrangement for everyone involved.

Van Marum, who had initially wanted the director's post to go to Reinwardt, changed his mind from the moment Reinwardt accepted to step into Brugmans' role. Although Van Marum could not help noting that Temminck "had been very selfish," he recognized that the burden of a double assignment as professor and director for Reinwardt would be too much to be practical.[13] Both Van Marum and Reinwardt saw the advantages of merging 's Lands Kabinet with the Leiden University Cabinet. Eventually Van Marum wrote to Minister Falck, acknowledged the benefits of a national museum for the Netherlands, and expressed his wish to contribute in any way possible.[14]

Despite Van Marum's positive attitude, the creation of a national museum was potentially detrimental to the collection of the Hollandsche Maatschappij in Haarlem, which was still under Van Marum's rule. He realized that, as the new Rijks Museum had secured the necessary funds and facilities for an expansion, it would soon surpass the Haarlem collection, so that visitors to the Leiden museum would not consider the Haarlem collection worth a visit. Van Marum decided not to compete with Leiden. Instead, his thoughts turned to giving the Haarlem cabinet a new focus that would make it useful as well as unique. He suggested that the cabinet concentrate its efforts on the indigenous fauna of the Netherlands, a subject that had always interested the Hollandsche Maatschappij. The collection would be reorganized and expanded with a collection of Dutch fish – prepared by Temminck's pupil and former doorman, Reindert Draak – and a collection of insects, a generous donation from

---

11 Van Hoorn to Temminck, 24 August 1821, Naturalis Biodiversity Center Archives, Archive C.J. Temminck; Holthuis, *Rijksmuseum*, 27.

12 Holthuis, 27.

13 Van Marum to Reinwardt, 8 September 1820 and 7 July 1821, NHA, Archive Martinus van Marum 529-20b; Van Heiningen, *Correspondence of C. G. C. Reinwardt*, 341, 434.

14 Van Marum to Falck, 8 September 1820; Weiss, "The Masses and the Muses," 22.

professor Van Swinderen.[15] Although the new policy was very well conceived, it was not implemented with enough energy to achieve Van Marum's goals, and the cabinet remained incomplete.[16]

While everyone was vying and campaigning for the post of director of 's Rijks Museum, Reinwardt was in Java, scarcely receiving information in letters from friends and officials. He remained on the sidelines of the discussions. After Temminck's appointment Reinwardt confessed that he was satisfied with the outcome: "It is a true pleasure to me that Mr. Temm[inck] has been appointed Museum Director and that I can stay out of the war against Moths and Larder beetles."[17] The museum's director was not only head of the institution, but also in charge of the safety and curation of the objects, fulfilling the double role of administrator and collection manager. Reinwardt was unwilling to take on these roles on top of his professorship, but he did resent Temminck's negligence of informing him about his appointment.[18] Although Reinwardt accepted the professorship in Leiden, he was worried about finishing his research and publications, i.e. seeing the results of his stay in the colonies.[19] He was allowed to remain in the Dutch East Indies a while longer to complete his projects, and he finally returned to the Netherlands in 1822.

2 Temminck's Directorate

By the end of 1820 Temminck was settled as director of 's Rijks Museum van Natuurlijke Historie, a wish he had harbored for more than six years. He finally occupied an official position, something he had coveted for over a decade. The museum's directorship would have repercussions on his scientific work, but at this stage Temminck did not seem to realize this, or if he did, he never expressed any concern.

After its establishment the museum was housed at the Hof van Zessen on the Rapenburg in Leiden and shared space with the Museum of Antiquities and the scientific instruments collection of the Koninklijk Instituut (the

15 Sliggers, "Naturaliënkabinet van de Hollandsche Maatschappij," 67.

16 Bierens de Haan, *Geschiedenis van een verdwenen Haarlemsch museum*, 50–52; Sliggers, 68.

17 Reinwardt to Van Marum, 1 January 1822, NHA, Archive Martinus van Marum 529-20b; Koolmees, "Caspar Georg Reinwardt," 50. Reinwardt referred here to common pests affecting natural history specimens, i.e. the larvae of moths and larder beetles that have a particular taste for dried plant and animals.

18 Reinwardt to Van Marum, 25 March 1822, NHA, Archive Martinus van Marum 529-20b; Van Heiningen, *Correspondence of C. G. C. Reinwardt*, 448.

19 Reinwardt to G. A. G. Ph. Baron van der Capellen tot Berkenwoude, 6 January 1820, in De Vriese, *Ons streven naar waren roem*, chap. 5.

Physisch Kabinet). Part of the Stadholder's collection was already there (it had been returned to the Netherlands from Paris by Brugmans), and so were other collections from the Museum of Antiquities. Despite Temminck's continuous complaints to the ministry about the lack of space and the building's shortcomings, it took eighty-five years and two consecutive directors for 's Rijks Museum to occupy a building of its own; this was on the Raamsteeg and opened in 1905. Even Temminck's private collection remained in his house in Amsterdam for a decade, and afterward remained unpacked in the basement at the Hof van Zessen. This was Temminck's way of pushing the ministry towards giving the museum more space, which, understandably, irritated Falck. Nevertheless, the collection expanded, and the building became ever more cramped and insufficient. Between 1824 and 1855, when the collections outgrew the building, a series of renovations and expansions were undertaken. An additional floor was added, and a whole row of houses was bought and demolished to make room for a new wing to house the collections. Temminck did not find a place in the museum for his magnificent collection of birds that was to his satisfaction until 1838.[20]

The collection remained in the expanded Hof van Zessen and was open to the public on Mondays and Wednesdays, from eleven o'clock in the morning until two o'clock in the afternoon. Judging from a newspaper advertisement for the museum opening, Temminck thought it was too early and too inconvenient for him to allow the public to visit. His announcement in the national papers seemed to be intended to manage the public's expectations rather than extend an invitation: "The Director of 's Rijks Museum van Natuurlijke Historie in Leiden announces that, although the galleries are not yet completed and that not all the objects in store can be displayed, those who wish to visit it *in this condition* will be given the opportunity, and entrance tickets will be available at the Papengracht[.]"[21] Visitors were not supposed to arrive in high hopes. Apparently, Temminck had a very clear idea of what made a museum complete and ready for opening.

The museum was primarily a research institute, and it remained so for more than a century. Interested and curious naturalists were very welcome to work on the collections at any time. Students from Leiden University, however, were only granted access to the museum's collections after obtaining a ticket from the museum administrator. They could visit the museum for four days a week,

20 Gijzen, "'s Rijks Museum," 249–56; Kortenbach, "Het Hof van Zessen"; Susanna, "Levensschets van Temminck," 58.

21 Temminck, "Van de Directeur van 's Rijks Museum," in *'s Gravenhaagsche Courant*, May 8, 1822 (emphasis original).

FIGURE 14 Rapenburg 28. Photograph by Jan Goedeljee [before 1882]. 's Rijks Museum van Natuurlijke Historie was housed in the Hof van Zessen at the Rapenburg, Leiden
COLLECTION ERFGOED LEIDEN EN OMSTREKEN, KFO20070090

three hours a day. This was hardly a welcoming arrangement, especially considering that students had access to the university's anatomical collections every day except Sundays, at any time, without prior arrangement.[22]

The royal decree of 1820 did not regulate the museum's organization – it only arranged the appointment of its director. Temminck selected the scientific and technical staff, and the Board of Trustees of the University had the right to veto his decisions. Temminck surrounded himself with like-minded naturalists. The administration of 's Rijks Museum was entrusted to Leonardus Franciscus Thijssen, who had traveled to Paris with Brugmans to reclaim the Stadholder's cabinet. When Thijssen passed away in August 1824, Joannes Andreas Susanna, his friend and a clerk at the Leiden court, succeeded him.[23] Susanna then worked with Temminck at the museum for thirty-three years. The museum records and surviving correspondence suggest that Susanna was

22 Vrolik, "Levensbericht van Coenraad Jacob Temminck," 75; Huistra, "Preparations on the Move," 17.

23 Details of the administration of the Rijks Museum under Temminck's directorship can be found in Gijzen, "'s Rijks Museum"; and Holthuis, *Rijksmuseum*. Biographical details for the staff Leiden Museum members can be found in Holthuis, *Rijksmuseum*, 18–38.

an influential figure. He took the helm whenever Temminck traveled abroad, and during the final years of his life Temminck asked Susanna and the curator of vertebrates, Hermann Schlegel, to act as directors of the museum. Susanna's life was strongly entwined with the history of the museum: he married Thijssen's widow, and one of his daughters married a [later] curator of the museum Jan Adrianus Herklots.

When the museum was founded in 1820 there were three scientific departments: that for vertebrates, invertebrates, and mineralogy (without fossils, as those were regarded as zoological objects). More than twenty years later a department of entomology was added. A curator led each of the departments, assisted by technical staff. The scientific personnel consisted of men Temminck had met personally or that had been recommended to him. He entrusted the Department of Vertebrates to Heinrich Boie, with whom he had corresponded ever since Boie had published some critical remarks on the *Manuel d'ornithologie*. Boie left his post at the University of Heidelberg's zoological collection to come to Leiden.[24] Working at the new 's Rijks Museum was apparently a very promising prospect for a young zoologist.

The collection of vertebrates was the most impressive of all, and also the largest. Beside the mounted specimens it also contained a substantial number of skulls and skeletons. Reinforcements came in 1822: Heinrich Christian Macklot was appointed as curator of the osteological collection, which had grown dramatically in just two years.[25] Of German origin, Macklot was trained as a pharmacist and had studied physics and medicine from 1818 to 1822 in Heidelberg, at Boie's alma mater, where he obtained his doctorate. He came to Leiden on Boie's recommendation.[26] With Macklot the department of vertebrates now had two curators, but only for a very short period. Both Boie and Macklot were appointed as members of the Natuurkundige Commissie voor Nederlands-Indië in 1823 – a scientific committee to explore the natural resources in the Dutch territories in 'East India' [Indonesia] – and Temminck had to look for their replacements. He turned to his network of acquaintances and colleagues.

The director of the Naturhistorisches Museum in Vienna, Carl Freiherr von Schreiber, had a suggestion: the naturalist and ornithologist Hermann Schlegel. Schlegel had been in contact with – and learning from – the ornithologists Christian Ludwig Brehm and Johann F. Naumann for years. By the time

24 Susanna, *Levensschets van Hendrik Boie*, 25.

25 Temminck, Report to the Ministry of Public Education, National Industry, and the Colonies on the Year 1821, 29 August 1822, Naturalis Biodiversity Center Archives, Jaarverslagen.

26 Holthuis, *Rijksmuseum*, 29.

Temminck wrote to Schreiber looking for an assistant to replace Boie, Schlegel had been working at the museum in Vienna for a year.[27] In 1825, satisfied with Schlegel's references, Temminck appointed him curator of the Department of Vertebrates. Schlegel eventually succeeded Temminck as the director of 's Rijks Museum in 1858. He was the only member of Temminck's curatorial staff who did not hold a university degree.

The invertebrate collection remained without a curator until 1823, when Temminck finally created the much-needed position. For this post he chose Jan van der Hoeven, a young naturalist with a doctorate from Leiden University. Van der Hoeven's dissertation bore the promising title "Dissertatio philosophica inauguralis de sceleto piscium." However, Van der Hoeven's appointment proved to be a misstep. He left the museum after only one year in the curatorial position to study medicine. A versatile and accomplished man, Van der Hoeven worked on comparative anatomy, medicine, entomology, anthropology, education, and even politics. He became a professor at the University of Leiden in 1826.

Temminck found a replacement for Van der Hoeven in the paleontologist Willem de Haan, who had also obtained a doctorate from Leiden University, and who would quickly become an authority on insects and crustaceans. Unfortunately, a spinal disease paralyzed him partially, and he had to retire in 1846. Jan Adrianus Herklots (Susanna's son-in-law) took over the invertebrates department. He too was a former student at Leiden University, and he had written a thesis on the crustaceans from the African Gold Coast [in present Ghana].

The realm of insects, which were part of the invertebrate department, was first taken care of by Anthony Johannes d'Ailly, the chemist from Amsterdam who had assisted at 's Lands Kabinet and continued to do so at 's Rijks Museum, but without an official appointment as curator.[28] In 1854 a separate Department for Entomology was established. The curator for the new department did not have a university degree in science: Samuel Constantinus Snellen van Vollenhoven was a lawyer by occupation and an entomologist by avocation, and an excellent one at that. A frequent visitor to 's Rijks Museum's insect collection, Snellen van Vollenhoven was the president of the Dutch Entomological Society and the founder of the Society's journal, the *Tijdschrift voor entomologie*, which was dedicated exclusively to insects. An excellent draftsman, he also provided many of the beautiful illustrations included in the

27 Schlegel, "Levensschets van Hermann Schlegel," 539–40.

28 Temminck rewarded d'Ailly for his efforts and assistance with a gift in the form of a collection of insects from Java made up from duplicates from the Leiden collection (Holthuis, *Rijksmuseum*, 31).

journal. By the time he was appointed as curator, his name was already very well-known in the entomological world – and it still is.[29]

Paulus Godfried van Hoorn (Brugmans' former assistant director at the Leiden University Cabinet) settled for a position as curator of mineralogy. He did so only grudgingly. After all, he had unsuccessfully contended for the directorship of the museum. Once appointed to curator, he vanished for weeks at a time. He was finally – and rather shamefully – forced to resign in 1837, after no fewer than seventeen years of conflict. He was replaced by Elte Martens Beima, an astronomer with a degree from Leiden University.

Plants ranked in second place. Although botanical collections were officially in the care of 's Rijks Museum, no botany curators were appointed. The museum collections, even though they included Reinwardt's plants from Java, consisted mainly of zoological specimens and this, together with Temminck's preference for zoology, might explain this lack of enthusiasm for the botanical collections. To remedy the situation, a royal decree was issued in 1829 for the foundation of 's Rijks Herbarium, which was located in Brussels with the renowned botanist Carl Ludwig Blume as its director. Blume had been Reinwardt's student and his successor, from 1822 until 1826, as director of the Botanical Garden in Java. After a long period of living and working in the Dutch East Indies, Blume possessed a considerable plant collection, including many specimens from the Natuurkundige Commissie, which went on to enrich 's Rijks Herbarium.[30]

Temminck chose his technical staff just as carefully as the curators. Cornelius Overdijk, supervised by De Haan, prepared the invertebrates until he, too, was appointed a member of the Natuurkundige Commissie in 1838 (curiously, it seems that he was not replaced until the 1920s). Hendrik Verstraten took care of the osteological preparations, and Jacobus Thomas ter Meer led the taxidermists' team. He had learned the art from his father, Hermanus Hendrikus ter Meer, a resilient and inventive man who, besides making mechanical puppets, also worked as an art teacher, firefighter, and watchman in Leiden, and doing some taxidermy work on the side. Some specimens prepared by him still survive in the Leiden collection, including a mandrill and a few baboon species.[31]

Jacobus Thomas ter Meer was truly gifted, so much so that he soon earned a good reputation thanks to his life-like mounts of birds and mammals. He

29 Krikken et al., "Samuel Constant Snellen van Vollenhoven"; Schilthuizen, "Early Entomological Art."

30 After the secession of the southern provinces and the establishment of the independence of the Kingdom of Belgium in 1830, the Herbarium was taken to the Hortus Botanicus in Leiden and remained there until 1839, when the institute moved to its own building at the Rapenburg. Smit, "Rijksherbarium," 5–11; Weber, "Hybrid Ambitions," 195–200.

31 Grondhout, "De Ter Meer's."

worked for 's Rijks Museum for fifty-seven years and passed on his artistic and technical knowledge to his sons Franciscus Hendrikus and Hermanus Hendrikus, who both also worked at the Leiden museum for several decades. Hermanus Hendrikus, in turn, taught the art of taxidermy to his son Herman Hendrik ter Meer, who became one of the best in his trade.[32]

Temminck was an expert taxidermist himself. He had first learned the craft from Levaillant and Naumann (his friend and author of a well-known manual on taxidermy) and later, by experimentation, continually refined his mounting technique for specimens.[33] He even developed a new way to mount fish, which were traditionally cleaned through a cut across their belly. This procedure damaged the underside of the fish and its fins, which are important for purposes of classification. Temminck proposed to cut the fish along one side only, from head to tail, and it was gutted and mounted so that one side and the ventral aspect remained intact. The procedure became known as 'Temminck's method.'[34]

Life-like mounts, however beautiful, were not meant to please the eye but to preserve the main characteristics needed to classify the animal. The museum's taxidermists had been practicing and improving their techniques to achieve this. The Ter Meers stopped stuffing the skins with hay and hemp and started making mannequins of wood or bundles of tied-up hay that mimicked the original form of the animal in question, making a frame over which the skin was stretched. These were the first steps towards developing the modern technique known as 'dermoplasty,' which involves making a customized frame that is precisely the right size and form to shape the animal skin.

In November 1835, Jacob ter Meer made a wooden model for a recently acquired skin of an African elephant, purchased from the Maison Verreaux, a family business dealing in natural history. Temminck was so pleased with the results that he decided to exhibit the model (without the skin) to the general public to demonstrate Ter Meer's extraordinary skill and expertise. Temminck even placed advertisements in the major Dutch newspapers inviting the public to see the model.[35] These advances also deserved mention in Temminck's report to the ministry:

32 Shufeldt, "Taxidermical Methods in the Leyden Museum," 1024; Völkel, *Herman H. ter Meer*. Herman ter Meer left the Leiden museum in 1907 to work at the Zoologisches Institut of the University of Leipzig, frustrated because only a handful of his excellent products were ever seen on display by the general public.

33 Naumann, *Taxidermie*.

34 Temminck, *Voorschrift*, 13; see also Gijzen, "'s Rijks Museum," 195.

35 Temminck, "Nieuwstijdingen," in *Middelburgsche Courant*, November 7, 1835; Temminck, "Binnenlandsche berigten," in *Nederlandsche Staatscourant*, November 4, 1835.

FIGURE 15 A Japanese Soldierfish prepared by Temminck's method, cut along one side from head to tail (RMNH.PISC.D.243)
NATURALIS BIODIVERSITY CENTER. PHOTOGRAPH BY ESTHER DONDORP

FIGURE 16 The uncut side of the Japanese Soldierfish, varnished and furnished with a glass eye for display (RMNH.PISC.D.243)
NATURALIS BIODIVERSITY CENTER. PHOTOGRAPH BY ESTHER DONDORP

> With this [use of wooden models] one can better restore the natural shape of the animal; every superficial muscle and bone can be made visible, and in this way, it is not necessary to pierce the skin; as the skin can be wrapped like a robe around the accurately built shape. The splendid final result justifies all the effort that goes into this process.[36]

However, Temminck employed the talented Ter Meers – four generations of them – at the museum not to please the visiting public, but mainly because their excellent mounts facilitated the morphological study of the specimens, which was so vital to systematics.

During his directorate Temminck ensured that the museum was filled with capable men with a keen interest in the collections in their care as well as in natural history and, particularly, zoology. Remarkably, eight of his nine curators had studied at a university, six of them had earned a doctor of philosophy degree, and almost half had studied at Leiden University. At the time anyone interested in pursuing a career in natural history either held a medical or sciences degree, or had enough wealth to own a private cabinet, or was a collector and dealer of natural specimens.[37] Indeed, two of Temminck's curators were medical graduates, and five of them had earned a doctorate in the natural sciences. Temminck sourced two of his curators from German institutes: Boie came from Heidelberg Universitätsmuseum and Schlegel from the Naturhistorisches Museum in Vienna. The most productive of them eventually became members of the Natuurkundige Commissie (Boie and Macklot) or remained at the museum for the remainder of their careers, or for the rest of their lives, whichever occurred first. Every one of the curators specialized in systematics, except Beima, naturally, as he was in charge of the geological collections. However, Temminck's decisions on staff had to be sanctioned by the University of Leiden and the Ministry of Home Affairs.[38]

Answering to both the minister and the board of trustees was extremely tiresome for Temminck. Every decision, whether it concerned new appointments, the acquisition of new specimens, or any other expense, needed to be approved by both. But in April 1822 Minister Falck signed a royal decree granting 's Rijks Museum the right to manage its own budget, effectively making it independent from the Leiden University budget and somewhat freeing Temminck's

---

36 Temminck, Report to the Ministry of Home Affairs on the Year 1828, 10 May 1829, Naturalis Biodiversity Center Archives, Jaarverslagen.

37 Gibson, "Careering Naturalists," 210–211.

38 The Ministry of Education, National Industry, and Colonies was split in 1824 into the Ministry of National Industry and Colonies and the Ministry of Home Affairs, Education, and Water Management. The museum fell under the administration of the latter.

hand. He was still obliged to report to the ministry and the trustees, which continued to have the final decision.[39] The reason for this break may very well have been Temminck's complaints, perhaps aided by his long friendship with Falck. At that point, however, Temminck actually already acted quite autonomously and informed only Falck about his plans. He had appointed Macklot without informing the trustees, he planned to purchase a collection of African animals without asking for permission, and he had ordered new museum cabinets without informing the board; for the latter he received a fairly mild reprimand from Falck inviting him to send his proposition to the trustees.[40] After 1823 Temminck stopped sending annual reports on the museum to the Board of Trustees and answered only to the minister.[41] How exactly he managed to do this without repercussions remains a mystery.

Temminck's opinion on third-party interference in the management of scientific institutions becomes very clear in his report on 1824 to the ministry. In it Temminck lamented that the British, although having expanded their empire considerably, had contributed so little to natural history:

> One should expect that the natural sciences in the British Kingdom would bring forth the most splendid results. But far from taking advantage of such a favorable period, during which they reigned over the seas, they have achieved nothing significant. Old crippled institutions, petty conventions, inept people, and useless societies enveloped in old privileges were always an obstacle, and are now still in place hindering the progress of good institutions. Nothing significant can be done and accomplished in England for the natural sciences as long as an Archbishop of Canterbury, the Lord Mayor of London, and the Speaker of the House of Commons occupy the highest ranks at the British Museum; and [as long as] the president of the Royal Society has the right to appoint a chemist as Director of Zoology.[42]

---

39 Royal Decree no. 110, 12 April 1821, NL-HaNA, RMNH 3.12.17, inv.nr. 1; De Jonge, "Macht, machinaties en musea," 194.

40 Falck to Temminck, 4 April 1822, NL-HaNA, RMNH 3.12.17, inv.nr. 1.

41 Dreier, "Voor museum en vaderland," 85.

42 Temminck, Report to the Ministry of Home Affairs on the Year 1824, 3 October 1824, Naturalis Biodiversity Center Archives, Jaarverslagen. Temminck referred to the three principal trustees of the British Museum: the Archbishop of Canterbury Dr. Charles Manners-Sutton; the Lord Chancellor, or Lord Keeper, John Scott Earl of Eldon; and the Speaker of the House of Commons, Charles Manners-Sutton, 1st Viscount of Canterbury (the Archbishop's son). The total number of trustees was forty-one.

Temminck could not understand, and was almost offended by the fact that the British Museum's trustees were politicians and clergymen. Judging by a report of the trustees from 1823, Temminck was right in part at least: "In short, the whole Zoological and Botanical department of the Museum is disgraceful to the nation, and very discreditable to the Trustees, to whose charge it has been consigned."[43] For Temminck the situation across the Channel served as a welcome warning to the Ministry of Home Affairs, indicating what might happen if the management of zoological institutions was entrusted to individuals without proper natural historical background.

Despite his disagreements with the trustees, what most affected Temminck in his directorate was an economic crisis that sent the Netherlands into debt, and nearly bankruptcy, forcing the country to reassess its political and economic priorities. The problems started just a few years after the museum's foundation. Three simultaneous conflicts had severely damaged the country's finances: the Java War of 1825 to 1830, the Dutch intervention in the Padri Wars in West Sumatra from 1821 to 1837, and the Belgian Revolution of 1830 with its nine-year aftermath. Amid these severe economic and territorial challenges 's Rijks Museum suffered budgetary cuts, as did 's Rijks Herbarium and other scientific institutions. By the beginning of 1824 the museum had already missed several payments. Temminck complained in his most sarcastic style:

> Why does the government always leave us without a penny in January, February, and March? Do they believe that we, like wild animals, hibernate, or live from our hoarded supplies like hamsters? No advances and no payments on time; fine housekeeping is that! Now, every man for himself.[44]

A few months later, as Heinrich Macklot prepared for his journey to Java as the newest member of the Natuurkundige Commissie, the government did not send the necessary funds for equipment and supplies. Temminck decided to resolve this single-handedly, as he was used to doing. He asked Macklot to acquire, given the government's "indécision flegmatique," whatever equipment he thought he might need without waiting for approval, and offered to pay for it himself if the government failed to do so.[45]

---

43 Trustees of the British Museum, "Annual Report, 1822," 390.

44 Temminck to Leonardus Franciscus Thijssen, 28 February 1824, Naturalis Biodiversity Center Archives, Archive C. J. Temminck.

45 Temminck to Macklot, 14 July 1824.

The lean times had only just started. Despite the erection of the new wing for the collections and annual subsidies, investments never seemed to keep up with the museum's growth. Temminck's annual report of 1825 concerns the lack of funds more than the activities at the museum. Although he kept the cost of acquiring new specimens at a minimum by exchanging them with other museums and private collectors, the bills were piling up. The museum's resources could not cover the costs of staff, a smith, a carpenter, basic materials (including alcohol, labels, and ink), heating, and lighting. The most basic needs seemed to have become a luxury. The ministry's financial support was simply not enough to keep the museum running, let alone to expand it further.[46]

The situation did not really change much after 1825. From the mid-1820s to the end of his life, Temminck constantly struggled to find sufficient funds. In 1826 Susanna explained to Friederich Molenhauer – a former student from Heidelberg and Van Hoorn's assistant until 1823 – that the situation had forced Temminck to economize so that he was not buying any new material for the collection.[47] By 1835 Temminck had to pay for his cherished African elephant – the one acquired from the Verreaux brothers and skillfully prepared by Jacob ter Meer – in installments as it was too expensive for his annual budget. He paid 2500 French francs for it, roughly twice the annual salary of an average worker in France. This was a small fortune (the Maison Verreaux asked 650 francs for not one but three rhinoceros), and clearly Temminck thought the elephant was worth it.[48] He carefully omitted the purchase price from his report to the ministry, which was already fully submerged in the economic crisis.[49]

To make matters worse, the 1840s brought with them political change that made things even more difficult for Temminck, and for that matter, for all museums in the Netherlands. These new difficulties were not just economical, but there was also a palpable change in attitudes towards cultural and scientific institutions which was partly fueled – or embodied – by the liberal minister of home affairs and later the prime minister, Johan Rudolph Thorbecke. Thorbecke's derogatory monickers for 's Rijks Museum van Natuurlijke Historie

46 Temminck, Report to the Ministry of Home Affairs on the Year 1825, 31 December 1825, Naturalis Biodiversity Center Archives, Jaarverslagen.

47 Susanna to Moldenhauer, [1826], Naturalis Biodiversity Center Archives, Archive C. J. Temminck, NAT_ARC_TEM_01975.

48 Hes, "History of Animal Preparations," 1–2.

49 Temminck to the Ministry of Home Affairs, 30 September 1835, "Verslag van Temminck aan de Minister van Binnenlandse Zaken over zijn wetenschappelijke werkzamheden tijdens zijn verblijf in Parijs," Naturalis Biodiversity Center Archives, Archive C. J. Temminck, NL-LdnNAT_Temminck_49.

and 's Rijks Herbarium were, tellingly, "the monkey house" and "the hayloft."[50] Thorbecke's view of the Leiden museum was nowhere near Temminck's idea of a 'national monument.'

For Thorbecke science and culture were not a governmental matter. Although he acknowledged that the government was responsible for maintaining museums, he thought that art and science were not among its concerns. In practice he favored a model in which the government would not provide any financial support for art and science if the relevant sectors could do without it, or if they could be supported by private initiatives.[51] The annual government subsidies for 's Rijks Museum had shrunk from f5000 in 1830 to f3600 in 1850. The persistent financial drought brought Temminck to the brink of a nervous breakdown. He wrote to the minister of home affairs in 1851:

> 's Rijks Museum van Natuurlijke Historie still splendidly maintains its well-earned fame, a fame magnified by the deterioration of similar institutes, even those of great empires. As long as Heaven gives me life and strength, I hope to spare from such a sad fate the object to which I have devoted all my time and diligence for more than thirty years, and I will succeed in that effort as long as my strength is reinforced by the protection and support of the Government, which I have enjoyed, as I gratefully acknowledge, to the present day, and which I will surely not lack in future under the strong direction of Your Excellency.[52]

Apart from the worries brought on by the museum's administration, Temminck faced another, more alarming problem: his hands had started shaking. From around 1840 onward he suffered from tremors that made writing very difficult. In the spring of 1842 Temminck wrote to his wife from Paris complaining about his shaking hands, which he considered a sign of aging, but celebrating the fact that he was otherwise of healthy body and mind.[53] He was sixty-four years old. Notwithstanding the optimistic note, his handwriting deteriorated further until it was virtually illegible. Temminck was forced to stop using pen and ink and write with pencils instead. In 1846 he confessed his lack of strength and confidence to his friend Lichtenstein with some sadness: "Work is not so quickly accomplished as it once was, and yet there are more things to do. One

50 Schlegel, "Levensschets van Hermann Schlegel," 69.
51 Van Berkel, *De stem van de wetenschap*, vol. 1.
52 Temminck, Report to the Ministry of Home Affairs on the Years 1847–1850, 26 November 1851, Naturalis Biodiversity Center Archives, Jaarverslagen.
53 Temminck to Anna A. Smissaert, 2 April 1842, Temminck Family Archives, A. M. van Lynden-de Bruïne private collection, no. 6.

demands as much of oneself as before, but one does not satisfy these expectations as well as formerly. Many outlines for articles are lying about unfinished and will die off quietly."[54] But Temminck was not yet ready to retire.

In 1851 the ornithologist and nephew of Emperor Napoleon I, Charles Lucien Bonaparte, complimented Temminck in a charming letter: "Telling you that I am working on the Gallinaceous and Water Birds equals telling you how much I am thinking of you, because, if you are the Papa (not the Pope) of all birds, you are the Genius of the fowl and the gulls[.]"[55] Temminck cheerfully replied: "And long live the Gallinaceous birds, the pigeons, and the water birds! They return an old man to the time of his youth, and rejuvenate me by a quarter of a century."[56]

Nevertheless, Temminck was well aware of his waning strength. In 1854, at the age of seventy-six, he wrote to Prince Max von Wied-Neuwied that he was feeling like a "veteran in the decline of his career, but one that still has the hope of dying in action."[57] In the summer of 1857 Temminck passed on the scepter to Schlegel and Susanna. Six months later, on January 30, 1858, Temminck passed away, two months before his eightieth anniversary.[58]

Regardless of all the problems, drawbacks, arguments, and disappointments, Temminck had achieved his goal of putting the Netherlands on the map. Indeed, Temminck had managed to build a collection that ranked amongst the best of its kind in Europe. As William Swainson put it in 1840: "The most celebrated [national museum] in the world is that of France: next may be ranked the museums of Berlin, Vienna, Holland, Bavaria, Denmark, and Florence."[59] Charles Bonaparte went even further:

> The superiority of the Leyden Museum over any other is unquestionable, not perhaps so much on account of its containing a greater number of species than those of London, Paris, Philadelphia and Berlin, but for the freshness and perfection of the specimens, for the quantity of skeletons, and above all for the never-sufficiently-praised series of individuals of the various species of both sexes, in different ages, and from

54 Temminck to Lichtenstein, 27 July 1846; translated in Stresemann, *Ornithology*, 150.

55 Bonaparte to Temminck, 1 January 1851, Temminck Family Archives, A. M. van Lynden-de Bruïne private collection, no. 46; De Beaufort, "Temminck, uit zijne brieven geschetst," 38.

56 Temminck to Bonaparte, 8 January 1851, Temminck Family Archives, A. M. van Lynden-de Bruïne private collection, no. 47.

57 Temminck to Prins Max von Wied-Neuwied, March 1854. Temminck Family Archives, A. M. van Lynden-de Bruïne private collection, no. 48.

58 Susanna, "Levensschets van Temminck," 77.

59 Swainson, *Taxidermy*, 73.

> different localities and countries, which facilitate one's judgement, and show at once in most cases, especially with Mammalia, what is or is not a good species.[60]

3 Colonial Nature

Under Temminck's direction, the collection of 's Rijks Museum had grown dramatically and given prestige to both its director and the expanding Kingdom of the Netherlands. The museum's main suppliers of specimens were the Natuurkundige Commissie and Temminck's extensive network of collectors, museums, dealers, and civil servants, who all required as much charm, command, and determination as he could muster. While Temminck had full control over his network – but not the finances – the commission and the museum were separate but intertwined organizations, and their relationship reflected the economic and political upheavals of the country, as much as its colonial ambitions.

The three collections that were to form the founding collections of the Leiden museum (the Leiden University Cabinet, 's Lands Kabinet and Temminck's private collection) were impressive, but not nearly sufficient to bring the museum to the elevated level of those in the neighboring countries. The Dutch government was painfully aware of this shortcoming – even before 's Rijks Museum was established.

The earliest attempts at remedying this situation were made by Caspar Reinwardt while he was working in the Dutch East Indies. But although he had collected many natural history objects for 's Lands Kabinet, he was not supposed to occupy the official role of a naturalist abroad. Reinwardt had been sent to Java in 1815 as the 'director of agriculture, arts, and science in Java and its neighboring islands.' He had been handed a list of detailed instructions, including an endless list of questions to which the government needed answers, and promptly. Reinwardt was asked to calculate the extension of cultivated land, forests, and wilderness; survey waterbodies; draw up a survey of the different occupants of the region, including indigenous peoples, Chinese and European immigrants, and their offspring; survey existent coal, iron, tin, and sulfur mines; investigate whether it was possible to source wood for shipbuilding; find out how much rice, coffee, sugar, pepper, cardamom, and other spices were produced in an average year, and whether coconuts were used as a food, for producing oil, or for distilling alcoholic drinks; find out which

60 Bonaparte, "On the Lorine Genus of Parrots, Eclectus," 142–46.

animals were edible; check the quality of animal horns; and inquire what children learned at school – to list but a few of his tasks. It was an astounding list of one hundred and twenty tasks.[61] It was also an uncomfortable reflection of how little the Dutch government knew about these territories at the time after nearly two centuries of Dutch presence in the region.

Reinwardt, undaunted, took the mission on all the same. He advised the colonial government, undertook survey explorations throughout the archipelago, collected all kinds of natural objects, conducted botanical investigations, cultivated various potentially useful plants, and founded 's Lands Plantentuin (the National Botanical Garden) in Buitenzorg (now Bogor) in 1817. He even found the time to reorganize the school and public health systems, set up a medical committee for health care in Batavia, and reopen the military school in Semarang, among other things. Despite all his steady activity, this was no more than a drop in the ocean. In order to gain domain over the Dutch East Indies, the Dutch government needed more information, and above all, more control over the territories.

Due to a series of very unfortunate coincidences, four of Reinwardt's precious shipments of specimens were lost at sea.[62] Falck instructed that the material that did reach the Netherlands was to be kept at 's Lands Kabinet in Temminck's and Van Marum's care. Incidentally, Brugmans protested against this arrangement in 1817, when the issue of a national museum was still unresolved, arguing that the University of Leiden should also benefit from the material arriving from Java.[63]

Soon it became evident that Reinwardt would never be able to comply with all the instructions he received. Sending reinforcements to Java for scientific research in the field of natural history seemed a sensible idea, especially since France and Britain had been able to explore and describe the archipelago just a few years previously – this was a race for knowledge. During the French and British interregnum (i.e. from 1806 to the Treaty of London in 1814) the archipelago had been explored by naturalists including Alfred Duvaucel (Georges Cuvier's stepson) and Jean-Baptiste Leschenault de la Tour. Thomas Horsfield, Alexander Hare, and Joseph Arnold had also been active when Sir Thomas

61 De Vriese, *Reis naar het oostelijk gedeelte*, 37–48; see also Weber, "Hybrid Ambitions," 120–22.

62 De Vriese, *Reis naar het oostelijk gedeelte*; Weber, "Hybrid Ambitions," 140–41; Smit, "Rijksherbarium." Lists of some of the objects collected and sent by Reinwardt to 's Lands Kabinet – and later, to 's Rijks Museum – are kept in the Naturalis Biodiversity Archives, NAT_ARC_TEM_02275, NAT_ARC_TEM_02306.

63 Brugmans to the Trustees of Leiden University, 26 September 1817, UBL-AC2, 70; Dreier, "Voor museum en vaderland," 41.

Bingley Stamford Raffles was lieutenant-governor in Java from 1811 to 1816. Most of the collected specimens were sent to the Muséum in Paris and to Sir Joseph Banks in London. The results of their efforts were impressive, both in terms of the gathered collections and in the description of many new species. For instance, Georges Cuvier described some of Leschenault's birds that had been collected in Java, and Horsfield used these collections in the preparation of his *Zoological Researches in Java and the Neighbouring Islands*, which was published in 1824.[64]

When the Netherlands regained control of parts of the Sunda archipelago, the king wished his nation to be able to compete (economically and scientifically) with the neighboring countries, and it is conceivable that Temminck, via his friend Falck, emphasized the need for a larger collection of natural history objects. According to Holthuis, Temminck was behind the idea of formally organizing an expedition to Java.[65] In early 1819 the Dutch government had not only decided to send more explorers to Java, but it had also chosen the men for the job: Heinrich Kuhl from Hanau, and Johan Conrad van Hasselt from Doesburg, both students of Van Swinderen in Groningen at the time. Temminck knew them both well. While he was a guest of Temminck in Amsterdam, Kuhl received a letter from Minister Falck informing him that he had been chosen to travel to the Sunda Islands by order of the government. Kuhl was excited; the message felt "like an electric shock."[66]

Temminck and Kuhl had met a few years earlier in Berlin, while both were visiting Lichtenstein. Kuhl had also met Van Swinderen in Hanau and had made such a good impression on him that Van Swinderen invited him to study in Groningen. Kuhl had gladly accepted the offer, and Kuhl and Van Hasselt had then started to work with Van Swinderen and Temminck. Van Swinderen and Temminck were dazzled by the young naturalists' skills and intelligence and therefore recommended them to the king.[67] For their new adventure in Java, they would be trained in Paris, London, Groningen, and Leiden, all and under Temminck's instruction paid for by the Dutch government. This was a once-in-a-lifetime opportunity, one that made them the envy of their peers. Heinrich Boie wrote to Naumann: "At the moment Kuhl is receiving money from the Dutch Government, to last for a year from Michaelmas, so that he can go to Paris and London and then leave for the East Indies – lucky man! If only Heaven will preserve his cheerfulness and his health, his name will be eternally

64 Smit, "Rijksherbarium," 8; Jansen, "Ornithology of the Baudin Expedition (1800–1804)," 186.
65 Holthuis, *Rijksmuseum*, 21.
66 Kuhl to Van Swinderen, in Sirks, "Indisch natuuronderzoek," 101.
67 Greshoff, "Kuhl en Van Hasselt," 70–71; Klaver, *Inseparable Friends*, 9; Sirks, 101.

FIGURE 17 *Heinrich Kuhl*. Lithograph by Friedrich Fleischmann [no date]
STICHTING NATIONAAL MUSEUM VAN WERELDCULTUREN. COLL. NR. TM-3728-703

celebrated!"[68] Alas, Heaven did not comply. Kuhl died only nine months after arriving in Java, three days short of his twenty-fifth birthday.

By royal decree, Kuhl, Van Hasselt, the artist Gerrit Laurentius Keultjes, and the taxidermist and illustrator Gerrit van Raalten became the first four members of the Natuurkundige Commissie voor Nederlands Indië in May 1819. The royal decree stated that after Reinwardt's return, the king "wished to promote more and more the knowledge of the natural composition of the products of our possessions in the East Indies," and accordingly, Kuhl and Van Hasselt were "to expand the scientific knowledge of the products of Nature in those countries."[69]

Compared with Reinwardt's instructions, this was deceptively simple. Initially, the commission was not asked to succeed Reinwardt or to aid him in his duties. It was to succeed Horsfield, Diard, and Leschenault de la Tour, and its specific purpose was to collect, describe, and study the nature of the colonies. The royal decree clearly instructed that all collected material should be deposited in the yet-to-be-founded 's Rijks Museum (in July, Temminck and Falck were still discussing the terms).[70] However, this form of government-sponsored science was not only intended to serve 's Rijks Museum and Herbarium, but also to boost the economy of the Dutch Empire with these new "products of Nature." This would eventually lead to confrontations between the government and Temminck.

The museum finally was founded in August 1820, less than four months after the birth of the commission. Officially the explorers were employed by the minister of public education, national industry, and colonies, and received support from the colonial governor-general. However, Temminck interfered in practically everything from the beginning. He steered the Natuurkundige Commissie, as far as he could, instrumentalizing it in the interest of the museum and its practice of systematics. Unfortunately, the material from the commission (including specimens as well as manuscripts) arrived in the Netherlands only intermittently and with great difficulty. Often the commission shipped the expected crates of specimens with years, even decades, of delay. Some never arrived at Leiden but sunk to the bottom of the ocean.[71]

The arrival of manuscripts was a rare event. Most naturalists in Java kept their field notes with them and then passed them on to their successors. Regrettably,

68 Boie to Naumann, 6 March 1819, in Stresemann and Thomsen, "Ornithologen-Briefe aus den Jahren 1816 bis 1820," 15; translated in Stresemann, *Ornithology*, 128.

69 Royal Decree no. 10, 2 May 1820, NL-HaNA, Rijksmuseum van Natuurlijke Historie te Leiden (RMNH), nummer toegang 3.12.17, inv.nr. 1; Veth, "Overzicht van hetgeen," 21.

70 Royal Decree no. 10, 2 May 1820.

71 Greshoff, "Kuhl en Van Hasselt," 86; Veth, "Overzicht van hetgeen," chap. 8.

Java was not a safe place for precious manuscripts. At one point the governor-general, Baron Godert A. Gerard Philip van der Capellen, gave Van Raalten the order to make copies of Kuhl and Van Hasselt's field notes and drawings. These copies also remained in Java and were only found when the house where Diard had lived was demolished in 1838, more than fifteen years after Van Hasselt's death.[72] The specimens, however, had been sent to Temminck and had arrived at the Leiden museum with very little written information.

In order to centralize the management of the whole enterprise, a second commission was established in 1827. The minister appointed Reinwardt, Temminck, and Carl Ludwig Blume (who was now in the Netherlands after four years serving as the director of 's Lands Plantentuin) as supervisors. Among other things they were in charge of appointments and travel schedules. Importantly, they were to make sure that the commission shipped material to the Netherlands on time, and they were to organize the publication of discoveries that any returning explorers had made abroad.[73]

Since slow progress was made in the 1820s in researching the material collected in the Dutch East Indies, Blume, Reinwardt, and Temminck instigated the publication of a work in which at least some of the results would finally be published, mainly by the curators at 's Rijks Museum. To accomplish this, they needed government support, which they finally received in the form of Royal Decree no. 101, on February 10, 1839. The decree made the publication possible, but disappointingly, it also limited its range, determined it was to be written in Dutch, and put a limit to the available funds. So while the government funded this new enterprise, it also specified that it should consist of no more than twenty-six parts.[74] Temminck was the designated chief editor, while Reinwardt, Blume, and the zoologist Jan van der Hoeven formed the editorial board. Both the funding available and the proposed number of parts proved to be insufficient. With some ingenuity Temminck managed to secure three additional issues on top of the twenty-six planned and named them 24 *bis*, 24 *bis* continued, and 26 *bis*.[75]

The full title became *Verhandelingen over de natuurlijke geschiedenis der Nederlandsche overzeesche bezittingen. Uitgegeven op last van den Koning door C. J. Temminck* (Dissertations on the Natural History of the Netherlands Overseas Possessions. Published by Order of the King, by C. J. Temminck).

---

72 Greshoff, 86.

73 Veth, "Overzicht van hetgeen," 90–91.

74 Royal Decree no. 101, 10 February 1839, NL-HaNA, RMNH, 3.12.17, inv.nr. 1; Veth, 72; Husson and Holthuis, "Dates of Publication of 'Verhandelingen'," 18.

75 Royal Decree no. 57, 15 April 1843, NL-HaNA, RMNH, 3.12.17, inv.nr. 1; Husson and Holthuis, 20–21.

It took nine years to complete. By 1844 only three volumes had been published: *Zoologie* by Salomon Müller, Hermann Schlegel, and Willem de Haan, *Botanie* by Pieter Willem Korthals, and *Land- en Volkenkunde* (on geography and ethnography) by Salomon Müller.[76] Both Korthals and Müller had been members of the Natuurkundige Commissie. The twenty-nine issues appeared between 1839 and 1847.

The Natuurkundige Commissie fell directly under the control of the government and Temminck – at least during the first decade of the commission. In the 1830s the colonial government slowly took over the direction of the commission, issuing new instructions as soon as its members arrived in Java. There they depended on the governor-general of the colonies for their housing, payment, materials, and the organization of the expeditions. As the governorship was, it seems, not an easy post to hold for a long time, the commission's situation varied with each new governor. The governor-general the first commission encountered was Godert Alexander Gerard Philip van der Capellen. In 1826, after ten years in office, he was replaced by Hendrik Merkus de Kock. Kock left just a few months later and was succeeded by Leonard P. J. du Bus de Gisignies, who stayed until 1830. The following governors lasted an average of three years in the post, except for Jan Jacob Rochussen, who stayed for six years. During the commission's thirty years of activity, ten governors succeeded each other, each with particular views and different levels of influence.

The commission's members usually occupied their roles for even shorter periods. Most died in the field from tropical diseases, from accidents, syphilis, or even by murder, so that the rate of turnover was high. The first party was among the most unfortunate. Kuhl lived only nine months in the tropics until he died from a liver infection, Keultjes followed him two days later from an unidentified tropical disease, and Van Hasselt died of dysentery two years later.[77] Van Raalten died from tropical fevers in 1829, and malaria took Boie's life two years on, but Heinrich Christian Macklot, who proved to be more resistant to tropical diseases than his colleagues, traveled and worked for eight years until he was killed in Krawang amidst a Chinese insurrection during which Macklot's and Boie's manuscripts burned to ashes.[78] Some of the young

76 The ethnographical collections put together by the Natuurkundige Commissie were considered the collectors' private property, at least until 1837, when 's Rijks Ethnographisch Museum (the National Museum of Ethnography) was established in Leiden. Salomon Müller's collection, for instance, was purchased by the Ethnographisch Museum (now Museum Volkenkunde) in 1864 (Ter Keurs, "Agency, Prestige and Politics," 172).

77 Klaver, *Inseparable Friends*, 21, 255; Veth, "Overzicht van hetgeen," chap. 3–4, Weber, "Collecting Colonial Nature," 80, 87.

78 Veth, 56, 72.

FIGURE 18 *Berggezigten en ethnographische tafereelen uit de binnenlanden van Timor.* Engraving after a drawing by Pieter van Oort, ca. 1828–1829. A Timorese man shows a snake to members of the Natuurkundige Commissie. In Coenraad Jacob Temminck, ed., *Verhandelingen over de natuurlijke geschiedenis,* [vol. 3] (1839–1842): plate 29
NATURALIS BIODIVERSITY CENTER ARCHIVES, MMNAT01_PM _NNM001000871_029

naturalists may not have had a particularly strong constitution, and some may have been ill-prepared for life in the tropics. Some were simply unlucky.

In any case, this instability in both the colonial government and in the commission led to misunderstandings and lack of support, money, and, at some point, accommodation.[79] On top of everything else, colonial wars broke out. During the Java War (also known as the Diponegoro War), which raged between 1825 and 1830, things became utterly unpleasant for the colonial government, and the Natuurkundige Commissie was, understandably, relegated to a very low position on the scale of official priorities. Du Bus de Gisignies, who had taken up the post of governor-general in 1826, not only denied payment and facilities to the commission but even advised its members at some point to pack up and go home.[80] It was an endless row of disasters.

A decade after the establishment of the commission, the colonial government began to interfere not only in the finances and practicalities of daily life, but also in the selection of subjects with which it should concern itself. In 1830 the Dutch botanist Pieter Willem Korthals, who worked at 's Rijks Herbarium under the direction of Blume, joined the commission. Yet Korthals'

79 Sirks, "Indisch natuuronderzoek," chap. 4; Veth, 3–13.
80 Boie to Temminck, 11 November 1826; quoted in Veth, 115.

appointment had much more to do with the colonial government's desire to exploit Javanese resources than with the museum or the herbarium.

After the Java War the Dutch government abandoned its liberal policies, and with Johannes van den Bosch as a new governor at the helm, the colony was under pressure to become a more profitable – and more docile – dominion of the Dutch Empire. Among other things, the government implemented a new cultivation system (*Cultuurstelsel*) that was designed to increase the export of crops to Europe, and thereby also the economic profit to the Netherlands. Naturally, the colonial authorities were eager to evaluate the system's efficiency and wanted to explore how it might be implemented in remote areas.[81] It is in this context that Korthals was appointed a member of the commission. Daniel Jacob van Ewijck, a civil servant of education, art, and science at the Ministry of Home Affairs, interfered with the Natuurkundig Commissie directives by providing direct instructions for Korthals.[82] He conveyed in no uncertain terms that Korthals' botanical work should not focus entirely on systematic research, which was of no apparent use in Van Ewijck's view:

> His botanical investigations shall not, in any way, be restricted to the description of their morphology, nor to the determination and distinction of genera and species of plants, but, as Mr. Korthals' mission is meant to be of immediate benefit for the region, for the Netherlands, and for the relationship between the two, they shall have as their goal a more general, useful, and fruitful knowledge of plants.[83]

He did not include a word about the museum, the herbarium, or their collections. Unsurprisingly, Temminck resented Van Ewijck's interference. Temminck's papers include a bundle of seven sheets recording his thoughts and questions on these instructions, and he was not pleased with the meddling of the colonial government. He wrote, apparently agitatedly: "Why is Korthals not allowed to work according to his instructions [from Blume, Reinwardt, and Temminck]? Why are his reports concerning Economy or Agriculture rejected or still unanswered? Why this deliberate burden to engage in anything other than collecting? Are [we] to be of service to the Indian [colonial] Government?"[84] Apparently the colonial government thought that the

81 Manse, "Kennis is macht," 42; Van den Doel, *Het rijk van Insulinde*, 48–50.
82 Van Ewijck to Temminck, 20 August 1830, NL-HaNA, RMNH, 3.12.17, inv.nr. 4; Dreier, "Voor museum en vaderland," 56.
83 Van Ewijck to Temminck, 10 November 1830, NL-HaNA, RMNH, 3.12.17, inv.nr. 4; Dreier, 56.
84 Temminck's minutes about the instructions for the Natuurkundige Commissie [1836], Naturalis Biodiversity Center Archives, Archive C. J. Temminck, NAT_ARC_TEM_00453.

commission was, indeed, to be of service to Dutch expansionist plans. At this point Europeans had sent expeditions surveying, measuring, collecting from, exploring, and claiming other regions of the world for centuries. These expeditions were motivated by a mixture of colonial ambitions, economic interests, and scientific projects.[85]

In 1828, the Natuukundige Commissie was instructed to travel to New Guinea, which was hitherto unclaimed by European colonial powers, to survey the coast and find a suitable place for a settlement. The governor of the Moluccas, Pieter Merkus, was alarmed by the British presence in northern Australia: he feared that they posed a threat to the Dutch monopoly on spices, especially on cloves, nutmeg, and cinnamon. New Guinea was a protectorate of the sultanate of Tidore, but the sultan was prohibited from settling on the island without permission of the Dutch colonial government in Java. Merkus thought that this cunning arrangement gave the Dutch the right to claim New Guinea. After negotiations with Achmadul Mansur Sirajuddin Syah, the sultan of Tidore, the Dutch sailed from Timor on the *Iris* and the *Triton*. Onboard were soldiers, women, children, and livestock for the new settlement, and they were accompanied by the members of the commission.

The colonial government labeled the undertaking a scientific expedition, thereby disguising its true intentions in order to avoid a possible skirmish with the British. The commission members would assess the suitability of the climate for a Dutch settlement, note the presence of edible plants, test the soil fertility, and record the availability of drinking water. On July 4, 1828 the Dutch settled in Triton Bay (now Kaimana Kota in West Papua, Indonesia) and built a temporary wooden structure, the Fort DuBus. (Due to a series of unforeseen circumstances the settlement was abandoned seven years later.)[86] During their months at Triton Bay the naturalists collected an impressive selection of plants, animals, and ethnographic objects, all of which were new to Europeans. Temminck was particularly happy with this introduction to a fauna that was mostly unknown to him.

Towards 1830 things began to change. All the setbacks, losses of lives, the lack of funds, and the changes on the political stage had weakened the initial force of the Natuurkundige Commissie. The enterprise as Temminck had envisioned it (i.e. in the spirit of earlier explorations by the French and British) was being transformed into an instrument for the colonial government, which was pushing it towards a role that was not unlike that which Reinwardt had occupied

85 See, for example, Anderson, "Natural History and Scientific Voyage."

86 Mörzer Bruyns, *Met de Triton en Iris*; Modera, *Verhal van eene reize*; Müller, *Reizen en onderzoekingen*.

FIGURE 19 *Dendrolagus ursinus*. Tree-kangaroo from Triton Bay, New Guinea. Drawing by Pieter van Oort, August 1828. Design proof for plate 19 in Coenraad Jacob Temminck, ed., *Verhandelingen over de natuurlijke geschiedenis*, [vol. 3] (1839–1842)
NATURALIS BIODIVERSITY CENTER ARCHIVES, MMNAT01_PM _NNM001000872_039

as the director of agriculture, arts, and science in 1815. Later expeditions to Sumatra and Borneo were also a mixture of colonial service and natural history research. On them, naturalists studied soil types, crops and their possible profitability, made maps, and prospected for copper, gold, and coal mines – activities that were of little or no profit for the Leiden museum. Inevitably this created tension and a conflict of interest between the Dutch government, the colonial government, and both 's Rijks Herbarium and Rijks Museum.

The commission was terribly expensive: not only did the members receive fees and funds for the material they needed, but they were also accompanied by soldiers, Javanese assistants, translators, porters, hunters, cooks, and guides. Delicate political issues were at play: only part of Java was under Dutch control, and the naturalists had to negotiate access to certain areas with either the British or local authorities.[87] Worse of all, of the eighteen members of the commission only six survived. Infections, diseases, rhinoceroses, Chinese revolts all took their toll. Specimens were lost in shipwrecks, manuscripts burnt, expeditions canceled or interrupted, and most of the young men's ambitions were never realized. In 1850 the government decided to dissolve the Natuurkundige Commissie.[88] In its thirty years of activity the commission provided the Leiden museum with tens of thousands of zoological specimens and as large a number of pages of useful manuscripts, drawings, and diaries. The cornucopia of natural history specimens included birds, mammals, fishes, plants, all kinds of insects, starfish, corals, snails, spiders, frogs, snakes, seeds, seashells, and nearly every other kind of zoological object, and a wealth of plants and minerals besides. All now bear very distinctive labels recording the names of their collectors: "Kuhl&vHasselt," "Boié," "Diard," "S. Müller," "Macklot," or "Forsten."

4 A Worldwide Web of Collectors

The Natuurkundige Commissie was not the only supplier of specimens for the Leiden museum. Crates with precious cargo arrived from all corners of the world. By 1830 the museum employed no fewer than eight taxidermists and technical assistants who feverishly worked on this wealth of material. After all,

87 Weber, "Collecting Colonial Nature"; Van Wingerden, "Science on the Edge of Empire."

88 For more information on the Natuurkundige Commissie see Boeseman, "Collectors and Fish Collections"; Fransen, Holthuis, and Adema, "Type-Catalogue of the Decapod Crustacea," app. II; Ter Keurs, "Agency, Prestige and Politics"; Mees, "Vogelkundig onderzoek"; Mörzer Bruyns, *Met de Triton en Iris*; Sirks, "Indisch natuuronderzoek," chap. 4; Stresemann, *Ornithology*, chap. 8; Veth, "Overzicht van hetgeen," 3–13; Weber, "Collecting Colonial Nature."

Temminck had his mind set on expanding the collection in the shortest time and at the lowest cost possible.[89] This short-term objective was actually simply an extension of his strategy for his private cabinet, which he had followed ever since he inherited it from his father in around 1800. Temminck proved up to this challenge.

The growth of the collection was an immediate concern for Temminck. The early annual reports for the ministry reflect his constant preoccupation with growth. Advances in natural history depended directly on collections, and the competition was fierce. Besides, 's Rijks Museum was required by royal decree to act as repository and fill the universities' cabinets, which increased the need for more specimens. Temminck divided the collection into two groups: one for the museum and one of duplicates intended for the universities of Utrecht, Groningen, Ghent, Liège, and Leuven, as well as for exchange with other institutes.

Most of the incoming material was sent from the Dutch East Indies by Reinwardt and the Natuurkundige Commissie, but Temminck also needed specimens from other parts of the globe. As soon as he took office, he put everyone to work. He relied on two main sources: other institutes and individual collectors. He even asked the ministry to instruct anyone who was abroad in service of the government to collect for the museum: ship commanders, medical officers, and anyone posted to exotic places like Japan, Makassar, or China. Payment should match the value of the collected objects – very much like his father had done during the period of the VOC.[90] However, Temminck was not idle while he instructed the government to put civil servants at work. Besides keeping an eye on the Natuurkundige Commissie, he actively corresponded with anyone willing to shoot, stab, skin, pin, put in jars, and otherwise kill and preserve specimens for the Leiden museum.

To ensure a high quality for the specimens and to guide those who had the spirit to collect but not the knowledge to preserve, Temminck wrote a short manual on taxidermy and preservation, focusing mostly on the techniques necessary to ensure safe transport and prevent the specimens from decaying en route. This small booklet easily fit into a pocket or travel-bag, and its title left no room for interpretation: "Instructions on How to Handle Natural History Objects, to Send them and Preserve them from Decay: For the Benefit of 's Rijks Museum van Natuurlijke Historie in Leiden." In the space of only twenty-seven pages, Temminck concisely summarized the basic techniques for preserving

89 Temminck, Report to the Ministry of Public Education, National Industry, and the Colonies on the Year 1821, 29 August 1822, Naturalis Biodiversity Center Archives, Jaarverslagen.

90 Temminck, Report to the Ministry, 29 August 1822.

any kind of natural object, from mammals to worms, and even gave directions regarding the desired size of minerals. The closing remarks noted: "It is always useful in science to have more than one natural history specimen; moreover, all well-preserved objects are of equal importance for the Museum's collection, independently of their degree of beauty of color or shape."[91] Given Temminck's policy of relegating the less perfect or complete specimens for didactic collections, he may have added this final remark to make sure that all available specimens were sent to Leiden, leaving the final judgment of their value to him. After all, the chances that, for example, an ugly or discolored tropical fish might be new to European naturalists were reasonably high.

Soon Temminck had an extensive and occasionally distinguished network of collectors at his disposal. It included merchants, medical doctors like Philipp Franz von Siebold in Japan or Hubertus Benedictus van Horstok in South Africa, pharmacists like Hendrik Haagen Dieperink in Surinam, civil servants like Hendrik Severinus Pel in West Africa, and even consuls, like Samuel Willem Ruyssenaers in Egypt or Reinhard Frans Cornelis van Lansberge in Venezuela.[92] Temminck never missed a chance to direct these men's efforts towards his own goals and interests. Hendrik Pel, for example, had started working for Temminck in the museum's taxidermy department as a thirteen-year-old boy in 1832. Knowing Pel's earnest desire to travel, Temminck asked the minister to send the boy along with an expedition to the Kingdom of Ashanti (now a region in Ghana), which was planned for 1839. The minister turned this proposition down, possibly in light of the death toll within the Natuurkundige Commissie (this particular region of Africa was nicknamed "the white man's grave") and because of the volatile political situation in Ashanti.

Temminck and Pel insisted on the plan and devised a way for Pel to travel to West Africa. They proposed a dual post: half civil servant, half collector for the museum. The minister agreed, and in 1840 he sent Pel to the Gold Coast (the coast of the Gulf of Guinea, West Africa) as a naturalist and as commander of Fort Coenraadsburg in Elmina. He endured the harsh climate and the tropical diseases. In 1842 he was appointed 'agent for the recruitment of Negro soldiers' ("Agent van werving van neger rekruten") in Kumasi, the capital of the Ashanti Kingdom.[93] However, the Dutch government's attempt to recruit slaves for the

---

91 Temminck, *Voorschrift*, 28.

92 Temminck, Report to the Ministry of Home Affairs on the Years 1847–1850, 26 November 1851, Naturalis Biodiversity Center Archives, Jaarverslagen; Gijzen, "'s Rijks Museum," 97–114.

93 Pel, report, 2 Feruary 1842, "Register van uitgaande stukken van den agent van het Succursale-depôt te Coomassie," NL-HaNA, Kust van Guinea, 1.05.14, inv.nr. 1109; Holthuis, "Biografische notities," 10.

Royal Netherlands East Indies Army soon turned out to be a violation of the Anglo-Dutch Slave Trade Treaty of 1818, and British objections to this system led to its immediate cessation. Much to Temminck's disappointment, Pel could no longer travel to the interior of Ashanti, but he worked as a civil servant in several Dutch trade posts along the Gold Coast. There he managed to collect large numbers of animals for the museum with the help of two local hunters, visiting markets and paying Ashanti and Fante people for any specimens they brought to him.[94]

These are just a few examples of Temminck's legions of contacts. Agatha Gijzen listed over four hundred correspondents who wrote with the Leiden museum's director and its curators during the period of Temminck's directorship. Most of them sold or exchanged specimens while a few only provided information and scientific facts. The list is not exhaustive, but it gives a good impression of the museum's long reach around the globe.[95] There are no records of non-European collectors working for the Leiden museum, and although the Natuurkundige Commissie, Pel, and nearly every European in the tropics had local helpers, hunters, and assistants, these were rarely mentioned by name in field notes or letters. There were some notable exceptions. Pieter van Oort, for example, recorded details of the villages, their inhabitants, and their customs, and noted down the names of his guides, translators, assistants, and even his Javanese lovers.[96] This was an idiosyncrasy on his part rather than the result of official instructions or commonplace practice.

Occasionally professional collectors heard about Temminck's hunt for specimens. Seeing a chance to expand their business, they contacted him directly to offer their services. There are records of a particularly avid German collector named Joachim Brehm, a pharmacist stationed at the Cape of Good Hope, approaching Temminck about a possible transaction. Brehm had a botanical garden in Uitenhage, collected plants, and carried out experiments investigating their medicinal properties.[97] When he heard of Temminck's wishlist from a certain Mr. Kachelhoff, who also lived at the Cape of Good Hope and already sent materials to 's Rijks Museum, he wrote to Temminck informing him that he was able to provide some of the specimens on the list. Brehm asked what price Temminck usually paid for "[b]irds, four-legged animals, insects, butterflies,

94 Holthuis, "Biografische notities," 12–24; Gassó Miracle, "Het beestje bij zijn naam noemen."
95 Gijzen, "'s Rijks Museum," 300–305.
96 Van Oort diaries, Java, 1831–1832 (part II). Naturalis Biodiversity Center Archives, Natuurkundige Commissie arcchief, NNM001001118; Weber, "Collecting Colonial Nature," 85.
97 Glen and Germishuizen, "Brehm, Joachim," 107–8.

fishes and snakes, and perhaps also flower seeds and dried plants."[98] Brehm already had some specimens ready and nicely prepared: he had treated the birds with arsenic soap and the four-legged animals with "grass and salt," had placed snakes in spirit and pinned or papered the butterflies. Brehm made sure to mention that he also had a lion, the head of a sea cow, and soon he would also have a young elephant available. This was an offer Temminck could not resist, and so it was that Brehm became one of the many collectors for the museum. However, he proved to be somewhat unreliable: his prices were often too high, and the quality of his specimens too low.[99]

In addition to acquiring collections from those who worked for him individually, such as Brehm and Pel, Temminck also worked with professional dealers like the Deyrolle family (a business still in existence today), Gustav Adolph Frank in Leipzig and Amsterdam, the Maison Verreaux, or the German naturalist Eduard Rüppell. Sometimes collections for sale at auction or through other channels were interesting enough to set Temminck in motion. However, any purchase required permission from the board of trustees and the ministry, and more often than not, also extra funds, which frustrated Temminck to no end. Occasionally he bought a collection without waiting for authorization, as he did in 1822 with the French naturalist Pierre Antoine Delalande's collection (sometimes De Lalande, or De LaLande). Delalande had collected many specimens between 1818 and 1820 while traveling in South Africa, together with his nephew, Jules Verreaux, the oldest of the Verreaux brothers, who was only eleven years old when they set off.[100] Temminck simply went ahead with the purchase and wrote to Thijssen: "While waiting for an answer, I have already bought and received the specimens, but do not speak to anyone about what has happened but to Boie."[101]

Then again, Temminck did not need the trustees' permission to exchange specimens either with other museums, societies, universities, or academies, or with private collectors. This was a cheap and convenient way to obtain any species he thought was missing from the museum's collection, so that exchange became one of the museum's most secure sources of specimens. A gift economy flourished between natural history museums, private collectors, and professional dealers. As the Dutch economy and politics limited the museum's resources and the growth of its collection, Temminck had no choice but to be

98 Brehm to Temminck, 15 October 1821, Naturalis Biodiversity Center Archives, Archive C. J. Temminck, NAT_ARC_TEM_00179.

99 Gijzen, "'s Rijks Museum," 176.

100 Molina, "More Notes on the Verreaux Brothers," 31.

101 Temminck to Thijssen, 13 May 1822, Naturalis Biodiversity Center Archives, Archive C. J. Temminck.

resourceful. He actively corresponded with collectors, bargained with dealers, and exchanged specimens to help crystalize his vision: a world-class collection in a museum of scientific prominence.

Everything Temminck did – all of it from Leiden – reflected his ambitions, from the choice of staff to his instructions to colonial collectors. After 1830 Temminck is rarely seen outside the museum except for trips to other European museums and private collections. Temminck's directorship and research, and the museum's direction as a scientific institution were all centered around Leiden. The museum was, after all, Temminck's headquarters. From here he managed, researched, wrote, and quarreled until his health stopped him. And while the Leiden museum grew, the rest of the world found itself in a frenzy of activities, from exploration, fieldwork, and collecting to academic teaching and research. During this period natural history fragmented into diverse fields of inquiry, each belonging to a particular domain: museums, universities, learned societies, and nature itself. The tensions between the Leiden University and 's Rijks Museum, the Dutch colonial ambitions, and Temminck's status as a naturalist were interconnected. But these dynamics were by no means restricted to the Netherlands: European natural science was, after all, the product of European expansion, politics, and each country's form of governance and social structure.

CHAPTER 4

A Place for Systematics

Between the publication of Temminck's first monograph in 1813 and the dissolution of the Natuurkundige Commissie more than three decades later, Temminck directed all his efforts towards raising the Leiden museum to the level of its Parisian counterpart, and to fashioning himself into a director equal to the prestige of his Parisian peers. This was a critical and eventful period for Temminck. After the establishment of 's Rijks Museum van Natuurlijke Historie things seemed to settle down for a while, but this period of stability was short-lived. Around 1830 an economic crisis and policy changes slowed down the growth of the museum. This impeded Temminck's work and that of his curators as well as the advancement and status of systematics in the Netherlands. The existing awkward relationship between the museum and Leiden University did not help particularly. Temminck soon found ways to bypass or even ignore any restrictions imposed by the government. He succeeded in expanding the collections even during the crisis of the 1830s, and this was mainly thanks to his networking skills. His directorship reflected his views on science in general and on systematics in particular, and was inevitably intertwined with the struggles of national politics. Temminck's directorate is best understood when all these factors are considered together, as is his changing status within European scientific circles.

Temminck, of course, was neither unique nor isolated in his ambitions. His aspirations and worries were familiar to naturalists of all nationalities, who operated in different settings and, more importantly, on different subjects. Academic and colonial naturalists, for example, also collected natural history specimens for research, but aims and approaches varied enormously from place to place. Disagreements and clashes occurred, and each naturalist thrived in his particular habitat. In this landscape systematics occupied its own, very particular position, in the natural history cabinet, whether private or public.

1 The Museum's Scientific Output

The most obvious element within Temminck's intellectual landscape was 's Rijks Museum van Natuurlijke Historie in Leiden. The museum had been both a blessing and a curse for him. As its director, Temminck had secured a

 | DOI:10.1163/9789004441491_006

prominent scientific position and unlimited access to the Leiden and other European collections. Unfortunately, Temminck was also overwhelmed by administrative tasks, the campaigning for funding, and attempts to deal with the material flooding in from the Dutch East Indies. Moreover, the king's strategy to nationalize science backfired and resulted in the isolation of his scientists rather than the acquisition of international prestige. In particular, his notion that Dutch scientists should publish in Dutch created problems for both the museum and Temminck.

The king's instructions for the Natuurkundige Commissie commanded that "no description of the objects observed or sent, nor observations of natural phenomena will be published in a foreign language nor sent to foreign societies nor scholars, except with permission from the minister [of public education, national industry, and colonies] and after previous publication by one of the learned societies or journals of this Kingdom."[1] The king was determined to claim any discoveries made by the commission for his country, but the Dutch language was not commonly read outside the Low Countries. French was the scientific language of the time, so that this decision from article no. 3 of the royal decree did not help promote the commission or the museum. If Dutch discoveries could not be read, spread, and shared outside the Netherlands, how could the work of the commission and the museum be noted, let alone recognized? Confronted with this dilemma, Temminck decided to bypass the rather inconvenient article no. 3 whenever possible.

As specimens and manuscripts arrived from the colonies, Temminck, short of staff and with his work mounting, needed to decide on how to deal with the material. One of the first things he did was to publish excerpts from Kuhl and Van Hasselt's letters in two journals, the *Annales générales des sciences physiques*, which was published in French in Brussels, and the *Algemeene konst- en letterbode*, which was published in Dutch in Haarlem. Temminck later translated (and at times slightly edited) many of the letters first published here in Dutch for publication in the Parisian *Bulletin des sciences naturelles et de géologie*.[2]

This making public of the correspondence was an attempt to circulate Kuhl and Van Hasselt's discoveries as quickly as possible. Once it became clear that they and other naturalists were not returning home to publish their findings themselves (as most of them died in the tropics), Temminck decided

1 Royal Decree no. 251, 29 April 1820, art. 3, in Veth, "Overzicht van hetgeen," 22; Striekwold, "Naam maken," 217–18.

2 A complete list of the published correspondence from Kuhl and Van Hasselt can be found in Klaver, *Inseparable Friends*, 57–63.

to distribute the relevant material among experts who could publish them. He would deal with the mammals and the birds himself, and Heinrich Boie was in charge of the amphibians, while De Haan took care of the mollusks, together with a French specialist, the Baron André Étienne J. d'Audebert de Férussac.[3] However, there were no Dutch experts that could describe and classify the thousands of fishes and insects collected by Reinwardt, Kuhl, and Van Hasselt between 1815 and 1821. Temminck proposed that Achille Valenciennes and Georges Cuvier worked on the fishes from Southeast Asia. For the insects, Temminck had Jacob Wilhelm Sturm in mind, the celebrated entomologist and engraver from Nürnberg.[4]

Valenciennes, from the Zoological Laboratory of Reptiles and Fishes of the Muséum National d'Histoire Naturelle in Paris, was working with Cuvier on the ambitious project of the *Histoire naturelle des poissons* (published in 22 volumes between 1828 and 1849). Valenciennes visited the Leiden museum in 1824 and in 1827 to study the collection of fishes, many of which were previously undescribed.[5] There he prepared drawings of the specimens as well as copies from the sketches originally made by members of the Natuurkundige Commissie. He also selected some specimens to be sent on loan to Paris for further study. (Their return would prove to be rather tricky and forced both Temminck and Schlegel to go to Paris to retrieve the fishes).[6]

Naturally, Cuvier and Valenciennes published their results in French (as they were included in their *Histoire naturelle des poissons*), against the explicit wishes of the king. Invoking the infamous article no. 3, Daniel Jacob van Ewijck of the Ministry of Public Education, National Industry, and Colonies strongly reprimanded Temminck for this flagrant violation of the royal decree not once, but three times.[7] Unconcerned, Temminck proceeded with his original plans. Cuvier and Valenciennes described the fishes held in the Leiden museum. Temminck worked on the collection of mammals and published the *Monographies de mammalogie*, in French. The first volume appeared in Paris and Amsterdam in 1827, and the second volume appeared in Paris and

3 Temminck, Report to the Ministry of Public Education, National Industry, and the Colonies on the Year 1824, 3 December 1824, Naturalis Biodiversity Archives, Jaarverslagen.

4 Temminck, Report to the Ministry, 3 December 1824; Gijzen, "'s Rijks Museum," 40.

5 Roberts, "Freshwater Fishes of Java."

6 Temminck to the Ministry of Home Affairs, 30 September 1835, "Verslag van Temminck aan de Minister van Binnenlandse Zaken over zijn wetenschappelijke werkzamheden tijdens zijn verblijf in Parijs," Naturalis Biodiversity Center Archives, Archive C. J. Temminck, NL-LdnNAT_Temminck_49; Gijzen, "'s Rijks Museum," 224.

7 Van Ewijk to Temminck, 9 May 1825, 28 May 1825, and 24 December 1825, NL-HaNA, BiZa, 2.04.01, inv.nr. 4859; Striekwold, "Naam maken," 217.

Leiden in 1835. Temminck described more than eight hundred species of birds in the *Nouveau recueil de planches coloriées d'oiseaux* between 1820 and 1838. The members of the Natuurkundige Commissie had collected about a quarter of them.[8]

Temminck always chose to write his extensive treatises in French. He published only a handful of short articles in Dutch and later translated most of them into French to incorporate them into his more comprehensive publications.[9] Temminck was not only a devoted Francophile but, first and foremost, seriously committed to serving the international scientific community. Apparently, since Temminck was the author, it seems that the minister did not object to him publishing in French.

Almost all curators at the Leiden museum preferred publishing in French as well. Boie wrote an unpublished treatise on reptiles and amphibians (based mostly on Kuhl and Van Hasselt's notes and specimens) entitled *Erpétologie de Java* before leaving for Java as a member of the Natuurkundige Commissie in 1823. Boie's manuscript was never published. It had been sent to Brussels for printing, but it was lost during the Belgian Revolution in 1830. The manuscript, including the drawings that were to serve as the originals for the plates, was somehow eventually returned to the Leiden museum, and Schlegel published an excerpt of Boie's descriptions in 1826.[10] Willem de Haan wrote several essays on insects, the *Mémoires sur les métamorphoses des coléoptères*, the first of which appeared in 1835.[11] Between 1830 and 1850 Hermann Schlegel published mostly in German and French.[12] Temminck must have been displeased when the king commanded in 1839 that the *Verhandelingen over de Natuurlijke Geschiedenis der Nederlandsche Overzeesche Bezittingen* be published in Dutch: these were the results of decades of work by the Natuurkundige Commissie.[13]

---

8 Temminck to the minister of public education, national industry, and the colonies, 24 November 1836; quoted in Veth, "Overzicht van hetgeen," 84.

9 See Appendix, Bibliography of Coenraad Jacob Temminck's works.

10 Boie, "Erpétologie de Java," manuscript and plates [1822–1823], Naturalis Biodiversity Center Archives, Archive Natuurkundige Commissie; Schlegel, "Notice sur l'Erpétologie de l'ile de Java"; Holthuis, *Rijksmuseum*, 29.

11 See also Vrolik, "Levensberigt van Wilhem de Haan."

12 In the second half of the century, and especially after he had succeeded Temminck as director, Schlegel focused primarily on the Dutch fauna and wrote mostly in Dutch. Significantly, these works were not only intended for fellow naturalists, but also for students of natural history. See Hubrecht, "Hermann Schlegel"; Schlegel, "Levensschets van Hermann Schlegel."

13 Husson and Holthuis, "The Dates of Publication of 'Verhandelingen'"; Fransen, Holthuis, and Adema, "Type-Catalogue of Decapod Crustacea."

Outside of the Netherlands the long-expected *Verhandelingen* were met with a mixture of satisfaction and disappointment. The choice of language was criticized and its limited reach lamented. The British naturalist Hugh Edward Strickland wrote: "This superb work contains figures and descriptions of many new species from the remote islands of the Malay archipelago; and it is only to be regretted that so valuable a publication should be compiled in a language with which few men of science out of Holland are acquainted." In the same report Strickland noted a specific publication by Schlegel in the *Tijdschrift voor natuurlijke geschiedenis*, saying that "being unacquainted with the Dutch language, I regret my inability to give a summary."[14] It was not in the best interest of the Leiden museum and its scientists, nor that of the Dutch government, to insist on publications in Dutch.

As the collections grew, it soon became apparent that the museum was missing an essential element: a catalog of the Leiden collections. Thirty years after the establishment of the museum, Charles Bonaparte could not bear this lacuna any longer. In a footnote (of three pages) in an article on parrots, Bonaparte declared that "no book on Natural History, we shall never enough repeat it, would prove more effectual to the progress of science, more creditable to the nation, to the government, and to the able individuals willing to accomplish the labor, than the Catalogue of the Leyden Museum."[15] His tone was perhaps a touch theatrical, but this passage reflects the importance of museums' catalogs. A collection catalog was more than a numbered list of objects: it was an essential tool for systematics. For the realization of a complete classification of life forms, it is crucial to know which species have already been described and which form part of collections and are available for examination. Bonaparte urged his friend Schlegel – whom he expected not only to lead the enterprise but to illustrate it as well – to progress with the project, combining "the knowledge for which he has long been celebrated all over the world, with the skill of a first-rate draftsman."[16]

Temminck understood the need for a catalog of the collections as well as Bonaparte – cataloging had been his gateway into taxonomy, after all. In fact, he had started compiling a catalog sometime around the 1820s, entitled "Galerie zoögraphique du Musée des Pays-Bas." This was in French. Temminck was nothing if not determined to reach the international community. However, he

14 Strickland, "Report on the Progress," 188, 207.
15 Bonaparte, "On the Lorine Genus of Parrots, Eclectus," 144.
16 Bonaparte, 144.

only managed to list part of the mammals and the birds.[17] He stopped indexing because he could not keep up with the numerous specimens arriving in Leiden every month.[18] Schlegel, who accepted Bonaparte's challenge of 1850, took twelve years to draw up a painstaking inventory before publishing the first issue of the catalog, also in French, which listed only a handful of groups of birds. By then, Schlegel had succeeded Temminck as the museum director. The title of the catalog became: *Muséum d'histoire naturelle des Pays Bas: Revue méthodique et critique des collections déposées dans cet établissement*. He made sure that the government subscribed to sixty copies, paying ƒ300 annually for five years.[19] The project of publishing the catalog was passed on from director to director as some sort of revered family heirloom. Each of them added one or more volumes, but it was never finished. The collections grew faster than the curators were able to catalog. Fourteen volumes in total appeared between 1862 and 1908.

Despite its increasingly towering size and prestige the Leiden museum also lacked its own journal. This was not due to a lack of interest on Temminck's part. He had envisioned 's Rijks Museum as a Dutch version of French Muséum, and such a vision included, quite naturally, a museum journal. Temminck had already picked a title: *Annales du Musée d'histoire naturelle des Pays-Bas*; but it never came to life.[20] Curiously, the Dutch government did not support this idea, despite the additional prestige and financial support that a journal might bring. When the Leiden museum finally produced a journal in 1879, under Schlegel's editorship, it was an English publication, entitled *Notes from the Royal Zoological Museum of the Netherlands at Leyden*, or *Notes from the Leyden Museum* for short. By then, the language of science was no longer French nor German.[21]

Institutional publications, such as the collections catalog or editing a journal, fell under the director's responsibilities. But whether Temminck was involved in the research and publications of his staff, is doubtful. Most curators were either publishing in journals or self-publishing, i.e. paying for their

17 Temminck, "Galerie zoögraphique du Musée des Pays-Bas: Catalogue des mammifères exposés dans les galeries du Musée des Pays-Bas," [1825], Naturalis Biodiversity Center Archives, Archive C. J. Temminck, NL-LdnNAT_Temminck_27. Apparently he also cataloged the bird collection, or at least part of it, as mentioned in Finsch and Hartlaub, *Die Vögel Ost-Afrikas*, vol. 4, 162.

18 Snellen van Vollenhoven, "Coenraad Jacob Temminck," 68.

19 Schlegel, *Muséum d'histoire naturelle*; Holthuis, *Rijksmuseum*, 69; Dickinson and Pieters, "Some Bibliographic Findings."

20 Holthuis, 69; Veth, "Overzicht van hetgeen," 79.

21 For a study of the evolution of scientific language, see Gordin, *Scientific Babel*.

work to be published out of their own pockets, as Temminck did. Some curators occasionally received small subsidies from the government. Schlegel even took on the task of publishing a work entitled *Abhandlungen aus dem Gebiete der Zoologie und vergleichenden Anatomie*.[22] The enterprise did not last long. Only two issues were published, one on cetaceans in 1841 and one on falcons in 1843. The reason for the discontinuation of the work is unknown, but since Schlegel remained interested in this particular subject it seems likely that he ran out of funds.

De Haan, Schlegel, and Snellen van Vollenhoven were among the most productive of the museum curators. They worked mostly, although not exclusively, on systematics, each within his field of expertise. De Haan, for instance, organized the classification of jellyfish, made comparative studies of the anatomy of crustaceans and other arthropods, researched the mysteries of the metamorphosis of beetles, and described many new species of crustaceans and insects from the Dutch East Indies and Japan.[23] Snellen van Vollenhoven concentrated on insect classification and life histories, and from time to time he attended to pest management and other issues of immediate practical value.[24] Schlegel, an autodidact like Temminck, wrote handbooks on general zoology, did comparative studies of the venom glands of snakes, published a series of monographs on the comparative anatomy of whales and dolphins, and wrote a treatise on falconry, as well as a significant number of other works. However, he was mainly preoccupied with vertebrate systematics and the quest for a satisfying definition of the species concept.[25] He spent more than a decade studying the snakes at the Leiden museum before publishing a two-volume monograph in 1837, the *Essai sur la physionomie des serpens*, in which he revised the existing classification systems and the characters used to identify the different groups and species of snakes.[26]

A particular episode in the early years of the museum illustrates this process of specialization in systematics. After he had been the curator of invertebrates for a year Jan van der Hoeven decided to go back to Leiden University to study medicine. His wish to study was, however, not his primary reason for leaving: he wanted to pursue a career in comparative anatomy. Van der Hoeven followed Brugmans' school of thought (but not his lectures, as Brugmans had died before Van der Hoeven arrived at university). Immediately after he was

22 Holthuis (*Rijksmuseum*, 69) noted that Schlegel published his *Abhandlungen* "at his own risk," implying that he had no financial support from the government.

23 Vrolik, "Levensberigt van Wilhem de Haan."

24 Krikken et al., "Samuel Constant Snellen van Vollenhoven."

25 See also Schlegel, "Levensschets van Hermann Schlegel"; Smeenk, "Hermann Schlegel."

26 Schlegel, *Essai sur la physionomie des serpens*.

appointed curator in 1826, he requested a room to house a separate collection of anatomical preparations.

Unsurprisingly, Temminck refused. He had spent years convincing Falck that anatomy and physiology required separate collections in the universities, and that the museum should primarily hold specimens specially prepared for systematics. He had also undertaken every effort to keep the museum as independent as possible from the university, which he considered the appropriate place for comparative anatomy. Additionally, the presence of a second collection within the museum would inevitably result in the division of funds, space, and time between the two, which Temminck was not prepared to do. Therefore Van der Hoeven left the museum, frustrated with the apparent narrowmindedness of the museum's director.[27] Van der Hoeven returned in 1858 to claim the directorship of the museum, a post that, according to the regulations of the museum and Leiden University, was rightfully his as the professor of zoology. However, Schlegel claimed to be the rightful successor himself. The dispute was intense, to say the least. In the end, Van der Hoeven was appointed *opperdirecteur* (chief director) and Schlegel *directeur*. The situation was a disaster, and Van der Hoeven left the museum once more, in 1860, now even angrier than the first time.[28]

Other than the tangles of the museum's politics, perhaps the most significant issue for Temminck's ascent to a place among the most prestigious naturalists of Europe was the format and pace at which his scientific publications appeared.

2 Temminck's Podium

As natural history flourished and works in the field mushroomed, momentum was as important as content. A legion of European naturalists was studying and describing exotic species from their respective countries' colonies, and with this, natural history advanced at a vertiginous pace. But in Temminck's case, content was more important than momentum. He poured all his time and attention into extensive and detailed multi-volume works, published some time apart. In the space of forty-seven years Temminck published, both as author and editor, fourteen titles (a total of twenty-six volumes) and wrote sixteen journal articles; by contrast, Snellen van Vollenhoven published 184 titles,

27 This episode and the power struggles between Schlegel and Van der Hoeven are described in De Jonge, "Macht, machinaties en musea."

28 De Jonge; Holthuis, *Rijksmuseum*, 41–42.

most of them journal articles, across thirty-seven years. Five of Temminck's books were on birds, three on mammals (eighteen volumes in total), and one dealt with nearly everything there was to know about the Sunda archipelago, in three volumes.[29]

From 1839 to 1844 Temminck acted as editor of the *Verhandelingen*. During the 1830s he worked simultaneously on the second edition of the *Manuel d'ornithologie*, with Meiffren Laugier de Chartrouse on the *Nouveau recueil de planches coloriées d'oiseaux* (this actually took nineteen years), on the *Monographies de mammalogie*, which was issued between 1827 and 1842, and on the volume on mammals for Philipp Franz von Siebold's *Fauna japonica*. Published in five volumes between 1833 and 1850, this splendid work was based on the zoological collections put together by Siebold and Heinrich Bürger in Japan, a country with firmly closed borders and virtually unknown to Europeans.[30] The volume on the vertebrates was authored by Temminck and Schlegel, while Willem de Haan wrote that on the crustaceans. The government agreed to buy several copies as a means of financial support, but the publication was paid for mainly by Siebold himself, in spite of the obvious fact that the staff of the Leiden museum wrote all five volumes and that this should therefore have been financed by the ministry.[31] In fact, during Temminck's directorate, the only funding the museum received for scientific publications was in the form of the purchase of copies of the *Verhandelingen* and the *Fauna japonica*.

There is one title among Temminck's publications that seems out of place. In his biography of Temminck, Joannes Susanna remarked: "in 1846, we meet him on a path where we did not expect to find him."[32] For reasons that had more to do with politics than science, Temminck had decided to gather everything that was known about the Dutch colonies of the east in his three-volume encyclopedia *Coup-d'oeil général sur les possessions Néerlandaises dans l'Inde archipélagique* (1846–1849). It included descriptions of its geography, local history, civil administration, ethnography, agriculture, horticulture, cattle breeding, and of course, natural history. Temminck used information from a wealth of

29 See Appendix, Bibliography of Coenraad Jacob Temminck's works; Krikken et al., "Samuel Constant Snellen van Vollenhoven."

30 For more information on the exceptional life of Philipp F. von Siebold and his collections, see Forrer and Arlette, *Siebold en Japan*; Holthuis and Sakai, *Ph. F. von Siebold and Fauna japonica*; Tjon Sie Fat and Van Vliet, *Philipp von Siebold*; Ueno, "The Western Influence on Natural History in Japan"; Yamaguchi, *Von Siebold and Natural History of Japan.*

31 Holthuis, *Rijksmuseum*, 39.

32 Susanna, "Levensschets van Temminck," 65.

FIGURE 20 Japanese crested ibis, or *Nipponia nippon* (RMNH.AVES.195123), Philipp Franz von Siebold Collection
NATURALIS BIODIVERSITY CENTER. PHOTOGRAPH BY ESTHER DONDORP

FIGURE 21 *Ibis nippon* (Japanese crested ibis, now *Nipponia nippon*). Hand-colored lithograph after a drawing by Josef Wolf or Hermann Schlegel. In Philipp Franz von Siebold, ed., *Fauna japonica*, vol. 4 (1849): plate 71
IMAGE FROM THE BIODIVERSITY HERITAGE LIBRARY. DIGITIZED BY HARVARD UNIVERSITY, MUSEUM OF COMPARATIVE ZOOLOGY, ERNST MAYR LIBRARY

sources, including the reports from the Natuurkundige Commissie (especially those by Korthals and Müller), government archives, and published books.[33]

He wrote this eclectic encyclopedia for two purposes. First, Temminck wanted to prove wrong those who had accused the Dutch nation of keeping all this valuable knowledge to itself. Once more he made sure to reach his intended audience by writing in French instead of Dutch.[34] Sir Thomas Stamford Raffles, who, among many other things, had been lieutenant-governor of British Java from 1811 until 1816, had been extremely critical of the Dutch colonial government in his *History of Java*, as had been John Crawfurd, resident governor at the Court of Yogyakarta from 1811 to 1815.[35] Temminck rebutted their accusations of harshness and secrecy: "The Dutch nation and its government have often been accused of a *supposed* reserve in everything concerning our overseas possessions. I am committed to proving that this imputation has no foundation whatsoever."[36] Remarkably, Temminck never visited the Dutch 'overseas possessions.'

His second reason for publishing the *Coup-d'oeil* was to draw the Dutch government's attention to the benefits of further exploring the colonies. It was an unveiled plea to resurrect the Natuurkundige Commissie. He repeatedly stressed the need for expeditions to discover more about the natural history of the Sunda archipelago, an enterprise that was essential not only for the museum but also to boost the economy of the Dutch nation:

> The value of these expeditions of discovery is nowadays widely recognized; we recommend them from the double point of view of the progress they will produce in the sciences and the influence they can have on the future of our distant possessions; even the advocates of a strict economy and those who only look at a positive balance of expenses will be forced to accept that these expeditions, up to now always a meager investment by governments in the progress of the sciences, invariably have a purpose that is real, immediate and of public benefit.[37]

Despite his insistence, these government-sponsored expeditions akin to the Natuurkundige Commissie were too expensive, too dangerous, and too

33 Ulrich Gerard Lauts, a professor of Dutch language and literature at the Museum for Science and Arts in Brussels, published very positive reviews in the *Algemeene konst- en letterbode* in 1847 and 1849.

34 Temminck, "Avant-propos," in *Coup-d'oeil général*, vol. 1, xv–xviii.

35 Raffles, *A History of Java*, vol. 1; Crawfurd, *History of the Indian Archipelago*, vol. 2.

36 Temminck, "Avant-propos," in *Coup-d'oeil général*, vol. 1, xvi (emphasis original).

37 Temminck, *Coup-d'oeil général*, vol. 2, 424–25.

bureaucratic to be efficient. After Schlegel became director of 's Rijks Museum in 1858, he enrolled collectors who already lived abroad, such as the medical officer Heinrich Agathon Bernstein and the civil servant Carl Benjamin Hermann von Rosenberg, which proved to be a more successful strategy.[38]

In addition to these bulky monographs, Temminck wrote only sixteen journal articles in four and a half decades: ten about mammals, four about the geographical distribution of animals, and only two on birds. The first appeared in the *Annales générales des sciences physiques*, edited by members of the Académie Royale des Sciences et Belles-Lettres in Brussels. After the Belgian Revolution, which shook him to his core, Temminck turned to the Dutch journals *Bijdragen tot de natuurkundige wetenschappen* and *Tijdschrift voor natuurlijke geschiedenis en physiologie*. (In 1831 Temminck had nothing but harsh words and contempt for Brussels: "The enrichment of the Academies and of that ungrateful Brussels, of those mutinous cities of Belgium, whose memorial pillars still stand despite the superstitious Vandalism, will do service to the civilization of a stupid and fanatic people; at last the works of natural history, through which the arts were cultivated, were without hesitation put under legal protection by their rebellious hands."[39]) These latter articles were in Dutch. However, Temminck clearly preferred to write long and detailed books, which he revised continually. He spent a very long time – thirty-five years, to be precise – perfecting and polishing the *Manuel d'ornithologie*.

This single-mindedness, which initially earned him prestige as a thorough, comprehensive, and meticulous naturalist, tied him to the same subject for decades. Ultimately, this turned against him. An anonymous British reviewer initially praised "the truly scientific and valuable" *Manuel* for its sound descriptions and classification, which was, he stated, free of "the fashionable rage – that worse than Cholera-plague of modern science – for inventing new systems, and forging new names, more unintelligible and discordant than, erst, the tongues which broke upon the ear of ancient Babel." Temminck's monograph was, he concluded, and admirable textbook. But ten pages later, and somewhat unexpectedly, the same reviewer then referred to him as "the wary and tardigrade Dutchman."[40] A thirty-five-year gap between two editions of the same book was, it seems, too long. By contrast, in 1828 the ornithologist

38 Fransen, Holthuis, and Adema, "Type-Catalogue of Decapod Crustacea," 311.

39 Temminck, "Algemeen overzigt van de orde der Cheiroptera," 316, fn. After the Belgian revolution 's Rijks Herbarium was moved from Brussels to Leiden.

40 "A Retrospect of the Literature of British Ornithology," in *The Analyst*, 84, 94. "Tardigrade" here translates to as 'slow-paced'; sloths were known as *Tardigrada* at the time. Today the term applies to a very different group of animals, the minuscule, long-lived and endearing water bears, also known as moss piglets.

William Swainson apologized for a two-year delay in the publication of the second volume of the *Fauna Boreali-Americana*, which dealt with the birds collected by John Richardson, the surgeon and naturalist to Captain Sir John Franklin's second expedition.[41]

At some point Charles Bonaparte also grew annoyed with Temminck's tardiness. After entrusting some specimens of American bats to him, Bonaparte waited patiently for no fewer than twenty years for him to publish about them. When Temminck finally did so, the results were disappointing: "We may therefore conclude, that he has devoted very little time to this subject during these twenty years, for notwithstanding his immense materials, and the aid which he has received from every side, the work would have afforded us but little light if its defects had been less prominent."[42]

Ironically, while his early, hefty but precise books had catapulted Temminck into fame, his slow pace was eventually partly responsible for his decline. During Temminck's lifetime, and especially after 1840, journals became a more and more popular medium for naturalists to publish, communicate, and debate their findings. In 1845 Hugh Strickland listed no fewer than forty-two journals that were of interest to ornithologists, indicating that there were even "many others."[43] Journals were far more efficient vehicles than books for reporting news. In a period of constant discovery and fierce competition, naturalists could no longer afford to ponder a particular systematic arrangement for decades. When Temminck was finally ready to release a description of species new to him, he found that others had already described them – mostly in quickly written, short journal articles – which lead to disputes about plagiarism and priority. This tainted his reputation, and he never recovered from it.

His slow pace meant that he was losing touch with the speed (for popular groups like birds and mammals at least) of the developments in systematics. Darwin worked for nearly eight years on barnacles, and his monograph was warmly received.[44] Temminck's work was specialized, descriptive, bulky, and above all, meticulous:

> What this work [i.e. the *Manuel*] contains is based on the most rigorous examination of nature, without any kind of compilation; all species have been examined and often compared between them in all the cabinets in Europe; these may be the only merits of my work, and [this is] the only

41 Houston, "Hudson's Bay Company and its Collectors," 350–51.
42 Bonaparte, "Observations of the State of Zoology," 21–22.
43 Strickland, "Report on the Progress," 200.
44 Charles Darwin, *Monograph on the Sub-Class Cirripedia.*

difference that will distinguish it from the works of sedentary and library naturalists.[45]

Traveling through Europe, from cabinet to cabinet, examining all the specimens that needed to be examined is not only a conscientious way of working, but it is also an exceedingly slow one. After taking up the directorship of the museum, Temminck could no longer sustain this methodology. He did not have the time to roam through Europe freely for months and visit collections, as he had done for the *Manuel*. He was aware of this shortcoming and noted it in the introduction of the *Monographies de mammalogie*. He had to content himself with "provisional" descriptions and classifications for the species he had been unable to examine first-hand, a method with which Temminck was not particularly happy – his work felt unfinished and not entirely ready for publication.[46]

Temminck was well aware of an increasing impatience among his colleagues, and perhaps in himself as well, to see more works on zoology published at the Leiden museum, and especially the results of the Natuurkundige Commissie's efforts. As a last resort, he asked experts from the very museums he considered competitors to study and describe these new species. He had already done so in the 1820s for groups he was not familiar with; Cuvier and Valenciennes, for example, attended to the Japanese and Southeast Asian fish. But perhaps the most fruitful collaboration for Temminck was that with Gustav Hartlaub, a bird collector and naturalist connected with the Bremen Museum of Natural History. Faced with the extensive African collection gathered by Hendrik S. Pel (Temminck's pupil who enrolled as a civil servant in the 'Gold Coast'), Temminck decided to tackle only the mammals, and in 1850 passed on the birds to Bonaparte and Hartlaub.[47]

Now seventy-two, Temminck felt his health deteriorating and his energy diminishing, so he decided to ask Hartlaub for his assistance. Hartlaub came to Leiden several times to study Pel's collection from West Africa.[48] Temminck had already examined the specimens.[49] Altogether, Temminck identified and named two dozen new species, but he is not credited for any of them today. Hartlaub took over the task of formally describing Pel's birds in one book and two articles. He gave Temminck due credit for his discoveries by writing

45 Temminck, "Introduction," *Manuel d'ornithologie*, 2nd ed., vol. 1, xxvii–xxviii.
46 Temminck, "Discours préliminaire," in *Monographies de mammalogie*, ix.
47 Snellen van Vollenhoven, "Coenraad Jacob Temminck," 68.
48 Hartlaub, "Vorwort," in *System der Ornithologie Westafrica's*, vi.
49 Temminck, "Zoologische schetsen."

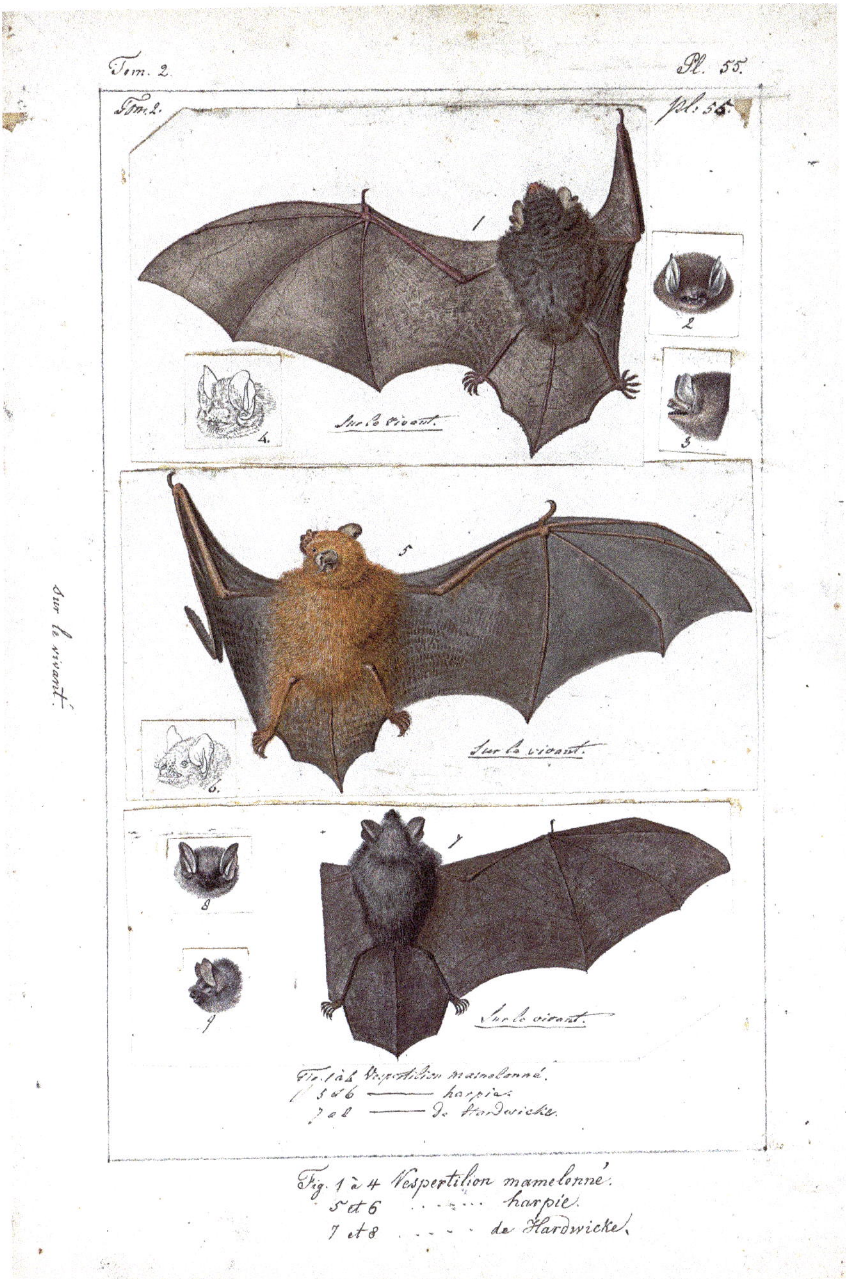

FIGURE 22 Preliminary design for plate 55 of Coenraad Jacob Temminck's *Monographies de mammalogie*, ca. 1830. Temminck used original drawings by the members of the Natuurkundige Commissie
NATURALIS BIODIVERSITY CENTER ARCHIVES, MMNAT01_PM_NNM001000766_002

"Temm." after each new species name coined by Temminck.[50] However, in a rather cruel twist of fate, the Rules of Zoological Nomenclature now dictate that Hartlaub should be cited as the species' author because the descriptions were probably Hartlaub's, although Temminck identified them as new and named them.[51] Of the twenty-eight new species described by Hartlaub based on Pel's specimens in 1855, Temminck had labeled twenty-five. Hartlaub is now deemed the sole author.

While Hartlaub and Bonaparte took care of the birds, Temminck published a volume on West African mammals in 1854, entitled *Esquisses zoologiques sur la Côte de Guinée*. It was, in Susanna's words, his "swan song."[52] Temminck had intended this to be the first of a series, the *Faune d'Afrique*.[53] Pel's collection was then studied over the years by several specialists, including Schlegel, Herklots (who wrote his doctoral dissertation on this material), and the ichthyologist Pieter Bleeker.[54] Each of them published on the collection independently, so that Temminck's envisaged series on the African fauna never appeared.

During Temminck's directorate from 1820 to 1850, the curators kept up a quicker pace than Temminck on average. The most striking difference between the curators' works and Temminck's is not the content, but the form of their work. They chose to publish mostly articles and short notes rather than monographs of several volumes, and thereby published more frequently. It seems that this allowed the curators to expand their range of subjects from the purely classificatory to comparative essays and more applied research. It is difficult to pinpoint the exact reasons for these differences. One reason may be that the curators were, on average, a quarter of a century younger than Temminck. This may not seem much of a difference, but the practice of systematics was evolving rapidly. Cuvier laid the grounds for comparative anatomy between 1800 and 1817, and it was embedded in nearly all classification systems by the 1820s.

Most curators developed their scientific ideas and skills in the 1810s and the 1820s, and thus during the period when Temminck's earliest works appeared. Short notes in professional journals accelerated communication, facilitated

50 Hartlaub, "Beschreibung einiger Vögelarten"; *System der Ornithologie Westafrica's*; "Systematisches Verzeichniss." See also Holthuis, "Biografische notities," 28–29.

51 See International Commission on Zoological Nomenclature, *International Code of Zoological Nomenclature*, art. 50. Dekker lists many of Hartlaub's types in "Type Specimens of Birds. Part 2."

52 Susanna, "Levensschets van Temminck," 68.

53 Temminck, "Avant-propos," in *Esquisses zoologiques*, xi.

54 For example, Bleeker, "Mémoire sur les poissons de la côte de Guinée"; Herklots, "Additamenta ad faunam carcinologicam Africae occidentalis"; Schlegel, "Over eenige nieuwe soorten van vergiftige slangen van de Goudkust."

debates, and gave momentum to systematics. Temminck began to publish on systematics when he had already passed the age of thirty, possibly as a consequence of his lack of formal training. By comparison, De Haan had started studying in Leiden at seventeen and wrote his earliest publications when he was twenty years old. There was a difference in temperament and taste between Temminck and his curators: Temminck was scrupulous and methodical to the point of being fastidious, but, perhaps more importantly, his curators had much more time available for research. Judging from the archives, Temminck took care of the management of the museum single-handedly. Moreover, from the mid-1840s onward, Temminck's trembling hands could not keep up with his mind. Because of this affliction, writing took much time and effort, and caused much frustration. Susanna took on the task of writing most of Temminck's letters and the annual reports to the ministry.

Just as Temminck's work was conspicuously absent from natural history journals, it seems that he did not give any public lectures either, and neither did he contribute in a noticeable way to the activities of the Dutch learned societies. He was a member of forty different societies and institutes, from the Imperial Society of Naturalists of Moscow to the Zoological Society in London and the American Philosophical Society in Philadelphia. It was an impressive, extensive network: Strickland's Report of 1844 listed thirty-four scientific societies that were in some way or another occupied with ornithology.[55] Temminck used these societies' platforms to publish a few articles and correspond with naturalists worldwide, but his contributions in the form of reports or lectures were minimal.

In the Netherlands, where it would have been relatively easy for Temminck to participate in meetings with his peers, it seems that he only did so very sporadically. In October 1838 Temminck read one of his papers before a meeting of the First Class of the Koninklijk Instituut van Wetenschappen (the same institute that had rejected his application for membership thirty years previously). In 1841 a second paper followed, but that was all.[56] Both papers were on Japanese mammals, were published in Dutch journals in 1838 and 1841 respectively, and were later translated into French.[57]

Meticulous morphological descriptions may not have lent themselves to being read in front of an audience and were better presented on paper, with

55 Strickland, "Report on the Progress," 200.

56 Temminck, "Kennis en verbreiding der zoogdieren van Japan"; "Over eenige geslachten van zoogdieren."

57 Temminck, "Description d'un nouveau genre de mammifères (Urotrichus)"; "Faune des Iles de la Sonde et de l'empire du Japon," in *Fauna japonica*, vol. 1.

illustrations, tables, and all kinds of comparisons and references. Nevertheless, some presentations on zoological or botanical descriptions were given. To give just one example, during the thirteenth meeting of the First Class of the Koninklijk Instituut, the members presented essays on subjects as diverse as the shape of the woolly mammoth's pelvis, the structure of the orangutan's carpal bones, the indigo plant, and the mammal fauna of Japan.[58] Talks specifically on systematics were rare, which may not have aided the field's visibility within the learned societies in the Netherlands.

In any event, Temminck was not a particularly engaged member of any society. Whether this was a consequence of his specialization in technical systematic research, of his avoidance of lively social interactions, or a combination of both, remains unclear. Remarkably, in his letters Temminck never discussed difficult issues or shared his discoveries with any of his correspondents. His network kept him informed of new publications and served mostly the exchange of information and specimens, but it was not his podium for debate. Perhaps he guarded his discoveries more zealously after his bitter experiences with Vieillot, Boitard, and Knip. Ultimately, Temminck's choice of language and platforms influenced his prominence and scientific authority. His focus always remained on zoological classification and his natural habitat, the natural history cabinet.

3 The Geography of Systematics

's Rijks Museum was the epicenter of Dutch systematics. Elsewhere academics and explorers approached natural history from different angles. Except in Groningen, anatomy and physiology had been traditionally the universities' business. Van Swinderen seemed to be the only Dutch professor with a taste for systematics. The universities had a long tradition of teaching natural history as applied to medicine, and since the 1815 reform of the education system they had favored comparative anatomy over systematics. Eventually this university tradition and both Brugmans' and Temminck's idiosyncrasies reinforced the divorce between the museum and academia.

Museums and universities had natural history collections prepared and ordered in such a specific way as to be useful to their intended users: anatomical preparations for the professors, stuffed specimens for the curators. The collections for anatomical and pathological studies were very different

58 Held on August 31, 1841 in Amsterdam, see Koninklijk-Nederlandsch Instituut van Wetenschappen Letterkunde en Schoone Kunsten, *Verslagen en mededeelingen*, vol. 4.

in composition from those needed for systematics. It was impossible to study the entire natural system, its variations, and the relations between groups and species without an extensive, complete collection, which was a representation of entire faunas. This is why Temminck reserved only a few examples of the "higher divisions" (i.e. mainly orders and families, the groups above genera in the hierarchy of living beings) for the southern universities.[59] The professors were more concerned with studying basic body plans than with cataloging living forms. Therefore, Temminck thought, a few objects as models would suffice.

Professor Brugmans particularly disliked systematics. He argued that it was pointless to divide nature into artificial units: "nature knows no system, no divisions, no classes nor species." His lessons were mostly on comparative anatomy, which he regarded as "the only solid foundation for the Systematic and Philosophic understanding of the Animal Kingdom."[60] Since the University of Leiden lacked the funds to expand the existing collection, Brugmans assembled a cabinet of natural history of his own, one that he could use for illustration in his lessons. His private cabinet reflected his views on nature: it consisted of anatomical samples, a few prepared specimens, a considerable collection of human skulls, and a pathology collection – one of the finest in Europe, according to Georges Cuvier.[61] Temminck had consulted Brugmans for "new anatomical facts" for his early monographs.[62]

That Temminck cared so little for the university collections did not imply that he dismissed comparative anatomy. For him, as for Brugmans, it was a subject that guided the Linnaean system. Both comparative anatomy and osteology were the "fundamental sciences," a point of departure for zoology. Temminck understood that comparative anatomy was useful for the general outlines of classification. However, once the major groups were delineated, additional information was required to refine the classifications. External characters, life histories, and geographical distributions allowed naturalists to perfect and polish the basic classification system.[63] For Temminck the two subjects were complementary. But they did require different types of collections for research.

---

59 Temminck, Report to the Ministry of Public Education, National Industry, and the Colonies on the Year 1820, 1 August 1821, Naturalis Biodiversity Center Archives, Jaarverslagen.

60 Quoted in De Jonge, *Het vergeten fenomeen*, 19.

61 De Jonge, 24. Leiden University purchased part of his collection after Brugmans' death in 1819, for the substantial sum of ƒ34,000.

62 Temminck, "Introduction," in *Pigeons et gallinacés*, vol. 1, 13.

63 Temminck, "Introduction a cette troisème partie," in *Manuel d'ornithologie*, 2nd ed., vol. 3, xv–xvii.

This discrepancy between the research fields at the University of Leiden and 's Rijks Museum meant that, outside of the competition for governmental funding and collection specimens, Temminck had little or no contact with students and professors in other disciplines, and no influence at all in higher education. This had greater consequences beyond Temminck's work. Whether or not universities practiced and taught systematics had an impact on the continuity of the field. Neither the universities nor the museum instructed future curators in systematics. Instead, zoological classification was – and still is – mainly a discipline learned through practice.

In Haarlem, Martinus van Marum would have agreed with Brugmans on his views on systematics. After all, he had been a student of the brilliant and famous physician Petrus Camper in Groningen. In his writings about the collection at the Hollandsche Maatschappij, he proposed enlarging the collection and turning it into a "more scientific collection of comparative anatomy, in which he [i.e. Van Marum] had developed a great interest and skills in comparative anatomy at the Anatomical School of Prof. Camper."[64] Apparently, the collection of prepared animals was of lesser scientific value in Van Marum's eyes.

Meanwhile, at the Athenaeum Illustre in Amsterdam, Reinwardt's lessons on natural history focused on chemistry and botany, mainly as applied sciences, and not on systematics. He had worked as a chemist's assistant in Amsterdam and had been introduced to the applied sciences through this work. Between 1800 and 1808 he was a professor of natural history and associate professor of chemistry and pharmacy at the University of Harderwijk. Botany was interesting to him as far as it served agriculture and medicine; classification in itself had never really caught his attention. Before departing for Java, Reinwardt focused mainly on economic botany and agriculture, and after he had spent six years in the colonies, his interest shifted towards processes and the interactions between organisms and their environment.[65] His approach to natural history was strictly Humboldtian and utilitarian and continued thus during his professorship in natural history at Leiden.

When Temminck heard about Reinwardt's growing interest in zoology, he tried to steer him towards anatomical research, advising him to leave aside "meticulous methodic comparisons." He hoped that in combining Reinwardt's anatomical research with his own on external characters, they could finally

64 "Van Marum's betrekking tot de Hollandsche Maatschappij der Wetenschappen, etc.," NHA, Archive Martinus van Marum 529-9; quoted in Sliggers, "Het naturaliënkabinet van de Hollandsche Maatschappij," 102.

65 Weber, "Hybrid Ambitions," 187–92.

"solve some puzzling issues in zoology."[66] Temminck suggested an association to Reinwardt through a division of their fields of study: if Reinwardt focused on comparative anatomy and Temminck continued with systematics, they could achieve a deeper understanding of these "puzzling" issues. Disappointingly, Temminck did not elaborate further on the nature of these issues. The fact that Temminck did not use terms like systematics or classification seems to testify to the uncertainties surrounding these emerging disciplines at the time.

While in this way the Leiden museum became the systematists' domain, the situation was different in other countries. In Germany, the Zoologische Museum was part of the University of Berlin, which had been founded in 1810. Nevertheless, its main focus was systematics. Karl Illiger was its first director, and following his death in 1813 he was succeeded by Hinrich Lichtenstein. Both assiduously corresponded and exchanged specimens with Temminck during their respective directorships. The Berlin museum also held an academic collection which was tied to the university's curriculum as well as a public institution.[67]

In France the Muséum National d'Histoire Naturelle differed from the Leiden museum in almost every possible respect. Between 1800 and 1850 the Muséum was an open, public space with zoological and botanical gardens and exhibition galleries. The museum combined academic teaching and research. Here, comparative anatomy and systematics shared avenues of research and offices, and different views of classification were proposed in the exhibition galleries and hotly debated in public lectures.[68] By 1832 the scientific staff of the Muséum was made up of titular professors occupying thirteen different chairs. Many of them were simultaneously employed in various remunerated posts in other institutions, mostly academic institutes and *écoles*. They were also involved in the French Academy of Sciences and were allowed to enjoy the benefits of having several salaries (*cumul*).[69] Besides the obvious financial advantage of multi-tasking, the professors had access to a broader public and a larger number of colleagues, students, and societies. The museum researchers were very well-known and present in the educational system.

Circumstances in Victorian Britain were different. While men of independent means placed their fortunes in the service of science, paid scientists

---

66 Temminck to Reinwardt, 25 November 1816, NL-HaNA, BiZa 1813–1848, 2.04.01, inv.nr. 4032B.

67 Ahrens, "Ornithological Collections of the Berlin Museum," 241–45; Nyhart, *Modern Nature*, 46; Stresemann, "Entwicklung der Vogelsammlung," 498.

68 Outram, "New Spaces in Natural History," 249–65; Limoges, "Development of the Muséum d'Histoire Naturelle."

69 Limoges, 215.

occupied a lower social status than the wealthy men.[70] Hugh Strickland was the epitome of the gentleman scientist. He was also one of the most prestigious systematists of the time. Of independent means, Strickland worked mostly on his private collection and also had access to other British natural history collections. He was well-connected and highly respected, but he was not employed by a museum or a university. He was closely associated with the University of Oxford, and his collection was bequeathed to the Museum of Zoology of the University of Cambridge after he died tragically at the age of forty-two, as a consequence of being hit by a train while examining geological strata along a railway.[71]

For the botanist Joseph Hooker, however, being a gentleman scientist was not an option because he lacked the necessary wealth and social status. After being rejected for the Chair of Botany at the University of Edinburgh in 1845, he decided not to pursue a university position further. At the university, botany was considered a minor subject within the medical field, not a discipline of its own. Hooker therefore declined a chair in botany at Glasgow University (the same chair his father William Hooker had held) and in 1846 accepted a job as a botanist for the Geological Survey of Great Britain instead. After almost a decade of traveling and publishing, he finally became the assistant director of the Royal Botanic Gardens in Kew in 1852, and then its director in 1865. Hooker's career was as much the result of his circumstances as a consequence of the universities' tradition of treating botany as part of the medical sciences.[72]

The conditions in which these naturalists worked varied greatly. The status of their disciplines, personal fortunes, and the culture in their respective countries all affected their positions and workplaces.[73] In the Netherlands this meant that systematics was located in the museum, and anatomy and medicine in the universities. However, there was another arena for scholars wishing to unravel the mysteries of living beings: nature itself.

Temminck's natural habitat was the natural history collection. He never engaged in any fieldwork other than going on shooting expeditions near his home in search of birds for his collection and his dinner. He never visited the Dutch colonies or any other tropical region where he could have admired his beloved pigeons and fowl alive and in their natural habitat. The exact reason for Temminck's apparently unadventurous disposition is unclear. Perhaps he

70 Allen, "Amateurs and Professionals," 15; Endersby, *Imperial Nature*, 8–12.

71 Jardine, *Memoirs of Hugh Edward Strickland*; Rookmaaker, *Calendar of the Scientific Correspondence*.

72 Endersby, *Imperial Nature*, 11.

73 For an overview of these and other factors see Bowler and Pickstone, *Modern Biological and Earth Sciences*.

did not feel physically fit for such fieldwork; perhaps the fate of so many members of the Natuurkundige Commissie discouraged him. Besides, managing the museum took up much of his time and also, in all probability, his energy.

However, Temminck did not complain about this lack of experience in the field. He never made it a priority for his research. His daily practice consisted not of observing and recording living animals but of methodically and fastidiously comparing as many specimens as he could lay his hands on, that is, mounted specimens. Comparing large series of specimens was far more appealing and rewarding for any systematist than a trip to the tropics, however exciting that might be, and this explains the desk-bound nature of Temminck's work in part.

Interestingly, the fact that he had never visited the Dutch East Indies seemed to Temminck an advantage rather than a shortcoming. He spent years completing his three-volume report on the Dutch East Indian colonies, the *Coup-d'oeil général sur les possessions Néerlandaises dans l'Inde archipélagique*. Temminck admitted that not having visited the islands himself made his account more matter-of-fact and less vivid than any account he would have written if he had been there. However, he reasoned that no-one could ever explore all of the archipelago islands and learn all there is to know about them. That would take several lifetimes. He could have never written the *Coup-d'oeil* from first-hand experience (that is, from experience in the field), but only by gathering information from written sources, libraries and collections.[74] Therefore, comprehensive surveys must be done from home. The same was true for systematics: a comparison and ranking of species were only possible with enough specimens at hand that covered morphological variation across a wide geographic range. While he acknowledged the importance of fieldwork (obviously for collecting, but also, for example, for gathering observations on the habitat, behavior, and distribution of species), Temminck believed that the goals of the naturalist in the field were simply different from his – classification could not be carried out in the field.

Temminck was not alone in this opinion. The first two members of the Natuurkundige Commissie, Heinrich Kuhl and Johan Conrad van Hasselt, feverishly collected specimens and information in Java with the hope of finishing their systematic work when they returned to the Netherlands, in 's Rijks Museum van Natuurlijke Historie. They had ambitious plans: they wanted to write a treatise on the fauna of Java using the material collected during their time in the field as well as materials from the Leiden collection, including

74 Temminck, "Avant-propos," *Coup-d'oeil général*, vol. 1, xxi–xxii.

specimens collected by Reinwardt.[75] As they were confident of their knowledge, they started carrying out part of the systematic work while still on Java. Kuhl and Van Hasselt wrote to Temminck: "Hardly a day goes by without making a new Genus or designating a new Species, and we still collect a great deal, which we cannot examine now and which we intend to research once we are in Europe."[76]

Kuhl's and Van Hasselt's zoological notes consisted mostly of descriptions of the species they collected, giving their scientific name, references (if known), localities, observations on behavior, description of external morphology, and extensive anatomical notes. Very often, a species' scientific name was followed by a *mihi*, or a *nobis*, the Latin equivalents of 'mine' and 'ours.' Placed after a species name, this indicated that the author of the manuscript was describing and naming the species in question for the first time. Kuhl and Van Hasselt were certain that they were dealing with species new to European science. This seems like a bold assumption, taking into account their isolation – news could take months to reach them – and that they had minimal access to scientific literature. They had no other books than those they had taken with them and an atlas of the animals and plants known at the time (which they had put together themselves), a compilation of plates from "the best natural history writers" representing "an almost complete *Systhema* [sic] *Natura* in plates."[77] Van Hasselt's library, for example, consisted only of Raffles' *History of Java* (1817), Sonnerat's *Voyage aux Indes orientales et à la Chine* (1782), and the first volume of Johannes van den Bosch's *Nederlandsche bezittingen in Azia, Amerika en Afrika* (1818).[78] This traveling library was hardly comprehensive and not very helpful for the preparation of their intended zoological treatise. They relied on the superb knowledge of the fauna and flora of Java they had acquired while training in the Netherlands, Berlin, London, and Paris. Their proficiency was even more impressive given their young ages: both Kuhl and Van Hasselt were in their early twenties at the time.

Temminck thought that Kuhl and Van Hasselt should postpone their plan of describing and classifying Java's fauna until they had returned to the museum,

---

75 Greshoff, "Kuhl en Van Hasselt," 66; Klaver, *Inseparable Friends*, 29; Veth, "Overzicht van hetgeen," 28.

76 Kuhl and Van Hasselt to Temminck, 18 July 1821; published in Kuhl and Van Hasselt, "Pjihorjavor," 99.

77 This was indicated in the will of Johan Conrad van Hasselt, 2 October 1823, Naturalis Biodiversity Center Archives, Archive Natuurkundige Commissie, MMNAT01_AF_NNM001001046_008. The atlas is not in the Naturalis Archives. It is possible that it remained in Bogor, or it may have been lost.

78 Will of Van Hasselt, 2 October 1823, Naturalis Biodiversity Archives, Archive Natuurkundige Commissie, MMNAT01_AF_NNM001001046_007.

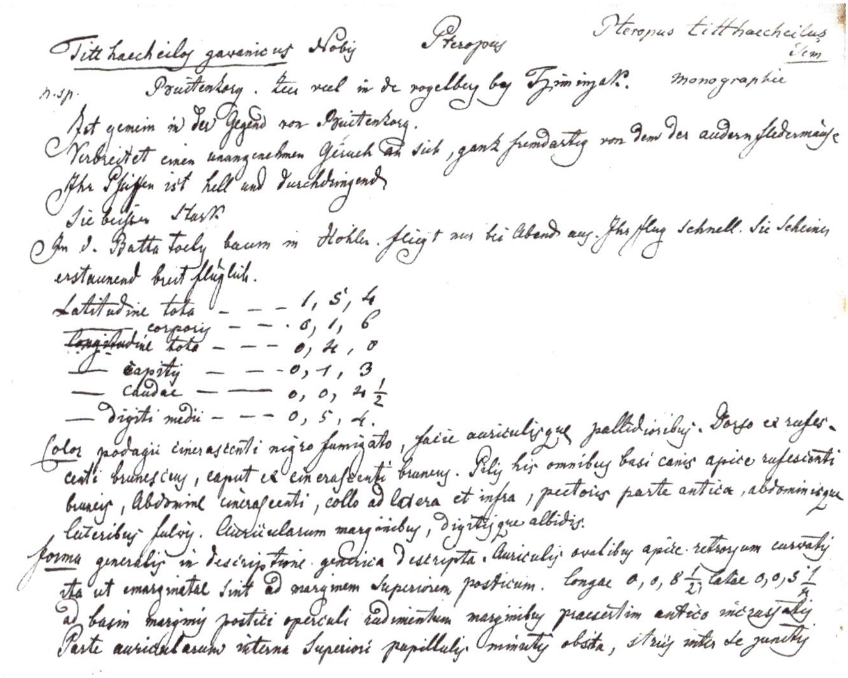

FIGURE 23 Description of a fruit bat (*Titthaecheilos javanicus*, now *Cynopterus titthaecheilus*) by Heinrich Kuhl and Johan van Hasselt (fieldnotes from Bogor ["Buitenzorg"], Heinrich Kuhl, Java, 1820–1821). Behind the species name, Kuhl and Van Hasselt noted "Nobis" ('ours') to indicate their authorship. After their death, Temminck described the species in his *Monographies de mammalogie* (1827) and named it *Pteropus tittaecheilus*, as noted in Temminck's hand in the upper right corner of the page.
NATURALIS BIODIVERSITY ARCHIVES, MMNAT01_PM_NNM001001033_013

an environment far better equipped for the comparative work that was (and still is) the primary practice of systematics. Unfortunately, they never had the chance, as both died much sooner than anticipated: Kuhl only nine months after his arrival in Java, and Van Hasselt two years later.

After Kuhl's untimely death, Temminck wrote to the minister of public education, national industry, and the colonies: "To make good field notes, to collect everything and to take the utmost care possible for the collections, as often as possible to dispatch shipments to the Motherland, this is the primary instruction for a naturalist in the tropical regions."[79] He had seen how Kuhl was trying to do everything at the same time and therefore left things unfinished, that

79 Temminck to the minister of public education, national industry, and the colonies, 21 March 1822; quoted in Veth, "Overzicht van hetgeen," 29.

his field notes were written hastily, almost stenographically, and that poorly preserved specimens had been lost. It was impossible to use one specimen for anatomical observations, describe it as a new species, make drawings, and then prepare it for shipping and to be preserved in a cabinet for systematic study. Kuhl had apparently found it difficult to focus on only one thing at a time, which made him, in Temminck's eyes, a poor collector: "I do not want to deny it, and the results show it, that Kuhl was not the right man to enlist for the expansion of a museum collection." Temminck lamented the fact that "he [i.e. Kuhl] could have done far more useful work in Europe, but he considered such a job to be too limited, however broad it may seem to others."[80]

Generally, young naturalists' time in the field was a stepping stone towards a career in systematics at home. Nearly all members of the commission received training in Leiden and Paris before going to the colonies.[81] The training prepared them on the one hand for the collection and preparation of specimens, and on the other for their comparison and arrangement using collections and literature. The first step was carried out in the field, the second in the natural history cabinets and libraries. Temminck expected the commission members to excel in both aspects, at least until it became painfully clear that their life expectancy was exceptionally short in the tropics.

Reinwardt was of an entirely different opinion. He fiercely contended that only field naturalists, not cabinet naturalists, could arrive at an accurate understanding of nature. He stressed the unity and connectedness of natural phenomena, and that therefore, in his opinion, it was only through work in the field that one could ever come to understand nature's organization.[82] Cabinets were nothing more than collections of lifeless objects removed from their original context.

These contrasting goals and methods in fieldwork and cabinet study had already been the source of tensions in Paris, decades before the birth of the Natuurkundige Commissie. In 1807 Georges Cuvier wrote a review of Humboldt's fieldwork report in which he drew a clear distinction between the "field naturalist" and the "sedentary naturalist." Cuvier noted that field naturalists were able to see the full wonders of nature through observing living beings in their environment. However, these observations were confined to the location of the traveler in the particular moment: "He [i.e. the field naturalist] is thus deprived of the possibility of comparing each being with those like it, of

80 Temminck to the minister of public education, national industry, and the colonies, 21 March 1822; quoted in Veth, 29.

81 Klaver, *Inseparable Friends*, 13.

82 Reinwardt, *Redevoering van C. G. C. Reinwardt.*

rigorously describing its characteristics, and is often deprived even of books which would tell him who had seen the same thing before him." Moreover, cabinet naturalists had enough time and material available for pondering all kinds of questions:

> The traveller can only travel one road; it is only really in one's study (*cabinet*) that one can roam freely through the universe, and for that, a different sort of courage is needed, courage which comes from unlimited devotion to the truth, courage which does not allow its possessor to leave a subject until, by observation, by a wide range of knowledge, and connected thought, he has illuminated it with every ray of light possible in a given state of knowledge.[83]

Cuvier's concluded that explorers and museum researchers pursue such different goals that they necessarily require different sets of skills to succeed, but that it was museum scientists who "led natural history towards the truth." En passant he also stated that the true "hero" is not the valiant field naturalist (who is much admired by the general public) but the tenacious and knowledgeable sedentary naturalist.[84] Cuvier's review seemed to especially aim at discrediting Humboldt's work and glorifying his own. Perhaps the mutual antipathy between Cuvier and Humboldt, and the Muséum politics, colored Cuvier's review.[85] Humboldt had very different views on the subject: he believed that it was his all-encompassing work, embracing several disciplines at the same time, that would lead man to understand nature.

Not all expeditions were equally useful for the museum curator. The fieldwork preliminary to systematic research consisted of survey voyages, explorations to inventory the existing species in particular regions of the world. This was a very different endeavor in its aims, methods, and organization than Humboldt's or Reinwardt's. As Robert Kohler puts it: "Systematics in the survey mode required an elaborate infrastructure of museums, with their dedicated study collections and elaborated exhibits, and of social networks of well-to-do patrons and local participants."[86] Systematics and its disciplinary counterparts, anatomy and physiology, relied on comparison, hence Cuvier's and Temminck's insistence on the necessity of working not in the field, and not

83 Cuvier, "Analyse d'un ouvrage, de M. Humboldt," in *Fonds Cuvier 3159*; quoted and translated in Outram, *Georges Cuvier*, 62–63.
84 Cuvier, "Analyse d'un ouvrage"; quoted and translated in Outram, *Georges Cuvier*, 62–63. See also Outram, "New Spaces in Natural History," 259–61.
85 See, for example, Coleman, *Georges Cuvier, Zoologist*; Outram, *Georges Cuvier*.
86 Kohler, "Reflections on the History of Systematics," 22.

in the colonies, but within a cabinet of natural history in Europe. Humboldt's methods were different because, to put it simply, he was practicing a different field of natural history altogether, the disciplines we would now call ecology and biogeography. His work was mainly a geographical and physical study of nature, based on observation and, above all, measurement. Obviously, these could only be made in the field. There were different kinds of field naturalists, just as there were different kinds of sedentary naturalists, which was a consequence of the breaking up of natural history into several disciplines. The different disciplines required different fieldwork methods just as they required different types of collections.

Unlike Cuvier and Humboldt, Temminck and Humboldt were friendly with each other. They corresponded sporadically, mostly to ask the other for advice on particular issues for their books. Temminck asked Humboldt for information one could only gain in the field, and Humboldt asked Temminck for facts only he had access to in the large collections. In 1822 Temminck sent Humboldt an engraving of the Andean condor by Nicolas Huet which was intended for the *Nouveau recueil de planches coloriées*. Huet had probably prepared a watercolor for the plate from a specimen in the Kaiserliche Hof-Naturalienkabinett in Vienna.[87] Aware of the limited usefulness of descriptions based on prepared specimens, Temminck asked Humboldt for his opinion on the plate's accuracy, as Humboldt had seen the bird during his travels and described it in 1811.[88] The Andean condor's head was difficult to draw from museum specimens, as its skin folds inevitably change form and color in the process of taxidermic preparation.[89] In a footnote to his description of the condor, Temminck reminded the reader of the problems one encountered studying prepared animals and the need for field observation to come to the aid of the sedentary naturalist:

> It is necessary to point out to those who compare our plate with that published by M. von Humboldt, that the drawing made by this scholar [i.e. Humboldt] and retouched by Barrabant [sic] pictures the head and neck drawn from a living individual; in the plate attached here, these same parts were drawn from a stuffed one. Before publishing it, I submitted my engraving to M. von Humboldt; this scholar finds that the occiput is

87 Raat, "Humboldt and Temminck."

88 Humboldt and Bonpland, "Essai sur l'histoire naturelle du Condor," in *Recueil d'observations de zoologie*, vol. 1, 26–30.

89 Humboldt to Temminck, [February–June 1822], Naturalis Biodiversity Center Archives, Archive C. J. Temminck, NAT_ARC_TEM_01613-15.

> raised too high, the back is too arched, and the transverse folds of the neck are not at all visible; all these defects are the result of a deficient preparation of the skin in the mounted specimen.[90]

Temminck and Laugier decided not to alter the original design by Huet but to include Humboldt's comments in their description, in a footnote. Not satisfied with this, Temminck wrote an *Addition à l'article du Catharte Condor* which comprised a description and an illustration of the female Andean condor and a detail of the male's head. This time, the original drawing for the plate was drawn after a living bird from the "Ménagerie du Jardin du Roi" (the Jardin des Plantes de Paris).[91] The differences between the drawing from the Vienna specimen and the one after a living condor are striking.

Occasionally Humboldt also asked Temminck for advice, appealing to his extensive knowledge of animals and their locations – the kind of knowledge one gathers from natural history collections. He approached Temminck while preparing the second volume of *Cosmos*, asking for examples of animals inhabiting separate regions for "the subject of zoological geography":

> Is there a bird that is known to be present everywhere, are there identical birds in the tropics of two continents, e.g., Africa and America? Are there identical mammals and birds in the northern and southern temperate regions? Ignorant as I am, I would like to be precise in the few pages I will dedicate to animal Geography. The sources I need are not easily accessible to me. I would like to have a few engaging examples for these questions. What you so easily can provide, is solid gold.[92]

Temminck was happy to help and attached some notes on the matter to his letter. He also planned to send Humboldt the first volume of the *Coup-d'oeil général*. Unfortunately, Temminck's notes do not seem to have survived, and Humboldt never published this zoological essay.[93]

---

90 Temminck and Meiffren Laugier de Chartrouse, "Catharte Condor," in *Nouveau recueil*, vol. 1, livraison 23, plate 133, text, [4], fn. 1.

91 Temminck and Meiffren Laugier de Chartrouse, "Addition à l'article du Catharte Condor," in *Nouveau recueil*, vol. 1, livraison 83, plate 494, text, [3].

92 Humboldt to Temminck, 16 August 1846, Naturalis Biodiversity Center Archives, NAT_ARC_TEM_01617.

93 Raat, "Humboldt and Temminck," 30. There are no copies of Temminck's notes on biogeography in the Naturalis Biodiversity Center Archives.

FIGURE 24 *Catharte* condor, *mâle*. Hand-colored engraving after a drawing by Nicolas Huet (after a prepared specimen of Andean Condor). In Coenraad Jacob Temminck and Guillaume Michel Jerôme Meiffren Laugier de Chartrouse, *Nouveau recueil de planches coloriées d'oiseaux*, vol. 1, livraison 23 (1822): plate 133

IMAGE FROM THE BIODIVERSITY HERITAGE LIBRARY. DIGITIZED BY THE SMITHSONIAN LIBRARIES

FIGURE 25 *Tête du* Catharte condor, *mâle*. Hand-colored engraving after a drawing by Nicolas Huet (after a living Andean Condor from the Jardin des Plantes in Paris). In Coenraad Jacob Temminck and Guillaume Michel Jerôme Meiffren Laugier de Chartrouse, *Nouveau recueil de planches coloriées d'oiseaux*, vol. 1, livraison 83 (1830): plate 494
IMAGE FROM THE BIODIVERSITY HERITAGE LIBRARY. DIGITIZED BY THE SMITHSONIAN LIBRARIES

It was not only zoologists who missed details when working from preserved specimens, but also botanists.[94] Plants are particularly susceptible to physical change after preparation: once dried, they are glued to sheets of paper and lose their original form and color. Often botanical specimens are missing important features like flowers, fruits, or roots. Surprisingly, what might seem an insurmountable stumbling block for botanists was actually incorporated in their classification systems from the beginning, which bypassed the issue.

Botanists like Joseph Hooker contended, as Temminck and Cuvier did, that the herbarium was the place for systematics. Hooker particularly disliked the proliferation of genera and species. He pointed out that colonial and local botanists, lost in detail, were prone to describe as separate species plants that were, in Hooker's view, mere varieties of the same species.[95] They lacked an overview of the whole range of species and varieties of the plants they studied. The issues concerning the classification of a particular fern from New Zealand served him as an example:

> In this (and in many similar cases) he [i.e. the colonial botanist] must bear in mind that I have examined many hundreds of specimens of the plant, gathered in all parts of the south temperate hemisphere, and have found, after a most laborious comparison, that I could not define its characters with sufficient comprehensiveness from a study of its New Zealand phases alone, nor understand the latter without examining those of Australia, South Africa, and South America.[96]

The classification systems of local floras highlighted the tension between the naturalist in the field and the botanist in the herbarium, or, as Jim Endersby puts it, between "the empire's center and periphery."[97] Collection building and system building were two different endeavors – although, of course, the second was impossible without the first. As natural history fragmented and expanded together with new disciplines and methods, certain activities became confined to certain places. Systematics flourished where extensive collections were available – mostly in museums.

By the middle of the century much had changed. It had been half a century since Temminck had inherited his father's cabinet around 1800. The works of Buffon and Brisson were now classic reference works, and Levaillant's

---

94 For a survey of the different practices of collecting and recording in the field see, for example, MacGregor, *Naturalists in the Field*.

95 Hooker, "Introductory Essay," in "Flora Novae-Zelandiae," xiii–xiv.

96 Hooker, xiii–xiv; Endersby, *Imperial Nature*, 155.

97 Endersby, 154.

fashionable books had been replaced by scholarly monographs and specialized journal articles. Naturalists and owners of private collections witnessed the birth of national museums in Berlin, Leiden, and Stockholm. These national institutes added many exotic, previously unknown specimens to their collections, as a result of the European imperialist expansion. The work undertaken in such collections – identifying, cataloging, and naming species – fueled systematics, and naturalists worked hard to include this ever-growing amount of species in their classification systems. At the same time, a network of like-minded naturalists and collectors – a small and intimate circle in continental Europe – became the foundation for a community of specialized systematists.

In the Netherlands zoological classification moved from private cabinets to the new Rijks Museum van Natuurlijke Historie, and remained instrumental to the universities' curriculum. Simultaneously, Temminck moved from an enthusiast collector to being the director of a national institute. Once at the helm of the new museum in Leiden, Temminck struggled to keep the books balanced and his scientific prestige intact. As the newborn Kingdom of the Netherlands strived to gain status in Europe, government-sponsored scientific expeditions were sent to the Dutch colonies to boost the natural sciences and, at the same time, the country's damaged economy. New specimens for the museum and herbarium arrived in large quantities and placed the Leiden collection amongst the best in Europe. Colonial and economic ambitions guided the country's scientific endeavors, while its naturalists worked relentlessly to compete with France, Germany, and later Britain.

Natural history advanced and matured during this period, which led to the specialization in both methods and location. Approaches varied from country to country, as their social, political, and economic circumstances determined where and by whom natural history was being practiced and taught. Despite their differences, all European naturalists faced similar issues and challenges. Those trying to classify the natural world struggled to keep up with an expanding natural world, the unstructured multiplication of literature, the entangled nomenclature of species, and a lack of standardization. After 1830 many naturalists dedicated themselves exclusively to searching for universally accepted methods of classification.

Between 1820 and 1850 Temminck also became increasingly entangled in issues regarding zoological classification. These included the search for an explanation for the seemingly inexplicable distribution of species around the globe, the perceived similarities between related groups, and most of all for a standard theory for classification. Concepts needed to be defined, patterns required explanations, and laws awaited being uncovered. Standardization became both a priority and the battleground whence systematics emerged as a scientific discipline of its own.

PART 2

Zoological Classification: 1800–1850

CHAPTER 5

Patterns, Laws, and Types

During the first half of the nineteenth century, European systematists were as devoted as Temminck to naming and classifying the natural world.[1] From the 1830s onward the number of specimens in collections and of natural history books in libraries was greater than ever before, partly as a consequence of European expansion and exploration. The problem was that despite centuries of classification efforts, no universal classification system or method to achieve it had been agreed upon. There were as many conflicting classification systems as there were systematists. To make matters worse, the new and exotic specimens arriving from the colonies could not easily be incorporated into the existing systems. Were the existing classifications deficient, or were they perhaps based on erroneous definitions of genus and species? Which characters were vital to classification, and which were superfluous? Who was to decide? The need for standardized approaches to and agreements on nomenclature became painfully evident. Classifications were, after all, much more than mere lists. They were theories about the natural order.

1 Geographical Patterns and the 'Type' Concept

Naturalists classified animals mainly according to their anatomical features (external and internal), but additional information about their habits and geographical distribution helped clarify the degree of relatedness between them. Naturally, 'relatedness' was interpreted at the time not as genetic kinship but as a degree of morphological similarity that placed particular groups of animals close to each other in the 'great catalog of life.' This was a deceptively simple concept. As more and more animals were discovered, their geographical distribution often seemed to contradict existing classifications and, more disturbingly to contemporary naturalists, the biblical narrative. How could there be marsupials in North America and Australia but nowhere in between? However could a lethargic animal like the sloth manage to travel all the way from the

1 This chapter is based in part on Gassó Miracle, "Significance of Temminck's Work on Biogeography: Early Nineteenth Century Natural History in Leiden, the Netherlands." *Journal of the History of Biology* 41, no. 4 (2008): 677–716. doi:10.1007/s10739-008-9158-8. Adapted by permission from Springer Nature.

 | DOI:10.1163/9789004441491_007

mountains of Ararat to South America after the biblical deluge? It was perplexing. The answer to these questions posed several scientific and philosophical problems. The main conundrums were the origin of species (including humankind), the species concept, the limits of species variation, the interpretation of the Bible, and whether nature is governed by universal laws or via direct intervention by God. These were not discussions for the faint of heart.

Temminck was among those who refined his classification systems with observations on the animals' geographical distribution. He insisted that distribution patterns were a subject "rich in facts and of consequence for science" and essential to any systematic arrangement.[2] He turned to the existing literature for species already described, but for his most recent acquisitions this information came from the notes of the Natuurkundige Commissie and from his crew of committed collectors.

Southeast Asian faunas proved to be the most intriguing of all. While working on the collections gathered by Reinwardt, the Natuurkundige Commissie, Philipp Franz von Siebold, and some others, Temminck began to ponder the differences between the faunas of Asia and the rest of the globe. He compared the animal population of the Sunda Islands with that of Japan, listed which species lived where, and noted the differences in species composition between islands. Then he contrasted this survey with the fauna of Central Europe and North America. Content with his findings, he published four essays on the subject between the late 1820s and the late 1830s.[3] He was convinced that he had uncovered a universal pattern in the geographical distribution of animals on the globe and called it a "general law on animal organization." According to his principle, there is a correlation between faunas from different regions of the globe:

> The zoological products of Japan [...] offer us new evidence of the geographical distribution of living beings: they help us establish this law of animal organization, [...] according to which there is a relation in organization, external form, and behavior between almost all animals that inhabit the same latitude, however far from each other the regions may be, where they live and freely disperse, and the extension of the seas

2 Temminck, "Kennis en verbreiding der zoogdieren van Japan," 293; "Aperçu général," in *Fauna japonica*, vol. 5, 8.

3 Temminck, "Aperçu général," in *Fauna japonica*, vol. 5; "Blik op de dierlijke bewoners"; "Faune des Iles de la Sonde et de l'empire du Japon," in *Fauna japonica*, vol. 1, "Kennis en verbreiding der zoogdieren van Japan."

> between regions has no influence on this, and neither has the enormous space of uninterrupted land that stretches between them.[4]

This was a generalization of the parallels that Temminck observed between mammals in Japan and those inhabiting Europe, Asia, and North America. The principle was drawn from observations of the fauna across the Northern Hemisphere, and Temminck assumed, generalizing rather confidently, that it was valid for the whole globe, even if it applied less strictly to the Southern Hemisphere. There was, he noted, a higher degree of diversification of forms in the tropical and subtropical regions. Unfortunately, Temminck did not attempt to explain this difference:

> This law applies in some respects to both hemispheres, but more strictly so to the Northern Hemisphere, where the species are exactly the same, while in the Southern Hemisphere, the groups and the species are distributed following a more diverse model, even though they very often present a great analogy with the animals that live in parallel latitudes.[5]

Temminck's law was born out of his recently acquired familiarity with the Japanese mammals. He observed that none of the species in Japan new to him were substantially different from those already known from central Europe and Asia. Indeed, none of the mammals in Japan had a fundamentally new morphology, and thus resembled European mammals in their anatomy. Except for a single species there were no monkeys in Japan, and this Japanese species was comparable in every way to the only species of monkey in Europe, the Gibraltar Monkey, also known as the Barbary Macaque. Temminck noted that both species of monkey lived in regions with similar climates, while monkey species similar to those found in the tropics – capuchin monkeys, for example – were not found in Europe, Japan, Central Asia, or North America.[6] These observations resulted in his universal principle, according to which animals from a particular region are alike in their "general organization" to animals from other areas on the same latitude:

---

4 Temminck, "Aperçu général," in *Fauna japonica*, vol. 5, 7. This passage is a literal translation of the Dutch text of the 1838 law in "Kennis en verbreiding der zoogdieren van Japan." Apparently, Temminck saw no reason to expand or alter the original text.

5 Temminck, 7.

6 Temminck, 4.

> Among the Japanese mammals, not a single species has been found that differs from the known basic forms (*types*) in any distinctive manner, such as those found in Australia, except for a few mammals that differ slightly from the known genera; all other mammals can be placed in the European and Asiatic systematic categories.[7]

These few lines encapsulate the theoretical foundations of Temminck's thinking, that is, the underlying ideas that guided his zoological research. The gist is that he found that the same basic forms or "types" inhabited similar latitudes. These types provided molds for living beings; they represented the morphological plan shared by animals belonging to the same group, located along the same latitude.

Temminck thus identified the type as the basic blueprint for a genus, and the species as varieties of that genus. The species was, therefore, the physical expression of the type in a particular region of the globe.[8] Species would then be defined by the principal characters of the genus they belonged to, that is, of the ideal type. For example, a bear is, in its general sense, a basic form of organization (a type, the blueprint for the genus *Ursus*), while the American Black Bear is the expression of the type in a particular region, North America (the species *Ursus americanus*, belonging to the genus *Ursus*).[9] The Polar Bear, or *Ursus maritimus*, is another species from the genus, a variation of the type 'bear.' Despite the characters that distinguish it from its American cousin, it is evident that both share the same basic design, or type. Simply put, a bear is a bear, regardless of its habitat or the color of its fur.

For Temminck, every type was definite, clearly separated from other types, and therefore relatively easy to define. Consequently, so were genera. Local deviations of a type produced a range of different species found together in a particular region. But unlike genera, species were difficult to define: there is no firm line that allows a clear separation of one species from the other.[10] Temminck warned against the perils of describing all variations of a type as separate species, which would obstruct the progress of science and complicate matters to no end. To identify each small variation, naturalists would have to use characters that were insignificant and only recognizable "with the scalpel in hand or with a magnifying glass before the eye," when in fact one could, "at

7 Temminck, "Kennis en verbreiding der zoogdieren van Japan," 287 (italics original).
8 Temminck, "Aperçu général," in *Fauna japonica*, vol. 5, 7–8.
9 Temminck, "Kennis en verbreiding der zoogdieren van Japan," 283-84.
10 Temminck, "Aperçu général," in *Fauna japonica*, vol. 5, 8.

FIGURE 26 *Innuus speciosus* (Japanese Macaque, now *Macaca fuscata*). Hand-colored engraving after a drawing by Josef Wolf or Hermann Schlegel. In Philipp Franz von Siebold, ed., *Fauna japonica*, vol. 5 (1842): plate 1
IMAGE FROM THE BIODIVERSITY HERITAGE LIBRARY. DIGITIZED BY HARVARD UNIVERSITY, MUSEUM OF COMPARATIVE ZOOLOGY, ERNST MAYR LIBRARY

first glance, recognize all the main characters that belong to the type of the true species."[11]

The difference in Temminck's thinking between the type and the species reflects his belief in the principle of continuity. What he observed in the collections was a continuum of forms – the oft-quoted principle *natura non facit saltus* (nature does not make jumps). When he described a new genus of mole from Japan, Temminck was confident that it "filled a gap in the animal series," proving that as naturalists advanced in their research, they would eventually describe all the "links of this uninterrupted chain, which closely links together the different groups of animals," both ancient and living.[12] A clean separation between species was nowhere to be seen:

> The observation of a multitude of individuals from the same species, native to different regions, clearly indicates how difficult it is sometimes to determine precisely the limits between analogous species, and to affirm that there is a specific separation or a well-traced demarcation for all the species we recognize as such. In sum, the existence of *Genera*, in its conventional meaning, seems a paradox to me: I am even far from accepting the existence of *species* as an axiom.[13]

Apparently Temminck was rather unsure what a species actually was in the greater scheme of life. Did it represent a true, existing natural category? Or was it the result of man-made divisions, an artificial category? In contrast to his faith in the existence of types, Temminck remained somewhat skeptical about species.

The species concept most widely accepted in Temminck's time was Buffon's, according to which species were reproductively isolated groups that produce fertile descendants.[14] Animals of the same species breeding always resulted in more animals of the same species, and those of different species produced no, or at least no fertile offspring that could continue the species. (For plants, the botanist John Ray had already defined species in these terms in 1686.)[15] Cuvier

---

11 Temminck, "Kennis en verbreiding der zoogdieren van Japan," 292–93.

12 Temminck, "Description d'un nouveau genre de mammifères (Urotrichus)," 1.

13 Temminck, "Introduction à cette troisième partie," in *Manuel d'ornithologie*, 2nd ed., vol. 3, xix (italics original).

14 The concept of species in biology has been extensively studied and discussed. See, for example, Atran et al., *Histoire du concept d'espèce*; Haffer, "The History of Species Concepts"; Stevens, *Development of Biological Systematics*; Wilkins, *Species: A History of the Idea*; Zachos, *Species Concepts in Biology*.

15 Ray, "Præfatio," *Historia plantarum*.

described the species as a "group of organisms descending from each other or common relatives, all resembling each other."[16] After introducing this definition, however, he quickly admitted to the impracticability of dealing with this species concept in day-to-day work. Jan van der Hoeven defined species in the same terms and added that each displayed a "character specificus," which was immutable and particular to each species. He called the characters by which individuals of the same species differed from each other "varietates."[17]

Interbreedability was, indeed, not a very useful definition for zoological classification. The belief in an interrupted chain of beings made the definition of species even more difficult. Natural historians took on a more pragmatic approach. They defined species by choosing a set of distinctive characteristics, mainly morphological, interspersed with information on behavior and geographical distribution. Charles Darwin noted: "Geographical distribution has often been used, though perhaps not quite logically, in classification, more especially in very large groups of allied forms. Temminck insists on the utility or even necessity of this practice in certain groups of birds; and it has been followed by several entomologists and botanists."[18] The resulting approach to species was similar to that to genera: species resembled an ideal morphological blueprint. This was a practical definition, not a philosophical one, but it allowed for the paradox that species were, in fact, morphologically recognizable but terribly difficult – if at all possible – to separate from other closely related species.[19]

After loosely defining a species as a local variation of genera, or types, Temminck proceeded to identify correlations between types, species, and habitat. The particular morphology of a species was under the influence of the climate of its habitat. For example, Temminck described two new species of bats from Japan which were both covered in thick fur, while similar species of the same genus from tropical forests were nearly hairless.[20] Similarly, a Snow Leopard living in the Altai Mountains had a denser, thicker coat than a Sumatran Leopard, which was covered with short and shiny hairs. The leopard living near the pole was, according to Temminck, "provided by nature" with very thick and long fur.[21] Nature was thus directly responsible for shaping the type into different species according to the climatic conditions of the places

16 Cuvier, *Règne animal*, vol. 1, 19.
17 Van der Hoeven, "Inleiding," in *Handboek der dierkunde*, 28–29.
18 Darwin, *On the Origin of Species*, 419.
19 Farber, "Buffon and the Concept of Species."
20 Temminck, "Kennis en verbreiding der zoogdieren van Japan," 282.
21 Temminck, 283.

in which the animals lived. What kind of natural forces he referred to, he did not say.

Local conditions could also influence species' form, which resulted in subtle alterations from the basic model. Species of the same type shared a basic morphology. Temminck observed that even when species undergo certain alterations and appear as slightly different varieties, "this does not mean that these varieties are divergent enough to isolate them from the types and to classify them as truly different species."[22] These altered forms, or varieties, differed only little from the original species, for example in size or coloration. Temminck clarified neither to what extent species might differ from the type, nor how varieties might differ from the species.

Temminck's writings about the type and the species concepts revealed his philosophical foundation of classification. Even though different species existed as a consequence of their environment, this was not a transformist argument, i.e. an argument that species adapt to the environment and change into new, different species. The scope of variation was limited, and there is no word about the transformation of species in a progressive, directed sense in any of Temminck's publications. The same applied to varieties within species. The different features of species and varieties were "provided by nature" from the beginning of time. Furry leopards belonged to cold areas, and Temminck never suggested that if a bald Sumatran Leopard decided to migrate to Mongolia, its descendants would grow a thick fur in their new, cold environment.

Temminck's typological thinking, with types as molds or models, was common during the eighteenth and early nineteenth centuries.[23] He might have derived it from reading Buffon, who had advanced this concept towards the end of the eighteenth century. Buffon had attempted, and succeeded for the most part, to present a global theory of the earth and the natural history of minerals, humans, quadrupeds, and birds in his ambitious *Histoire naturelle, générale et particulière* (1749–1804). His ideas evolved over time, sometimes even appearing to be contradictory across different volumes of the *Histoire naturelle*, but his observations had a notable influence on later naturalists' view of the natural world.[24]

Buffon described a "prototype" or archetype upon which each individual is modeled, but "which seems, in its actual expression, to be altered or improved

22 Temminck, "Aperçu général," in *Fauna japonica*, vol. 5, 8.

23 For an in-depth discussion of the morphological type concept and its evolution in the nineteenth century, see Witteveen, "Suppressing Synonymy with a Homonym."

24 For more information on Buffon's writings and how they evolved over time see, for example, Farber, "Buffon and the Concept of Species"; Roger, *Buffon: A Life in Natural History*.

by circumstances; therefore, with regard to certain characteristics, there appears to be an astonishing variation in the succession of individuals, and, at the same time, an admirable uniformity in the entire species."[25] All horses, for example, are always shaped after the "first horse," which is "the exterior model and the interior mold."[26] This interior mold is more than an ideal plan, as it acts both as an arranging force and as a model for all animal forms. The characters of the "prototype" were perpetuated generation after generation, and they changed as the species multiplied.

The species was, for Buffon, a historical continuity that changed over time.[27] Temminck's type, in contrast, was fixed and eternal. He decidedly rejected Buffon's view that all species belonging to a specific group (for example, an order) were descended from a single original source. Thereby he also rejected Buffon's idea of the interior mold as an arranging force:

> What a strange error it is to indicate the Bizet or the Wild Pigeon as the common ancestor of all different species of Pigeons on the surface of the earth; what are the reasons that have led M. Buffon to establish a law that nature denies in every individual? I will make these reasons known, and then it will be easy to prove that M. Buffon's supposition lacks any credibility. The lack of observations made of nature is the principal source of an error that results from too great a confidence in his creative genius; this confidence, fatal for science, perpetuates the aberrations of a great man, whose merits are justly respected and whose talents are honored.[28]

The differences between some groups of pigeons, Temminck continued, were so clear and definite that they should never have "escaped such a great genius."[29] The whole group of pigeons could not possibly have originated from just one original species. To him, Buffon's assumption that climate alters form so as to create new species was unfounded and entirely wrong. While nuances in color could be due to the environment, the climate did not have "the magical power" to affect coloration and size in any significant way.[30] Besides, not only the external morphology of these species of pigeons was distinct, their behavior and anatomy were different as well. Temminck did not

25 Buffon, "Quadrupèdes," in *Histoire naturelle*, 61–62.
26 Buffon, 62.
27 Farber, "Buffon and the Concept of Species," 263–66; Roger, *Buffon: A Life in Natural History*, 399–400.
28 Temminck, "Discours sur l'ordre des pigeons," in *Pigeons et gallinacés*, vol. 1, 16.
29 Temminck, 17.
30 Temminck, 23.

accept that animals were subject to any significant change under the influence of exterior circumstances and dismissed the idea of an interior mold. While Buffon thought of the prototype as the source for the whole group of pigeons – or any other living being – Temminck found that many types were necessary to explain the entire order of pigeons, the diversity of which was so great that it could not possibly have come into being only by the influence of the climate on one primordial form.

Temminck's type fits the category of Paul Farber's "morphological type-concept," which he defined as a "basic plan or type that could be discerned at various taxonomic levels."[31] The type concept that dominated pre-Darwinian natural history was that of George Cuvier. He believed in a natural, true, and permanent plan of the structure behind every form. There was a type for every level of organization, from his four *embranchements* (the four 'branches' into which Cuvier divided the animals: Vertebrata, Articulata, Mollusca, and Radiata) to the species. Deviation from the types had to be limited because, according to the principles of comparative anatomy that he had established, any significant change in the animal body left it unable to function correctly.[32] Therefore, animal morphology could not be as variable as Buffon and Lamarck proposed. Cuvier stated that there were essential characteristics of animals that resisted all influences, and that time had no more influence on animal form than climate or domestication had.[33]

Temminck shared with Cuvier not only the notion of type as the basic form of organization, but also the concept of variations of a species, as limited to size and color. Cuvier wrote:

> Thus the most superficial characteristics are the most variable; color depends much on sunlight; the thickness of the hair on heat; size on the abundance of nourishment [...]. Thus, although the wolf and the fox live from the torrid zone to the glacial zone, they hardly give evidence, in this

31 Farber, "Type-Concept in Zoology," 100–101. On the development of the morphological type in the eighteenth and nineteenth centuries and essentialism, see Hull, "Essentialism on Taxonomy," parts I and II; O'Hara, "Representations of the Natural System in the Nineteenth Century"; Winsor, "Creation of the Essentialism Story"; Witteveen, "Naming and Contingency."

32 For Cuvier's theories see, for example, Appel, *The Cuvier-Geoffroy Debate*; Coleman, *Georges Cuvier, Zoologist*; Outram, *Georges Cuvier: Vocation, Science, and Authority*; Rudwick, *Georges Cuvier*.

33 Cuvier, *Discours sur les révolutions*, 62.

> immense space, of another variety than a little more or less beauty in their fur.[34]

The question of the relationship between variation and climate had also caught the attention of a German zoologist named Constantin W. L. Gloger. In 1833 Gloger observed in his *Das Abändern der Vögel durch Einfluss des Klima's* that birds tended to be darker near the equator, where the environment is more humid.[35] In fact, this link between climate and coloration applied to all warm-blooded animals, and it is now known as 'Gloger's rule': humid climates are inhabited by darker varieties of a given species. Temminck did not mention this rule in any of his works on biogeography or when discussing variation. He may have been unfamiliar with it, as Gloger's publication of 1833 is not, according to the auction catalog for his library, among the books Temminck owned, even though other works by the same author are listed.[36]

Cuvier's colleague and intellectual rival Étienne Geoffroy Saint-Hilaire also based his theories on the idea of a morphological type or, more explicitly, on the unity of type. In contrast to Cuvier, Étienne Geoffroy Saint-Hilaire stressed form above function. For him, function was a consequence of form, and therefore an animal form could change without critically altering the animal's physiology. Ergo, species were not necessarily fixed in their morphology. Moreover, he believed that more complex forms had progressed from simple ones. He maintained that vestigial organs, homologies (that is, similar structures in different beings, such as a bat's wing and the human hand), fossils, and embryological stages indicated a common origin of the existing forms, which descended from one archetype. Cuvier vigorously opposed Geoffroy Saint-Hilaire's views.[37] Their argument ended in one of the most notorious debates in the history of biology, and one of the most heated. Although both scientists' views were widely known at the time, there is no reference to Geoffroy Saint-Hilaire's unity of plan in Temminck's writings.

Temminck could certainly never have agreed with one of its more polemical implications, transformism. The parallels between Cuvier's theories and Temminck's ideas were not coincidental. Temminck admired Cuvier and considered him one of the greatest minds in natural history. He frequently

34 Cuvier, 59–60.

35 Gloger, *Das Abändern der Vögel durch Einfluss des Klima's*, 11–24. Although this idea is now known as 'Gloger's rule,' Stresemann (*Ornithology*, 70) noted that Peter Simon Pallas had already proposed it two years earlier in his *Zoographia Rosso-Asiatica* (1831).

36 See the 1858 auction catalogue for books from Temminck's library, *Catalogue de la bibliothèque délaissé par C. J. Temminck*.

37 Appel, *The Cuvier-Geoffroy Debate*.

visited Cuvier in Paris and referred to him as "mon maître, et depuis lors mon ami" (my teacher, and since then my friend).[38] He manifested his esteem by dedicating the *Nouveau recueil de planches coloriées d'oiseaux* to Cuvier as "an expression of our most respectful devotion."[39] Cuvier's writings strongly influenced Temminck, who applied Cuvier's typological thinking to interpreting the patterns of animal distribution that he observed in species from Southeast Asia and Japan. It is even possible that Temminck developed his law to fill the gap left by Cuvier in his writings – a discussion of biogeography. He applied Cuvier's principles of comparative anatomy and his type concept to the subject of animal distribution.

2 Temminck's Law versus Buffon's Law

Like many of his contemporaries, Buffon was intrigued by the problem of geographical distribution patterns. He addressed the subject in his *Histoire naturelle*, starting with the proposition that animal organization resulted from external factors, that is, from the environment in which they originated. Each fauna belonged to a particular area of the world, and he expected that regions with similar environmental conditions would contain similar species.[40] But as time passed, the newly described species from tropical countries stubbornly contradicted Buffon's expectations. He was surprised to see that naturalists had, in fact, found different animals in the Old and New World, even under the same climatic conditions. Buffon explained these unexpected findings by introducing geographical barriers. When mammals encountered insurmountable physical barriers, species became isolated and unable to spread from one continent to the other. And with this theory in hand, Buffon changed his mind:

> All of the animals that, by their nature, are unable to support the climate of the north, even those that can support it but are unable to reproduce in it, are therefore confined on two or three sides by oceans that they cannot cross, and on another side by regions so cold as to be uninhabitable. Accordingly, there is no longer any reason to be astonished by the general fact, which first appeared very strange, and which no one before us even

---

38 Temminck, "Avant-propos," in *Coup-d'oeil général*, vol. 1, ix.

39 Temminck and Meiffren Laugier de Chartrouse, "A monsieur le Baron G. Cuvier," in *Nouveau recueil de planches coloriées*, vol. 1.

40 Buffon, "Histoire naturelle, générale et particulière, servant de suite à l'histoire des animaux quadrupèdes," in *Histoire naturelle*, vol. 3, supplement, 270. See also Mayr, *Growth of Biological Thought*, 441.

> suspected, that no species of the torrid zone of one continent is found in the other.[41]

He thus explained why different regions comprised different species of animals. For the realm of plants, the naturalists Alexander von Humboldt and Augustin Pyramus de Candolle arrived at the same conclusion by comparing the floras of different regions.[42] Humboldt combined his observations on vegetation with Buffon's passage and extended it to a principle that applied to the whole globe and all animal groups, including birds, insects, and reptiles.[43] The principle that different forms inhabited different regions is often referred to as 'Buffon's law.' It was, like Temminck's law, a bold generalization, mostly based on the comparison of only the mammals known to the respective naturalists in the Old and the New World. Nonetheless, by the 1820s, Buffon's law was widely acknowledged by the community of naturalists. The Scottish geologist Charles Lyell was charmed by it: "the French naturalist caught sight at once of a general law in the geographical distribution of beings, namely, the limitation of groups of distinct species to regions separated from the rest of the globe by certain barriers."[44]

Remarkably, Temminck's law is the exact opposite of Buffon's law. Even more surprisingly perhaps is that Temminck never even mentioned it. Was Temminck ignorant of Buffon's observations? Presumably not. Temminck was a well-read, well-informed naturalist. He owned no fewer than four different editions of Buffon's *Histoire naturelle*, which he often consulted.[45] Nearly every work of natural history that dealt with the patterns of geographical distribution included Buffon's observations. Temminck often corresponded and occasionally talked to Humboldt, Cuvier, Illiger, and several others. It is hard to imagine that he had never read Buffon's law or works that referred to it, never heard of or talked about it with his colleagues. Nevertheless, he did not allude to Buffon's observations in his discourse on animal distribution. This odd silence may have been the result of his reluctance to explain the contradiction between his and Buffon's law rather than negligence on Temminck's part.

How did Temminck and the other naturalists reach opposite conclusions from looking at the same faunas? They were using different yardsticks for the

41 Buffon, *Histoire naturelle*, vol. 9, 96; translated in Nelson, "From Candolle to Croizat," 275.

42 On the history and implications of Buffon's law see Nelson. A general history of biogeography can be found in Browne, *The Secular Ark*.

43 Humboldt, "Sur les lois que l'on observe dans la distribution des formes végétales," 234.

44 Lyell, *Principles of Geology*, vol. 2, 66; Nelson, "From Candolle to Croizat," 274.

45 See *Catalogue de la bibliothèque délaissé par C. J. Temminck*, 2–3; Temminck, *Catalogue systématique*.

same measures. Buffon and other naturalists highlighted the differences, not the parallels between Europe and America. They emphasized variation, while Temminck emphasized similarity. Where others focused on individual species, Temminck concentrated on the general organization plan of related species, i.e. on genera. In fact, all naturalists dealing with distribution patterns during the first half of the nineteenth century, like de Candolle and Humboldt, agreed with Temminck that species from different regions, in their most basic form, were analogous. De Candolle wrote:

> The same distance from the equator, or the same degree of latitude, produces rather a resemblance in the forms, – an agreement in the families and genera, – than the same species, chiefly because, besides this geographical latitude, the height above the surface of the sea, the temperature during the growing season, the soil and constitution of the mountains, even the degree of longitude, and several other circumstances, have an influence on vegetation.[46]

Likewise, the British physician and ethnologist James C. Prichard recognized that in regions with the same "physical conditions" on different continents, "[t]he vegetable tribes will present, in each [region] respectively, analogies of form and general character; but few, if any, of the same species will be found in localities thus separated."[47]

The similarities in the composition of families and genera at the same latitude was not in dispute. But apparently Temminck was alone in placing much more importance on the recurrence of these groups than on the diversity of species, which he related to the influence of the environment. It is not surprising, then, that these naturalists interpreted the variation of forms throughout the globe in two entirely different ways. They were looking at geographical distribution from different perspectives. While Temminck emphasized similar genera along the same latitude, Buffon, Humboldt, de Candolle, Prichard, Wallace, and others were surprised by the great diversity of species and varieties in regions with the same climate.

Temminck's law, unlike Buffon's, passed through the nineteenth century virtually unnoticed. Sir John Richardson, the Scottish naval surgeon and naturalist known for his two famous Arctic expeditions, devoted a few lines to it. He understood that the validity of Temminck's law depended directly on the

46 Candolle, *Elements of the Philosophy of Plants*, 266.

47 Prichard, *Researches into the Physical History of Man*, 50; Nelson, "From Candolle to Croizat," 271.

taxonomic level one was looking at: the genera or the species. In a report on North American zoology published in 1836, he noted:

> Temminck adduces circumstances which favour a modern opinion almost directly opposed to Buffon's; namely, that all the genera which people the earth [...] are to be found in the equatorial zone, or at least within the tropics; and that the genera are spread abroad by means of analogues or species possessing exactly similar generic characters, which range in the same parallels of latitude, through all the degrees of longitude, and that notwithstanding the barrier which a wide ocean may be supposed to interpose. The comprehensiveness of this law will evidently be modified by the number of generic divisions admitted by naturalists, and it will be scarcely tenable if the geographical groups of species be raised to generic rank as has been of late frequently done.[48]

Indeed, whether certain populations were regarded as a genus, a subgenus, or a species was a matter of constant dispute. As new – mostly exotic – species were described, the scope and limits of genera were under constant revision. Temminck's law depended precariously on ever-changing genera, a category that Temminck regarded as permanent.

For his law, Temminck chose to focus on the northern half of the globe (which was much less diverse than the tropics) even though he was surrounded daily by thousands upon thousands of tropical specimens at 's Rijks Museum. He confessed to finding the species of the Sunda Islands too diverse and still too little known to point to general patterns. These islands were, "due to their position and their fertile soils covered by vigorous vegetation, the most favorable place to study the animals of the tropical region of the Old World [...]; while there [i.e. in the temperate regions of the Ancient Continent], better than anywhere else, one can obtain firm evidence about the geographical distribution of the species."[49] By applying his law to both halves of the globe, Temminck extended the patterns he observed in the Northern to the Southern Hemisphere, even if with a certain reservation.

The presence of marsupials both in Sulawesi and in South America was a confirmation for Temminck that his law was valid beyond the Northern Hemisphere: "The *Phalangers* and the *Couscous*, marsupial animals [...], are for the Ancient Continent what the *Opossums* and the *Didelphes* are for the

48 Richardson, "Report on North American Zoology," 223–224.

49 Temminck, "Faune des Iles de la Sonde et de l'empire du Japon," in *Fauna japonica*, vol. 1, iv–v.

parallel regions in the New World."[50] By focusing on the unique morphology of marsupials instead of on individual species, Temminck inadvertently alluded to their common ancestry, while Buffon's law highlighted their evolution and diversification.

3 On the Origin and Immutability of Species

The distribution patterns of animals and plants had puzzled naturalists for centuries. Explanations ranged from divine design to more materialistic causes based on universal laws, in which God's role was reduced to creation. If God's intervention was behind all patterns in nature, one's theories had to consider both divine design and environmental factors (such as barriers to dispersion). However, if nature was a dynamic entity, always in flux, one also had to contemplate historical events. As Augustin de Candolle put it: "All of the theory of geographical botany rests on the particular idea one holds about the origin of living things and the permanence of species."[51]

Temminck's views on the difficult issues of the origin and the fixity of the species were shaped by his type and species concepts, and his enthusiasm for Georges Cuvier's theories. Temminck tried to explain, on the one hand, the source of the types, and on the other, how they came to be distributed across such distant regions as northern Europe, America, and Asia. The first issue seemed straightforward to him. Temminck simply stated that a "pouvoir créateur" (creative power) was directly accountable for the origin of the types.[52] This power was undoubtedly of divine nature: God used ideal types as basic molds for creating species. After all, for Temminck, nature was an expression of divine greatness and manufactured according to God's will, and natural history was "the science that leads man to read from the divine pages of the book of nature; that allows him to know the animated and inanimate parts of creation; that unnoticeably elevates his soul to worship the Almighty."[53]

The second issue, an explanation for the types' actual distribution was, however, a very different matter. According to Temminck, it could not be due to migration because animals disperse until they encounter a physical barrier. Mountain ranges or the Pacific and Atlantic oceans would, indeed, be

50 Temminck, *Coup-d'oeil général*, vol. 3, 113, 16.

51 Candolle, "Géographie Botanique," 417; translated in Nelson, "From Candolle to Croizat," 285.

52 Temminck, "Aperçu général," in *Fauna japonica*, vol. 5, 8; Temminck, "Kennis en verbreiding der zoogdieren van Japan," 291.

53 Temminck, "Discours préliminaire," in *Pigeons et gallinacés*, vol. 1, 6.

unsurmountable obstacles to traveling mammals. In this point he agreed with Buffon. There had to be another explanation for the remarkable uniformity of types across continents: "It is very probable that the elements that have exerted their influence on creation have produced the same types in the most easterly regions of Asia and North America, as under the climate of Europe."[54] Unfortunately, Temminck did not specify what precisely these elements were, nor did he explain the mechanism by which they could reproduce a type in a particular region on earth. However, he did hint at an alternative, less literal interpretation of the Genesis.

That God was the source of nature itself was not in dispute in Temminck's time, but the biblical accounts of Genesis and the Great Flood were problematic at best. This was a matter of simple physics. The sheer number of species known towards the 1850s did not agree with the space allocated for the animals on the Ark. Besides, animals and plants could not simply wander around the globe and across oceans. Species with disjunct distributions, i.e. living in widely separated regions but found nowhere in between, were particularly confusing. One example is the humble violet. Most species of violets live in the Northern Hemisphere, and some in such distant places as the Andes, Hawaii, or Australia. Other species thrive on the summits of both European and South American mountains. How had these species come to be in their habitats? And why are no individuals of these species in the areas in between? As naturalists questioned the biblical accuracy, they started to embrace an elegant alternative explanation: there had been not one, but several acts of divine creation in different parts of the world.

This theory of multiple creations was first proposed by the German naturalist Johann Friedrich Gmelin and later adopted and spread by the botanist Karl Ludwig Willdenow.[55] It explained the apparently confusing distribution of species while allowing for God's absolute government: every region had been populated with its own set of species through a new, unique act of creation. Once in place, species dispersed until they encountered a physical barrier. One could then expect to find the outlines of these centers of creation by studying the species' geographical distribution.

Temminck pinpointed several regions with "special creations," two of them on the Indo-Australian Archipelago. A mountain chain measuring nearly 3,000 meters at its highest point divides the island of Timor into two regions, each with a separate set of species. Temminck recognized a fauna of an Asiatic-type on north of the mountain range which was comparable to that of the Greater

54 Temminck, "Aperçu général," in *Fauna japonica*, vol. 5, 8.

55 Larson, "Not without a Plan," 458–59; Kinch, "Geographical Distribution," 96–98.

Sunda Islands and the mainland, and a different fauna on the southern part of the island. The species there "belong to a Fauna that seems different, or at least transitory between the species characteristic of the Moluccas and the Australian groups, which appear to belong to a very particular creation."[56]

Similarly, he noted a line running through Timor, Sulawesi, and the Moluccas, which separates the Greater Sunda Islands' fauna from the Australian fauna. In Sulawesi, there were no elephants, rhinos, and tapirs "anymore," while marsupials and incubator birds could be found on Sulawesi for "the first time" in the archipelago: "The *Phalangers* and the *Couscous*, marsupial animals that are not found anywhere else in the Sunda Islands, are here [in Sulawesi] the first representatives of this group."[57] Notably, "first time" and "anymore" indicated the geographical direction Temminck was following in describing these island faunas, from west to east, and not a timeline. Temminck saw clear shifts from one fauna to another when moving through different latitudes, which agreed with his law and with the idea of multiple creations.

As Cuvier had demonstrated, fossils were also of separate creations, and in their case, of now-extinct animals. In the *Monographies de mammalogie* Temminck listed the number of known species according to different authors, including forty-two fossil species, that is, remnants of animals that were part of a creation prior to the present one. Concerning extinction, Temminck accepted Cuvier's catastrophism, which he described as the "new and illuminating ideas about the catastrophes that have changed the surface of the globe."[58] Catastrophism is the theory that the earth and all life forms on it had been shaped by violent, sudden events. This theory explained the extinction of the animals found in the fossil record but not in present faunas. When speaking of the chain of beings, Temminck included both extinct and living species.[59]

Temminck did not address the question of the fixity, or unchanging nature, of species directly, but his writings reveal that he believed in it. Temminck was certain that there were limitations to the amount of variation a species could undergo, namely changes in coloration and size that were due to the

---

56 Temminck, *Coup-d'oeil général*, vol. 3, 168, Temminck, "Blik op de dierlijke bewoners," 74–75.

57 Temminck, *Coup-d'oeil général*, vol. 3, 111–13. This shift would later be described by Alfred Russel Wallace and become known as "Wallace's line." Wallace's line runs, differently from Temminck's, between Bali and Lombok instead of the Lesser Sunda Islands and Timor. See also Camerini, "Evolution, Biogeography, and Maps."

58 Temminck and Meiffren Laugier de Chartrouse, "A monsieur le Baron G. Cuvier," in *Nouveau recueil de planches coloriées*, vol. 1, [i]; Temminck, "Discours préliminaire," in *Monographies de mammalogie*, vol. 1, ix.

59 Temminck, "Description d'un nouveau genre de mammifères (Urotrichus)," 1.

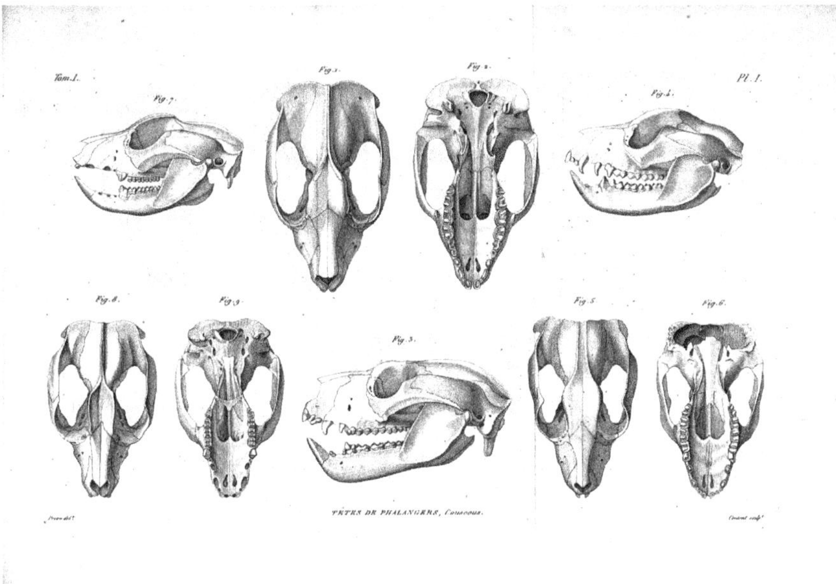

FIGURE 27 *Têtes de Phalangers*, Couscous. Engraving after a drawing by Jean-Gabriel Prêtre, showing the differences between the skulls of three different species of possums from the Moluccas and New Guinea. In Coenraad Jacob Temminck, *Monographies de mammalogie*, vol. 1 (1827): plate 1
IMAGE FROM THE BIODIVERSITY HERITAGE LIBRARY. DIGITIZED BY THE MUSÉE CANTONAL DE ZOOLOGIE, LAUSANNE

environment in which the animals lived. Varieties were not species in change, but slight deviations from the original form.[60] Temminck's definition of the type as an ideal morphological plan that was distributed by God along particular latitudes was inextricably linked with a static concept of the animal form like the one Cuvier had described.

The fauna of the Sunda Islands confirmed to Temminck that species were, indeed, immutable. Sumatra's fauna is different from that of the Asian continent and that of Java, its neighboring island, even though the short distance between them might make one expect otherwise.[61] This successfully contradicted the idea "adopted by some erudites" that the Sunda Islands were formerly attached to the Indian subcontinent.[62] These 'erudites' were none other than the French naturalist René-Primevère Lesson and the surgeon Prosper

60 Temminck, "Kennis en verbreiding der zoogdieren van Japan," 292.
61 Temminck, *Coup-d'oeil général*, vol. 1, 85.
62 Temminck, *Coup-d'oeil général*, vol. 2, 93. See also Temminck, "Faune des Iles de la Sonde et de l'empire du Japon," in *Fauna japonica*, vol. 1, x.

Garnot, who had traveled with Louis Isidore Duperrey on *La Coquille* during its memorable circumnavigation of the globe from 1822 to 1825. Based on their observations, they proposed that Sumatra had been physically linked to the mainland at some point.[63] However, Temminck reasoned that if the islands had ever been connected to each other or to Asia, then animal populations would have happily walked, crawled, swum, or flown between them, and following the separation of the islands, the same species would still inhabit Sumatra, Java, and the mainland. An argument of this kind can only be valid if one believes in the fixity of the species: otherwise, Sumatra and Java might have been connected and shared the same species, and after their detachment, these species could have changed into new ones, leaving both islands with different species. Therefore, if animals can change with time, a difference in species composition is a meaningless argument for rejecting a former link between the two islands. (The fact that continents move, clash, and break off again would not be generally accepted by the scientific community until the late 1950s.)

There was no place for the transmutation of species in Temminck's logic, but there was in Buffon's. He claimed that species changed from generation to generation due to the influence of the environment and Buffon's interior mold, and that their history was linked to that of the earth's surface.[64] In sharp contrast to Buffon, Temminck stressed uniformity: in space (the same types are found along the same latitude) and in time (species are immutable).

The immutability of the species was embedded in Temminck's definition of the type. Alternative definitions led to alternative ideas, as illustrated by Wallace's essay "On the Tendency of Varieties to Depart Indefinitely from the Original Type." It was the second of the now legendary papers by Darwin and Wallace in which they first proposed the theory of evolution by natural selection, and was read at the meeting of the Linnean Society of London on July 1, 1858. After presenting his arguments to prove that species evolved by what we now call natural selection, Wallace concluded:

> We believe we have now shown that there is a tendency in nature to the continued progression of certain classes of *varieties* further and further from the original type – a progression to which there appears no reason to assign any definite limits – and that the same principle which produces

---

63 Duperrey, *Voyage autour du monde*, 19; quoted in Temminck, "Faune des Iles de la Sonde et de l'empire du Japon," x–xi.

64 See also Farber, "Buffon and the Concept of Species," 284.

> this result in a state of nature will also explain why domestic varieties have a tendency to revert to the original type.[65]

For Wallace, variation from the original type was not only potentially unlimited, but it also allowed for the stronger or best-suited varieties to survive and replace the others. His type was not fixed, and certainly not immutable, while Temminck's was both.

In sum, there were two main and opposing currents during the first half of the nineteenth century. At one end of the spectrum, some naturalists believed that God designed nature carefully and distributed animals and plants intentionally on the globe in their current distribution. His intervention was the sole cause of the patterns they observed. Disjunct distributions proved that there had been multiple centers of creation. At the other end of the spectrum, some saw nature governed by general laws which had become effective after God's creation. The geography of plants and animals was a consequence of migration and extinction: dispersion, physical barriers, and changing geographies and species were the main mechanisms at play.[66] Temminck belonged to the first group, those who believed that God's plan was behind the structure of nature. Disjunct distributions were proof of God's influence in assigning certain types to specific regions. After 1860 some naturalists like Richard Owen and Louis Agassiz embraced the idea of a divine plan, as they saw evidence of a willful designer everywhere in nature, while Charles Darwin, Alfred Russel Wallace, and Charles Lyell found the answers to the problems posed by the geographical distribution of the species in evolution through natural selection.

4 After Temminck's Law

Temminck's general law was not as successful as he must have expected. There are hardly any references to it the scientific literature, with the exception – and perhaps only this one – of the note by John Richardson quoted above, which emphasizes the role of genera in Temminck's law. Richardson, as it turned out, was right. As systematics was in constant change, especially at the genus and species levels, his remarks turned out to be prophetic: with the continual changing and splitting of genera, Temminck's type concept began to destabilize. It was not even a well-defined concept. The botanist Joseph Hooker

65 Wallace, "On the Tendency of Varieties to Depart Indefinitely from the Original Type," 62 (emphasis original).

66 Kinch, "Geographical Distribution," 117–19.

cautioned his readers about the dangers of linking a subjective concept – in his case, the species – to the patterns in geographical distribution in 1853:

> It is natural to suppose that an observer who believes species to be arbitrary divisions of a genus, dependent on the naturalist's choice of characters, will adopt widely different conclusions as to their limits and origin, from one who regards them as distinct creation; and he who denies that a plant which grows spontaneously in England and New Zealand can have originated from one common parent, will reason differently on the subject of migration and dispersion from him who holds an opposite view.[67]

Temminck failed to address issues in an ordered manner that were vital for this law on biogeography: the genus, the species, and the origin of both. Temminck's law appears without further context in a text that was, in fact, an introduction to the description of new mammals. He defined the type-concept in a few separate articles and the introduction of the *Manuel d'ornithologie*. He did not dedicate any sections or chapters dedicated exclusively to a detailed theory of distribution. Temminck's rationale is scattered throughout his works. Since he did not construct a logical, comprehensive theory for his observations, it is not surprising that other naturalists ignored his law – if they were acquainted with it at all – as a possible explanation for the geographical distribution of animals and related problems. One has to distill Temminck's concepts from his writings by laboriously sifting through his works and interpreting them. Besides, typological thinking began to lose popularity among naturalists during the first half of the nineteenth century.

Temminck also left several other questions unanswered, such as the significance of variation, God's role in creation, and the meaning of disjunct distributions. He came across some peculiar patterns that he labeled "important" and "striking," and yet he did not attempt to explain them.[68] For instance, after studying the fauna of the Moluccas, he found that the differences in species among islands with the same climate and in close proximity to each other were most remarkable. He limited himself to listing which species lived where, and even when he did make comparisons between islands, he offered no explanation.[69]

Similarly, Temminck was amazed by the fact that the summit crater of Mount Sumbing, a volcano in central Sumatra, was populated by flora akin

67 Hooker, "Introductory Essay," in "Flora Novae-Zelandiae," vol. 2, i–ii.
68 Temminck, "Blik op de dierlijke bewoners," 67.
69 Temminck, *Coup-d'oeil général*, vol. 3, 235–36.

to that in the Alps. "The vegetation on its top carries all the characteristics of the alpine plants of Europe," including species of *Vaccinium* (a genus of dwarf shrubs that includes cranberries and blueberries) and *Rhododendron* (azaleas and related plants).[70] Why were the same types of plants inhabiting the mountains of central Europe, Sumatra, and South America? Did these mountains have similar climates and soil? Did God create these species in situ in the process of multiple creations? Again, Temminck offered no answers.

Explanations for Buffon's law came mostly from observations of climate, geography, soil, and other physical features. However, most naturalists agreed that there had to be something else, something involving the history of the species – migrations, dispersals, transmutation, and other events of the past.[71] His failure to quote and contest Buffon's law in an articulated way, his own typological thinking, and the lack of causal explanations – historical or otherwise – are the greatest shortcomings of Temminck's theories on animal distribution.

That being said, works like the *Faune des Iles de la Sonde et de l'empire du Japon* and the three-volume *Coup-d'oeil général* proved extremely useful to others, for the simple but significant reason that they included a complete and current overview of the faunas of the Sunda Islands and Japan. The account of the species living on some particular islands, but not on others, the report of species unique to Sulawesi and the Moluccas, and the transition from the Asian to the Australian fauna were all relevant observations at the time, even though Temminck was certainly not the first to address them. Half a century earlier Eberhard von Zimmermann, a professor of natural sciences at the Collegium Carolinum in Braunschweig, published what could be considered the first work on mammal zoogeography. In his *Specimen zoologiae geographicae quadrupedum domicilia et migrationes sistens* (1777) he compared the faunas of Asia, the Sunda Islands, and Australia – the very regions that had so intrigued Temminck.[72] Zimmermann also thought that the dispersal of animals from a single location was impossible, and favored the idea of multiple creations: species emerged (or had been created by God) in the regions that they currently inhabit. A map in the *Specimen zoologiae geographicae* depicted the groups of mammals that populate each region of the earth, and can be considered the first zoogeographic map to be published. In this map Zimmermann depicted the mammals of Asia as belonging to a different fauna than the one of New Guinea and Australia. The mammals of Timor and the Moluccas were

---

70 Temminck, *Coup-d'oeil général*, vol. 2, 82.

71 Nelson, "From Candolle to Croizat," 286.

72 Zimmermann, *Specimen zoologiae geographicae*. Temminck was at least acquainted with this work, as it is listed in the *Catalogue de la bibliothèque délaissé par C. J. Temminck*.

FIGURE 28 *Tabula mundi geographico zoologica sistens Quadrupedes hucusque notos sedibus suis adscriptos.* Hand-colored engraving after a drawing by August Wilhelm Knoch. In Eberhard August Wilhelm von Zimmermann, *Specimen zoologiae geographicae quadrupedum domicilia et migrationes sistens* (1777)
PRINCETON UNIVERSITY LIBRARY, HISTORIC MAPS COLLECTION, HMC01.6382

not assigned to either group; possibly he had little to no information about the mammals from those islands in 1777.

With the expansion of the European exploration of Asia, descriptions of Asian fauna multiplied. Naturalists like Charles Lyell and Thomas Horsfield were equally baffled by it.[73] Even the East India Company's hydrographer James Horsburgh took note of animal and plant distributions in his *Indian Directory*, a navigational aid for the seas from the 'East Indies,' China, and Australia to Africa and South America. In the 1840s, French explorers published several accounts of the Asian fauna based on information gathered during the voyages of the *Uranie* and the *Physicienne* (1817–1820), the *Astrolabe* (1826–1829), and the *Coquille* (1822–1825); onboard of the last traveled Lesson and Garnot, among others.[74] Yet Temminck had additional information to offer, facts that were new for many others.

Of particular interest were his newly described species and detailed comparisons with the Sunda Islands, as well as his details on the Japanese fauna, which was previously mostly unknown in Europe. These records were either entirely new or had previously been published in Dutch or German in journals with limited distribution.[75] Dutch zoological and botanical treatises, like Siebold's *Fauna* and *Flora japonica* or Temminck's *Coup d'oeil* and *Verhandelingen over de natuurlijke geschiedenis der Nederlandsche overzeesche bezittingen*, provided essential elements for the construction of new theories, including the idea of evolution by natural selection. William Swainson, Andrew Murray, Jan Baptist Jozef van Doren, Charles Darwin, and Alfred Russel Wallace had all consulted Temminck's works.[76] When looking for information about the Southeast Asian fauna, Darwin made use of Temminck's writings, among others. In 1839 he drew a table comparing the mammals from Sumatra, Java, Borneo, the Moluccas, and other islands aided by Temminck's *Coup d'oeil*, and later processed this

73 Lyell, *Principles of Geology*; Horsburgh, *India Directory*; Horsfield *Zoological Researches*.

74 Dumont d'Urville, *Voyage de la corvette l'Astrolabe*; Duperrey, *Voyage autour du monde*; Freycinet, *Voyage autour du monde*.

75 See, for instance, Temminck, "Kennis en verbreiding der zoogdieren van Japan" and "Blik op de dierlijke bewoners"; Müller, "Aanteekeningen," and "Ueber den Character der Thierwelt."

76 Murray, *Geographical Distribution of Mammals*, 116, 376; Swainson, *Treatise on the Geography*, 25, 357; Van Doren, *Mémoire sur les quadrumanes et les chéiroptères*. Wallace referred to Temminck's *Monographies de mammalogie* and the *Verhandelingen over de natuurlijke geschiedenis der Nederlandsche overzeesche bezittingen*, see Wallace, *Malay Archipelago*, "On the Habits of the Orang-Utan"; and "On the Orang-Utan." For Darwin's use of the *Coup-d'oeil*, see Camerini, "Evolution, Biogeography, and Maps."

information to answer the very questions Temminck had left unanswered in his writings.[77]

Perhaps the main reason for Temminck's unfinished biogeography is that, even though Temminck recognized its interest, he simply considered the field secondary to his central project. He was devoted to the description and classification of animals, following the path he had first taken at the turn of the century by cataloging his father's cabinet. After formulating his law, he remarked:

> It would be important to add to this rough outline a revision of the geographical distribution of animals, and study how the species are distributed on the globe along the longitude; but this subject, even though it seems so important, would take us too far away from the purpose of this essay, that is, the description of the mammals of Japan.[78]

Temminck's main preoccupation was, and continued to be, zoological classification. The concepts of type, genus, and species, which were the very foundation of Temminck's law, were also the essence of his systematic research, which was both praised and criticized. This, naturally, led to some drawn-out and, at times, heated disputes. Temminck's discussions with his peers reflect the main issues that concerned them all: discovering the natural system, standardizing nomenclature, and defining key concepts.

77 Camerini, 711–12.

78 Temminck, "Aperçu général," in *Fauna japonica*, vol. 5, 8–9.

CHAPTER 6

Systematics Wars

Considering the importance of Temminck's more philosophical views on nature and how best to comprehend it – which were the very foundation of his animal classifications – it is perplexing how little effort he put into explaining them. They only become clearer in two contexts: through his law on the geographical distribution of animals, and in his rebuttals of other naturalists' criticisms of his zoological classification methods. Temminck's discussions with his peers revolved mostly around the rules of zoological nomenclature, the establishment of genera, and the best method for arriving at a natural system of classification. These technical issues may not seem particularly interesting to the non-specialist, but they were – and still are – at the very core of how naturalists understood the natural world and their search for universal laws in the way Newton did. Therefore, these debates dealt with much more profound issues than just the naming of species and their grouping into categories.[1]

Particularly interesting are Temminck's disputes with the French ornithologist Louis Pierre Vieillot and the Irish zoologist, politician, and the founder and secretary of the Zoological Society of London, Nicholas Aylward Vigors. Their disagreements between 1800 and 1840 illustrate the three main questions that pushed the development of systematics further. First, the primary goal of classification was to find a natural system, one that reflected the real organization in nature. Second, there was a hindering lack of agreement on how to reach that goal. Although most naturalists shared the ambition to find a natural system, their approaches varied according to their scientific and social backgrounds, and beliefs. This resulted in a variety of systems of classification, some disarray in the nomenclature, and a number of arguments. Third, as more and more naturalists participated in these discussions, natural history changed from being the dominion of a few learned men into the field of expertise of many, both scholars and enthusiasts. In short, the pursuit of standardization (of rules, practices, nomenclature, forms of communication, and much else) was the main occupation – and preoccupation – of naturalists at the time.

---

1 This chapter is based in part on Gassó Miracle, "On Whose Authority? Temminck's Debates on Zoological Classification and Nomenclature: 1820–1850." *Journal of the History of Biology* 44, no. 3 (2011): 445–481. doi:10.1007/s10739-010-9265-1. Adapted by permission from Springer Nature.

 | DOI:10.1163/9789004441491_008

1 Temminck's Debates

Achieving standardization was only possible through debate and consensus. Temminck and Vieillot had been arguing ever since the latter had copied descriptions of some new species from Temminck's *Catalogue systématique* and assigned Latin binomials to them – something that the inexperienced Temminck had neglected to do. The arguments between Temminck and Vieillot concerned mainly two issues: the validity of the names coined by each of them, and the validity of their classification systems as a whole.

Regarding nomenclature, each accused the other of appropriating names from other works. In 1816 Vieillot wrote a book of seventy pages entitled *Analyse d'une nouvelle ornithologie élémentaire*, in which he proposed a whole new classification system and described numerous new species, of which quite a few had already been described before him by other naturalists.[2] In a flagrant act of plagiarism, he used these existing names and descriptions without giving credit to the original authors. Enraged, Temminck was quick to respond to what he considered an attack on his credibility and published the *Observations sur la classification méthodique des oiseaux, et remarques sur l'Analyse d'une nouvelle ornithologie élémentaire par L. P. Vieillot* in 1817. This was, in effect, a declaration of war.

Vieillot answered a year later in his entry on ornithology in the *Nouveau dictionnaire d'histoire naturelle*. Vieillot not only criticized Temminck's system, but he also accused him of plagiarism. Much to Temminck's surprise, Vieillot never mentioned the *Observations* of 1817.[3] Temminck wondered, with some sarcasm, if Vieillot only mentioned works published up to 1815 in order to avoid referring to Vieillot's own *Analyse* from 1816 and Temminck's *Observations*.[4] However, his skepticism may have been misplaced, since the second edition of the *Nouveau dictionnaire* had been commissioned several years earlier, and Vieillot may have already submitted his article to the editors before Temminck's *Observations* were published. Then again, Howard Saunders suggested that Vieillot may have modified his 1818 classification because of Temminck's criticisms on the *Analyse*.[5]

Vieillot's main criticism of Temminck's classification system was that it seemed to be nothing more than a compilation of previously extant ornithological systems. From his comparison of Temminck's groups with those of other

2 Vieillot, *Analyse*; Oehser, "Louis Jean Pierre Vieillot"; Stresemann, *Ornithology*, 122–23.
3 Vieillot, "Ornithologie," 96–98.
4 Temminck, "Introduction," in *Manuel d'ornithologie*, 2nd ed., vol. 1, xv.
5 Saunders, "Preface," in *Vieillot's Analyse*, iv.

naturalists, Vieillot concluded that Temminck's work was simply not original. Temminck's bitter answer to the *Dictionnaire* appeared in the introduction to the second edition of the *Manuel d'ornithologie* in 1820. Temminck defended himself and then attacked Vieillot's free use of his fellow-ornithologists' discoveries as well as his system of classification. Their animosity continued for the rest of their lives.

From 1825 onward Temminck engaged in yet another dispute, this time with Nicholas Vigors, who published an article questioning Temminck's classifications and in particular his concept of genus. Temminck and Vigors argued on a more technical than personal level. For some reason Temminck did not respond to Vigors' article directly, but in the second edition of the *Manuel*, where he also replied to Vieillot. Not only did Temminck's answer appear a decade after Vigors' article, but Temminck did not mention Vigors at all. Perhaps the 1825 article came to Temminck's attention rather late, or perhaps he deliberately chose to include his rebuttal to both Vigors and Vieillot in the *Manuel*. In any event, these exchanges lay bare these naturalists' different concepts of the natural world and of natural history.

2 Nomenclatural Chaos

In the period during which these disputes happened, both the natural history collections and the literature on the subject were growing rapidly. A bewildering amount of works dedicated to European and exotic birds appeared, many iconographic, others strictly descriptive, and some were accounts of adventurous *naturalistes voyageurs* (naturalists explorers).[6] One of the most immediate and obvious problems to appear in natural history was the multiplication of scientific names. As more naturalists dedicated themselves to studying local and tropical faunas, it became difficult to keep track of discoveries and access all publications. More often than not, the same bird was described independently by different naturalists who were unaware of each other's work, and consequently, the same species received two or more names. Additional confusion was caused by the practice of substituting older names with new ones that seemed more appropriate. On top of this, classification systems were continually changing. Knowledge increased, and groupings were under constant revision. Genera were split into two or more new ones, species moved from genus to genus, and the suprageneric levels were just as unstable. To the dismay of most naturalists, specific names changed with every review.

6 Farber, *Emergence of Ornithology*, chap. 5.

Temminck decided to put some order into this confusion. He began by listing all scientific names of birds known to him, then proceeded with validating each one, untangling the synonymy. (In systematics, names fall into a hierarchy, and only one may apply to a species. The discarded names then become synonyms).[7] He decided which name was to be preferred in each case on the basis of either priority or suitability: "I have spared neither effort nor means to bring the *General Index* [of the *Manuel*] to the level of the current knowledge in ornithology, that is, I have purged it as much as possible of the double, triple and often quadruple names which clutter the works of Gmelin, Latham and more recent publications."[8]

Temminck's "general index" was a clear reminder of the absence of a canon for good nomenclatural practice. In the early nineteenth century most binomials in zoology followed the precepts set by Linnaeus and amended by Johann Christian Fabricius in 1778.[9] Later, the Swiss botanist Augustin Pyramus de Candolle published his version of the rules for botanical nomenclature in 1813.[10] Also, every natural historian was familiar with these rules of nomenclature, although there had been no explicit agreement on universal rules and exceptions. Naturalists held different opinions and had different interpretations of the rules of nomenclature, which generated as much confusion as irritation. To make matters worse, almost all of them applied these rules inconsistently; some, including Temminck, adjusted their theories along the way, found justifiable exceptions, and broke the rules as they went along.

The rule of priority proved to be the most straightforward and, paradoxically, one of the most debated rules. It was generally agreed that the first scientific name ever given to a species or genus should be preserved for the sake of clarity and stability. Linnaeus wrote: "A generic name, that is worthy to be retained, ought not to be changed for any other, though more apt and proper. […] Such innovations ought by no means to take place, because new names [that are] more fit might be every day invented without end."[11] The German botanist and pharmacist Carl Ludwig Willdenow proposed the same law of priority in 1792 in his *Grundriss der Kräuterkunde zu Vorlesungen*. Any name given to the same species later was considered a junior synonym and therefore invalid. The rule of priority was widely acknowledged and frequently broken. Many, including Temminck, allowed for exceptions. Naturalists often proposed

7 Temminck, "Avant-propos," in *Manuel d'ornithologie*, ix.

8 Temminck, "Introduction," in *Manuel d'ornithologie*, 2nd ed., vol. 1, xviii–xix.

9 Linnaeus, *Systema naturae* 1st and subsequent editions; *Philosophia botanica*; *Species plantarum*; Fabricius, *Philosophia entomologica*.

10 Candolle, *Théorie élémentaire de la botanique*. See also Melville, *Towards Stability*; La Vergata, "Au nom de l'espèce."

11 Linnaeus, *Philosophia botanica*, 196.

new names which they considered to be more suitable: a better description of a species' specific character, behavior, or habitat. This was not a mere technicality: by changing a name, a naturalist invalidated the work of the author who had coined the first name and undermined his authority.

Therefore Temminck met Vieillot's plagiarism with anger and cynicism: "it is hardly imaginable that our savant believed in good faith that his plunder would go unnoticed, and that his new classification and nomenclature would be quoted as the fruits of his own discoveries, and of the conceptions of his genius."[12] In Temminck's eyes this was an unforgivable crime, "an insolent pillage, consisting of appropriating the research and the nomenclature published by other modern authors, who are nowhere mentioned, altering their ideas by changing the meaning of the words or their orthography."[13] The British zoologist William Swainson echoed Temminck's complaints: "M. Temminck has publicly protested against these plagiarisms, and others have spoken of them in terms of severe censure."[14] In fact, Vieillot was universally reprimanded for appropriating other ornithologists' descriptions.

Vieillot in turn pointed out that Temminck had happily replaced existing names for new ones and hence disregarding the work of others. He also thought, as Swainson did, that Temminck was relying too heavily on the main groups established by his predecessors, Latham and Illiger.[15] Vieillot regarded the *Manuel* as unoriginal: "This is the compilation that the author claims to be the fruit of his efforts and the model to be followed by modern ornithologists. Does this not remind us of the fable of the Jay?"[16] Temminck was infuriated: "Rest reassured, overly scrupulous Vieillot! I, in turn, will not evoke other allegories of the good La Fontaine."[17] Temminck's and Vieillot's disputes around names were a bitter contest for credibility. Temminck wrote:

> Some of my readers have undoubtedly already done justice to the conduct of M. Vieillot, and perhaps they will blame me for responding to puerilities; but being attacked, as I am, by a censor who seeks fame, more by his literary pretensions than by scientific works, I will in turn display my competence by employing other weapons than his.[18]

---

12 Temminck, *Observations sur la classification*, 55–56.

13 Temminck, 7.

14 Swainson, *Treatise on the Geography*, 184.

15 Latham, *Index ornithologicus*; Illiger, *Prodromus*.

16 Vieillot, "Ornithologie," 97–98. Here Vieillot referred to the La Fontaine's fable "The Jay Dressed in Peacock's Plumes" (Le geai paré des plumes du paon) in *Fables*, vol. 4, fable 9.

17 Temminck, "Introduction," in *Manuel d'ornithologie*, 2nd ed., vol. 1, xv.

18 Temminck, xi–xii.

That said, Temminck presented his counterarguments to Vieillot's approach to zoological nomenclature in the introduction to the *Manuel*.

Vieillot sometimes replaced existing names with new ones by translating Latin names into Greek, suppressing a few letters or altering the orthography, and then presenting the new names as his own. For Temminck it was not only a grave impropriety not to quote the author of the original name, but it was also against the fundamental principles of nomenclature and the practice of science itself. Since Linnaeus, scientific names were by universal agreement either Latin or Greek, and following the grammatical rules of their source language. However, if a binomial was a hybrid of words from both languages, or either grammatically or orthographically incorrect, it was common practice to replace it. This was particularly inconvenient to Temminck, who had no academic background and knew neither of these classical languages.[19] For him, the alternative nomenclature slowed down the progress of zoology and destroyed the work done by previous naturalists. Even though an early species name may not be in Greek or Latin, Temminck said, it was always better to preserve it, provided it had been widely recognized and accepted:

> In my view, it is preferable to keep the ancient name of a species that has been accepted around the world (even though the name might be barbarous in its composition, not having Greek or Latin roots), rather than substitute it with another name that, although it might be better chosen or grammatically more correct, it will likely cause misunderstandings, because nothing is worse for the development of the natural sciences, and in particular for the natural history of birds, than all these different opinions about the names of genera and species.[20]

Naturally, he here referred to opinions other than his own. When Temminck decided to change a scientific name, he said, it was to correct misidentifications and mistakes, such as when variations due to sexual dimorphism or molting stages had been described as new species.[21]

There were, however, exceptions for which Temminck allowed the replacement of a scientific name with a new one. One of these concerned geography. Animals often received names denoting the geographical origin of the species shortly after their discovery, such as *javanicus* or *borneensis*. Without yet having an accurate picture of the species' actual distribution, some of these names

19 Susanna, *Levenschets van Temminck*, 74, fn. 4.
20 Temminck, "Avant-propos," in *Manuel d'ornithologie*, xi.
21 Temminck, xi–xiii.

turned out to be erroneous. In these cases, Temminck felt that a correction was necessary. During an age of continuous discovery and with such large areas still to be surveyed, naturalists should therefore avoid "following this excessively defective method [...] of giving animals the names of countries" to prevent any confusion about their actual geographic distribution, "since it is rare that a species is confined within the limits of our geographical divisions."[22]

Of course, other naturalists had different opinions on the matter. Frédéric Cuvier – Georges' younger brother, a zoologist and the keeper of the Ménagerie du Jardin des Plantes – thought that naming animals after countries had an obvious advantage: a geographic epithet recorded the place where a particular specimen had been first discovered.[23] One example for this is the Sunda Leopard Cat, a small wild cat that inhabits the islands of the Philippines, Sumatra, Borneo, Java, and Bali. Anselme Demarest had named it *Felis javanensis* after a specimen from Java in 1816. By 1824 it had become clear that the cat's distribution reached beyond Java, so Temminck decided to change its name into *Felis minuta*.[24] The American naturalist Thomas Horsfield felt that the change was unjustified; the cat did, indeed, live in Java, even if not only there. He responded to Temminck's "inconsiderate" change by quoting a passage from the *Histoire naturelle des mammifères* by Geoffroy Saint-Hilaire and Frédéric Cuvier:

> Men have objected, apparently with reason, to the practice of giving names of countries to specimens that may have a wider distribution; but considering that these names are given when these organisms are discovered for the first time, when their distribution is still unknown; and as everybody knows, as it is easy to agree that these designations are not absolute, one may recognize that of all names that mean something, names of countries are the ones with fewer disadvantages [than others].[25]

However, if a species name was completely misleading (for example, Cuvier's name *javanensis* for a bear that is found only on Sumatra, and nowhere else), Horsfield accepted that the name was erroneously applied and could not be retained.[26]

22 Temminck, *Monographies de mammalogie*, vol. 1, 132.
23 F. Cuvier, "Nouveau recueil," 645.
24 Temminck, *Monographies de mammalogie*, vol. 1, 130–132.
25 Geoffroy Saint-Hilaire and Cuvier, "Ours de Sibérie," in *Histoire naturelle des mammifères*, livraison 42; quoted in Horsfield, "Description of the *Helarctos euryspilus*," 230, fn.
26 Horsfield, 229, fn.

Putting aside the bruised egos, the nomenclatural chaos impeded progress. Any museum catalog, systematic work, and description of species needed to include the different names available for every species if naturalists were to succeed in communicating meaningfully. In 1824 Horsfield attempted to put things into order when he cataloged the Javanese birds at the Museum of the Honourable East India Company. He listed the species names as accepted in 1820 in one column, and another column beside it listed the changes made after 1820 "by the rejection of *topical* names, the introduction of several names employed in the Nouveau Recueil de Planches coloriees d'Oiseaux, published by M. M. Temminck and Laugier, and by other necessary alterations in Nomenclature."[27] Horsfield noted with frustration that "the employment of names taken from the countries where animals are found or have been discovered, is a constant theme of discussion and declamation with Continental naturalists."[28]

Not all British naturalists agreed on this matter. William Swainson had quite a different view than Horsfield. He believed that names of species could, occasionally, be derived from their geographic range, "provided it is peculiar," but he warned the reader that such names should be avoided when all the species of the same genus inhabit one and the same country. In this case, he added, the original names should be canceled.[29]

Changes were sometimes needed to correct misidentifications, and sometimes to restore older names by applying the law of priority. But on average, the nomenclatural practices of the 1820s still depended very much on the author's personal views. An anonymous reviewer of Temminck's *Monographies de mammalogie* urged his readers to be cautious:

> But the propriety or impropriety of employing such names [of countries] does not fall fairly into discussion in this instance; such an enquiry, how indispensable soever previous to designating a species, becomes too late when the name has been applied. It must then be of necessity retained, unless it conveys a decidedly false impression; and even in this case, considerable hesitation would be experienced by every naturalist, who felt unwilling to increase the disarray already too prevalent in synonymy.[30]

27 Horsfield, "General Catalogue of Javanese Birds," [1] (emphasis original).
28 Horsfield, "Description of the Helarctos euryspilus," 229, fn.
29 Swainson, *Natural History and Classification of Birds*, vol. 1, 240.
30 "Analytical Notices of Books," 532.

If deciding how to apply the rules of nomenclature to topical names was complicated, the question of how to proceed with double descriptions and priority was a conundrum. Often species that had already been described were renamed and described again, sometimes multiple times by different naturalists within a few years. Questions arose about the reasons to draw up another description: had the author been unaware of previous publications? Or did he intend to replace it with a better description and a more suitable name? It was also possible that the first author had misidentified a species, for example, by considering a female a new species because the specimen was so different from the male. Authors usually explained their reasons for replacing existing names, and naturalists who failed to do so would be rebuked by their peers.

Some naturalists thought they should always strictly follow the law of priority, without exceptions. George Robert Gray, the zoologist and keeper of the Zoology Branch of the British Museum in London, referred to this in 1842 as "the right of priority." He wrote: "In accordance with established practice, however, I regard the names first given, notwithstanding the anathema which has been pronounced against them by some of our lawgivers, as sacred, and have quoted the subsequent series only as their synonyma."[31] While taking up the monumental task of cataloging all genera of birds ever described, Gray concluded, understandably, that the complications generated by the multitude of synonyms impeded the progress of science. He then listed all the names known to him in his *List of the Genera of Birds*.

The first version of Gray's list appeared in 1840 and then he left it to the "judgement of each individual who chooses to pursue the subject, to select those which he considers tenable, and to erase the rest." To this he immediately added, quoting Vigors, that "no individual possesses the right of dictating what are, and what are not, good generic groups, or of erasing from the list of genera those of his fellow Naturalists which may not exactly square with his own particular view."[32] Temminck vehemently disagreed and found himself being criticized for intentionally renaming species already named by others.

In some cases, Temminck seemed unaware of the double descriptions, or in any event he offered no explanation at all. In other cases, his replacements were deliberate. He mostly followed this practice for the following reasons: if he considered the species part of a different group than the one to which it had been previously assigned, or if the original description seemed defective to him in some way. For example, Temminck decided to draw up a new description of a particular species of hornbill, the Rhinoceros Hornbill, that Levaillant had

31 Gray, "Preface," in *Appendix*, vii, iv.

32 Gray, "Preface," in *A List of the Genera of Birds*, vii. Gray gave no reference to Vigor's quote.

FIGURE 29 Original watercolor of a Rhinoceros Hornbill, possibly by Pieter van Oort, ca. 1830. Temminck scribbled in pencil on the lower left-hand corner that this drawing should be used for plate 546 of his *Nouveau recueil de planches enluminés*, "Calao *à casque en croissant*."
NATURALIS BIODIVERSITY CENTER ARCHIVES, NNM001000157_001

named *Calao rhinoceros* in 1801, arguing that Levaillant had based his description on an artifact – a specimen made up from parts that belonged to different species. Apparently Temminck had obtained that particular specimen for the collection of the Leiden museum. Then, using illustrations drawn from a living specimen (in all likelihood by Pieter van Oort), he placed this bird into another genus and renamed it *Buceros lunatus* (now *Buceros rhinoceros*).[33]

With time additional descriptions appeared in Temminck's works, duplicating existing descriptions by other naturalists. The reviewers of Temminck's great iconographic work, the *Nouveau recueil de planches coloriés d'oiseaux*, pointed this out on several occasions. When the earliest installments were published, the authors received many positive comments for the exactitude of the figures, the accurate descriptions of the species, and the synonym lists

33 Levaillant, *Histoire naturelle d'une partie d'oiseaux nouveaux*, 26, pl. 13; Temminck and Meiffren Laugier de Chartrouse, "Calao à casque en croissant," in *Nouveau recueil*, vol. 2, livraison 92, plate 546, text, [1].

FIGURE 30 Calao *à casque en croissant* (Rhinoceros Hornbill). Hand-colored engraving by Jean-Gabriel Prêtre after drawings by (possibly) Pieter van Oort. In Coenraad Jacob Temminck and Guillaume Michel Jerôme Meiffren Laugier de Chartrouse, *Nouveau recueil de planches coloriées d'oiseaux*, vol. 2, livraison 92 (1838): plate 546
IMAGE FROM THE BIODIVERSITY HERITAGE LIBRARY. DIGITIZED BY THE SMITHSONIAN LIBRARIES

that accompanied every name.[34] A review in the *Annales générales des sciences physiques* from 1820 reads:

> We have delayed telling our readers about this important collection, and we are now most certain of the praise it should receive. The sixth installment that has just been published equals, if not surpasses, the installments that preceded it in beauty and perfection. After all, this is what one should expect from the reputation of its authors and from the authority these scholars have enjoyed for a long time.[35]

A year later, the reviewer stated: "This beautiful enterprise continues with admirable zeal, and its authors receive everywhere the satisfactory reviews they are entitled to expect from the many enthusiasts of the natural history of birds."[36] But in 1840, by the time the work was finally finished, it had also received mixed reviews. The quality of the final installments did not live up to expectations. Temminck had introduced many new names for species that had been described before. The *Annals and Magazine of Natural History* published a review that praised the work for the precise descriptions of the species and the accuracy of the plates, but criticized Temminck's nomenclatural practices:

> [T]he greatest blemish to the work is the passing over the discoveries of contemporary writers, with whose works M. Temminck should have been conversant; thus from the commencement of the series many birds are given under new denominations which had previously been described by other naturalists. A difference of opinion may have existed, but prior and contemporary labourers in the same field could scarcely all be unworthy of notice; even in the concluding Livraisons this is apparent.[37]

Charles Lucien Bonaparte detected a similar trend in Temminck's *Monographies de mammalogie*, in particular, in his treatise on a genus of vesper bats named *Vespertilio*. Bonaparte listed more than half a dozen of Temminck's species names that resulted from misidentifications or synonymization, or were otherwise invalid.[38]

---

34 "Nouveau recueil," in *Annales générales*, vol. 7; "Nouveau recueil," in *Annales générales*, vol. 8.

35 "Nouveau recueil," in *Annales générales*, vol. 7, 361.

36 "Nouveau recueil," in *Annales générales*, vol. 8, 395.

37 "Bibliographical Notices," 267.

38 Bonaparte, "Observations of the State of Zoology," 22.

Amidst this chaos, the introduction of order was one of the main concerns of early nineteenth-century zoologists. Hugh Strickland was particularly worried about this issue: "Any one who will take the trouble to examine the various papers on ornithological nomenclature in the *Analyst*, will find that specific names are as variable as the London fashions." For him, the solution lay in a strict application of the rule of priority: "Then will the nomenclature of Linnæus, Cuvier, and Temminck triumph over the crude inventions of a host of anonymous scribblers."[39] In an attempt to create some order, Strickland published a set of twenty-two rules for zoological nomenclature. Despite his best efforts, even this set of rules left plenty of room for interpretation and creativity. Strickland allowed for exceptions to the principle of priority (without indicating the cases to which this applied), and rule number six was so ambiguous that it invited controversy: "A name may be expunged whose meaning is false, as applied to the object or group which it represents. [...] Yet, in some few cases, where a name, though false, does not really mislead, it should be retained, if long established."[40] Swainson's 1837 rules were generally badly received by other naturalists or, if approved of, regarded as guidelines and recommendations.

Everyone agreed that, to attain the desired stability, naturalists should follow specific rules of nomenclature, and these rules must be accepted and adopted universally. But agreeing on the rules themselves was a very different matter. There was no recognized authority to settle the issue, a reflection of the fact that nomenclature and classification were still developing fields. Nicholas Vigors, convinced that nomenclature was a tool for classification rather than its foundation, wrote in 1825:

> Nomenclature itself is variable. From its very nature it cannot remain stationary: it must be enlarged to suit the increasing bulk of materials which it is meant to regulate; it must be altered to meet the more accurate information of every day. But the principles which direct its art never vary. Standing on the philosophical basis upon which they were placed by Linnæus, they are suited to every change of the science; in every alteration of views, in every modification of knowledge, they remain the same, unchanged, unchangeable. Ought it not to be the undeviating principle therefore to which we should adhere, and not the inconstant name?[41]

---

39 Strickland, "On the Inexpedience of Altering Established Terms," 128.

40 Strickland, "Rules for Zoological Nomenclature," 174.

41 Vigors, "Observations on the Nomenclature," 190.

The problem was that naturalists could not agree on the question which "principles direct the art of nomenclature." It would not be until 1842 that rules for zoological nomenclature were formalized by an established scientific organization, the British Association for the Advancement of Science.[42] The consequences of this went far beyond the stabilization of nomenclature: the rules marked a shift of the locus of authority from individual practitioners to a self-governing collective of naturalists. But there was yet another issue that ruffled feathers, for which a compromise was out of the question: the genus concept.

3 Establishing Genera

Since the works of Latham and Illiger had been published, naturalists generally agreed that the bulky Linnaean genera needed dividing into smaller units to accommodate the incredible number of new birds discovered since 1800. In Linnaeus' *Systema naturae* of 1766, seventy-eight genera were listed, which contained over nine hundred species. The subsequent, revised editions did not help to clarify the nomenclature. Temminck mentioned the "misérable compilation" by Johann Friedrich Gmelin, who "had the talent of making the thirteenth edition of Linnaeus the most indigestible book there is: therefore, those who insist on slavishly following it cannot fail to fall into the greatest errors."[43] In 1790 Latham's *Index ornithologicus* listed almost three thousand bird species in a little over one hundred genera. In 1816 Vieillot listed over two hundred and seventy genera in the *Analyse d'une nouvelle ornithologie élémentaire*. In the second edition of the *Manuel d'ornithologie* (1820–1840), Temminck cataloged almost two hundred genera of birds from Europe only. By 1844, Gray had described eight hundred and fifteen genera and listed over two thousand four hundred generic names known to him at that time.

Clearly, naturalists disagreed on what a genus was. They held widely divergent views, ranging from the most conservative, apprehensive about changing the Linnaean system, to the most innovative, routinely dividing large genera into multiple smaller ones. (Incidentally, these two radically opposed positions are still a matter of some controversy today; their advocates are known, colloquially, as 'lumpers' and 'splitters'). Species were moved from group to group, shedding their generic names with every rearrangement. Modifications reflected each naturalist's theoretical premises. Gray sketched out the situation in 1841: "an author will not hesitate to state, that he cannot adopt the

42 Rookmaaker, "Early Endeavours by Hugh Edwin Strickland."

43 Temminck, "Avant-propos," in *Manuel d'ornithologie*, xxxii.

genera of certain Ornithologists, because they are not what he is disposed to consider 'natural divisions'; he therefore proposes his own divisions, and designates them by his own names."[44]

Nicholas Vigors who, besides being a renowned naturalist, was one of the leading reformists in Britain, advocated a change in nomenclature as well as in methods of classification. In 1824 he wrote a passionate letter about Temminck's classification of birds as applied in his published works between 1813 and 1820.[45] The letter was addressed to the editors of *The Zoological Journal*, who published it in 1824, entitled "Some Observations on the Nomenclature of Ornithology; Particularly with Reference to the Admission of new Genera." Temminck responded in 1835, in the third volume of the *Manuel*'s second edition. Temminck and Vigors disagreed on three main questions. First, was a genus better than the old Linnaean sub-generic category, the 'section'? Second, where should naturalists place those species that fell in between genera? Third, which were the best characters to define a genus?

The debates about these three points illustrate perfectly the chaotic state of zoological classification at the time. The first disagreement between Temminck and Vigors was about the use of genera versus the use of sections. The section was an informal systematic category that helped cluster groups of species within the same genus, and it did not assign a Latin name nor an author name to it. Sections were particularly useful for large genera containing many species, breaking them up into smaller units, and were used both in botany and in zoology. In this way, no additional generic names were needed to place these species into separate groups. Temminck was very critical of the proliferation of genera that had taken place in ornithology between 1800 and 1820. He particularly disliked those descriptions that, in his eyes, were not supported by a careful comparison between forms, varieties, and related species from different geographic regions.

Temminck had already clashed with Vieillot on this point. In his view, Vieillot had divided existing systems and created so many tribes, subfamilies, genera, and sections that "he is one step away from making the species disappear and thus making ornithology comprise only genera and sections."[46] It was not only Vieillot's system that Temminck criticized here but also the general practice of splitting groups into smaller ones. This practice, he thought, complicated the systems and impeded the advance of science. For Temminck,

---

44 Gray, "Preface," in *List of the Genera of Birds*, 2nd ed., v.

45 This "Vigorian reform" and the situation of nomenclature and classification in Britain have been described by McOuat in "Cataloguing Power" and "Species, Rules and Meaning."

46 Temminck, *Observations sur la classification*, 6–7.

additional genera did nothing but create confusion, especially if they included species that were difficult to place within the existing system:

> But what fate awaits the *anomalous species*, so numerous within the class of birds? Minutiae will not always help to dissipate the uncertainty; in this case, the species is destined to be shifted from one genus to another, the source of many errors and multiple synonyms; rendering naturalists less likely to understand each other [...] When in doubt, to avoid this problem, a *new genus* is made. This procedure is the order of the day, and it indulgently boosts the self-esteem.[47]

Accordingly, he kept the genera large and synonymized many new names with the old ones. To make everything more manageable, to Vigors's dismay, Temminck used sections.

Vigors preferred a different approach. Naturalists generally had two options when a large genus seemed to include smaller units. One – Vigors' choice – was to elevate the genus to a family and then create new genera to accommodate these clusters, the other – Temminck's favorite – to keep the original genus name and then to refer to sections as sub-generic categories in which to group similar species within the existing genus. For Vigors the chief fault of this procedure was that it complicated nomenclature and sacrificed the brevity of Linnaean binomials. With the use of sections, genera became too big to be useful. The placement of a species within the system, in a genus containing dozens, sometimes hundreds, of species, became an impossible exercise. Therefore, Vigors concluded, sections were not in accordance with the spirit of zoological classification:

> Now if it should occur, that, owing to the increase of materials or of our knowledge respecting them, the terms assigned to certain groups should become too vague and limited in their application, with reference either to the magnitude of the group itself, or the variety of forms that enter into it, – as, for instance, where an original genus receives such a numerous accession of species, and such an addition of new forms, as renders the name, as generick, inapplicable to them all; – it becomes a question whether we should make an alteration in the terms of the nomenclature itself, or violate the principles which have been established for its direction.[48]

---

47 Temminck, "Introduction troisième partie," in *Manuel d'ornithologie*, 2nd ed., vol. 3, xxviii–xxix (emphasis original).

48 Vigors, "Observations on the Nomenclature," 190.

For Vigors it was natural that names changed to adapt to the growth of knowledge. The principles of classification, by contrast, were immutable: "There are no terms more definite in their meaning and application than those of genus and species."[49] Vigors considered Temminck's attempts to avoid creating new genera pointless. Even worse, while new names did not create confusion in science, the use of sections did. If sections were so distinct as to be separated from one another and the adjoining genera, then, Vigors said, "I cannot conceive that the subdivisions so characterized can be considered ought else than genera."[50] In this situation, he saw no reason to refrain from describing a genus:

> The follower of the modern views in science discovers, in an extensive family, a distinctly characterized group; he calls it a genus, and designates it by a scientifick name. The adherent to the Linnean nomenclature acknowledges the same distinct group; he calls it a section of the Linnean genus, and designates it by an unscientifick name. The process is precisely the same; the terms made use of alone are different.[51]

Sections and genera were synonyms for Vigors, but certainly not for Temminck. Temminck saw them as two very distinct categories with different meanings and, consequently, different functions. Temminck's sections were primarily useful for research: one could change them without fundamentally altering the classification system or the nomenclature. Sections could even be regarded as temporary if necessary. To Temminck, sections were always preferable to redundant genera, because they "may be adopted, modified or removed entirely without affecting the systematic order, and without hindering, by a series of new names, the understanding between naturalists throughout the world."[52] He concluded that sections, if made with precision, were preferable to new genera: "the first can depend on one's viewpoint and it is comparable to the rules of propriety in society; the second should serve as a universal rule."[53]

Temminck saw yet another advantage in the use of sections instead of genera: they were not followed by the author's name, as genera were. Sections deflated the ambitions of the naturalist describing new genera only to see his name written next to that of a new genus: "to flatter his self-esteem, he usually rubs the sponge over the name of his predecessor and says, by means of MIHI or by NOBIS in capital letters: *A moi la cargaison!* [The cargo is mine!], and

49 Vigors, 199.
50 Vigors, 199.
51 Vigors, 189.
52 Temminck, "Introduction troisième partie," in *Manuel d'ornithologie*, 2nd ed., vol. 3, xxx.
53 Temminck, xxxi.

very often gives priority to the genus."[54] Ironically, Vigors wrote of Temminck: "if it is an innovation to create and characterize a new genus, he [Temminck] is himself one of the greatest innovators of the present day. I have too high an opinion of the talents of that distinguished naturalist, and too high a respect for his character, to suppose that he would be influenced by any motives of prejudice or personal importance."[55]

Temminck's use of sections instead of genera followed his understanding of what a genus actually ought to be: a permanent category in the system. After all, the genera were embodiments of the types. He made an essential distinction between the sections as convenient classification tools and actual classificatory groups, the genera. Vigors and Temminck argued about the uses of sections and genera, but since they defined these differently, they were talking at cross-purposes.

The second issue leading to quarrels was the creation of new genera in which to include intermediate forms, i.e. species showing characters of two or more different genera. Temminck considered many new genera superfluous because, even when there were recognizable groupings of species within a large genus, the differences between them were so gradual that one could not discern where one group ended and the next started. On the particular case of finches, he wrote:

> It is a vain endeavor to invent double and even triple new names to form groups that are strictly methodical [i.e. for classification purposes only]. I have very carefully compared more than one hundred exotic species with our native ones; the result of this examination has confirmed my view that there is a gradual passage, without any demarcation, from one species to the other.[56]

It was often the case, Temminck said, that two separate genera had been described by taking species from the two extremes of the morphological continuum of one single genus, without knowledge of the intermediate forms. Seeing the differences between these two groups of species, one might indeed believe that they belong to different genera. But a few of the intermediate or transitional species, from the middle of the gradation, would prove that they belong to the same genus.[57] Temminck always preferred including related spe-

54 Temminck, xxxii.
55 Vigors, "Observations on the Nomenclature," 194.
56 Temminck, *Manuel d'ornithologie*, 2nd ed., vol. 1, 342–43.
57 Temminck, "Introduction troisième partie," in *Manuel d'ornithologie*, 2nd ed., vol. 3, xiv.

cies in one genus. He argued that, without a clear boundary between groups, it was impossible to define a new genus with enough precision for the definition to have any diagnostic value. Any new genus should exhibit distinct characters and have clear-cut boundaries with other genera, or not be described at all. Consequently, Temminck was happy to accept new genera as valid only when an author was able to back up a description with sufficient comparative research, and if he defined generic characters with enough precision.[58]

Vigors shared Temminck's view on this point: absolute divisions did not exist in nature. However, this realization led him to a very different approach to classification. For Vigors, every group did, indeed, gradually pass into another, in an uninterrupted chain of affinities that connected all living forms. If this was what Temminck considered to be a genus, then "we must admit one group, one genus alone, in Ornithology, which we may denominate the genus *Avis*."[59] The only way to deal with this apparent unity, Vigors wrote, was to admit the artificial nature of classification: divisions were arbitrary and man-made, because nature showed none. Boundaries were set around a group of species as they more or less approached the type species, the one exhibiting the defining characters of the genus.

Vigors' type-concept was the tool that allowed him to outline his generic groups. He did this by comparing the morphological differences of a species from the types most closely resembling it before deciding in which genus to place the species: "Upon these typical eminences I plant those banners of distinction round which corresponding species may congregate as they more or less approach the types of each."[60] He compared this method to the process of naming hills in an undulating landscape: "And thus it is that I conceive my groups to be at once separated and united; separate at their typical elevations, but united at their basal extremes."[61]

Temminck was not convinced by Vigors' approach. If a genus was composed of those species that most resemble its type, every genus must necessarily have a well-defined type. And that was what the new systems with their vaguely defined genera lacked, in Temminck's opinion. Every type was unique and separate from other types; each species always exhibited the type's main characters.[62] Therefore, only when the type was well defined was it possible to

58 Temminck, xxii; Temminck, "Description d'un nouveau genre de mammifères (Urotrichus)," 1.
59 Vigors, "Observations on the Nomenclature," 195.
60 Vigors, 196.
61 Vigors, 197.
62 Farber, "Type-Concept in Zoology"; McOuat, "Species, Rules and Meaning," 511–12; Witteveen, "Naming and Contingency."

allocate species to their type and to set the limits of the genus. Genera might be, as Vigors had said, artificial, but Temminck believed that natural or true genera could be discovered by ascribing the species to its right type (provided one knew its main characters). This method would, in due course, close the gap between the artificial genus and the natural entity.

For Temminck the first step was to define the characters of the type as accurately as possible before describing any new genus. On this aspect, William Swainson remarked: "Our author's [Temminck's] forte, indeed, like that of Illiger, is detail; but he seems, unfortunately for his own fame, to have imbibed the ancient notion that no genus is to be retained, if the links by which it is connected to another are discovered."[63] Nevertheless, Temminck remained determined to oppose the splitting of genera. After all, these rather dry technicalities embodied nothing less than his understanding of nature's organization as God's design.

The third issue in dispute, which was just as controversial as the other two, was how to differentiate a genus and its type. A type, said Temminck, should be defined by a few essential characters, intrinsic only to the form it represents, leaving no room for mistake. As types were the blueprints for genera, they should be clearly outlined by a small number of fundamental traits. A neat and concise definition was essential. Without it, Temminck argued, the description of any new genera (those made for the intermediate forms) must necessarily include a long list of minor or auxiliary characters on top of those already used for the types of the existing genera: "a good diagnosis is not as easy to make as one might think. Naturalists do not always know how to clearly define a species using a few words. A long description full of minute details is certainly the most confusing."[64]

For Temminck, any group exhibiting morphological characters also present in other genera could not hold its own as an actual category. If a genus included these intermediate forms, naturalists must use many "minute characters" as part of interminable and complex descriptions. This had already been a significant point in Temminck's criticism of Vieillot's system. Vieillot had described tribes and families using a countless number of characters, undermining the validity of his system, or so Temminck thought.[65] Ill-defined types only served to confuse classification and obscure the true types within the system.

---

63 Swainson, *Treatise on the Geography*, 184–85.

64 Temminck, "Introduction troisième partie," in *Manuel d'ornithologie*, 2nd ed., vol. 3, xxiv–xxv.

65 Temminck, *Observations sur la classification*, 12.

The key to this problem was finding one right set of characters. Temminck pleaded for the use of a set of carefully chosen features – the same for each genus – to make comparisons between groups possible. In ornithology, these characters should include not only the external characteristics such as the feet, the beak, the plumage, or the wings, but also skeletal features and traits from the species' geographic distribution and behavior. Minor characters such as variations in color or size should not be included, as these only clouded the natural relationship between species and genera; also, none of the main features should be left out. That would only hinder the comparison between groups. There were key characters and minor characters used to define a genus, and not all of them should be weighed equally in its classification:

> When one tries to establish a new subdivision within an existing genus, it is essential to know, by examining nature, all or at least most of the known species related to those one wishes to isolate. It is also necessary to examine the series of specimens with greater or smaller affinity to those that might be included in the new genus. Furthermore, it is imperative to know if the way of life, nutrition, and anything related to behavior and habitat of these species are markedly different from their congeners. It is important to distinguish mere varieties in a natural series of species from characters of a higher order, which are indicative of a different organization and behavior: generally, the eye of the experienced naturalist is better able to identify these accurately than those first and foremost devoted to the theory of natural sciences.[66]

Temminck listed the characters to be used for the definition and arrangement of new genera in the *Manuel d'ornithologie*. Any additional characters were superfluous and ineffective.

Additionally, and perhaps more importantly, only those naturalists who had enough empirical knowledge and practical experience were able to determine which characters were to be used in classification. Naturalists who worked on the theory of nature by contemplation or speculation were not capable of distinguishing variations from genuine diagnostic characters. There was a good reason why that Temminck's publications were long and slow in progress. Temminck's pride in his extensive studies exasperated Vigors: "With considerable, and perhaps justifiable, self-complacency, he adds, that it is this

66 Temminck, "Introduction troisième partie," in *Manuel d'ornithologie*, 2nd ed., vol. 3, x.

knowledge which draws the line of distinction between the true expounder of nature, and the pseudo-naturalist of the cabinet and the library."[67]

The question of what precisely a defining character was became a critical point of dispute, as it carried with it the possibility to create boundaries between taxa (a taxonomic group of any rank, such as a species, genus, or family, which is arranged in a hierarchy within a classification system). Temminck had already pointed out Vieillot's inconsistent choice of characters: "he makes use of distinctive features with a certain indifference, taking them from here and there; this approach is highly defective, as it makes it impossible to find equivalent traits suitable for comparison between genera."[68] In order to understand how the natural groups of birds relate to each other – and thus to understand nature's patterns – it was necessary first to accurately define these groups. For this, Temminck turned to Cuvier's principles.

Cuvier's works on fossils and comparative anatomy had led him to postulate a starting point for the classification of animals. He defined the higher taxa based on his principles of the correlation of parts and the subordination of characters. Simply put, organs essential for an animal to live, such as the respiratory and digestive organs, are deeply interdependent and the most important in classification. He grouped animals according to the degree of similarity in these vital organs first, arranging the smaller groups using the less critical anatomical features.

While popular, Cuvier's approach was not new. Botanists, who often seem to be one step ahead of zoologists for some reason, had already developed a similar method in the eighteenth century. Antoine-Laurent Jussieu outlined his method for plant classification in his *Genera plantarum* in 1789. He relied on the principle of continuity and the subordination of characters to establish natural groupings, which reflected relationships based on similarities. The groupings were to be made by synthesis, moving from small to large taxa (groups of related species constituted genera, related genera, families, and so forth). Jussieu's method was a success, but zoologists only seem to have adopted it after Georges Cuvier developed his particular version of the method for the animal kingdom in his *Règne animal* from 1817. However, as Cuvier advanced the description of the four *embranchements* (the four main groups in the animal kingdom at the highest level of classification), he stated that these were definitely isolated from each other, and thus breaking with the principle of

67 Vigors, "Observations on the Nomenclature," 193.
68 Temminck, *Observations sur la classification*, 41.

continuity. He also favored a 'downward' approach, defining larger groups first and then working his way down to increasingly smaller groups.[69]

Rather ironically, Temminck followed Cuvier in his general outlines for zoological classification while retaining the idea of continuity between species at the lowest rank of the system. Temminck did not explain this further, but his writings suggest that he saw nature not as a single, uninterrupted chain of similar forms, but a reticulated entity in which types were positioned at the nodes of each radiation. He saw nature as a structured whole with distinct higher [i.e. genera, families, and orders] taxa that presented sufficient variation at the lower levels to allow for continuity. Higher taxa and types were definite and isolated; the variations within species were so diverse that they formed a continuum. Types – and genera – were, therefore, definable by the selection of the right set of characters.

Temminck chose the characters to delimit his genera after Cuvier. But this was just the starting point:

> It is certain that comparative anatomy and the study of the osteology of animals must be considered as basic sciences and as the point of departure for the zoologist who wishes to build the methodical edifice; but it is not exclusively with the insights provided by these two studies that he must establish the second order groups.[70]

For example, the main features he used to define the families of pigeons and fowl were the beak and the feet (essential for feeding and locomotion) and the shape of the trachea (critical for respiration and reproduction, as the males use vocalization to attract the females). He then delimited each genus by a combination of characters taken from its skeletal features, the shape of the tail and wings, body size, habitat, and nesting behavior. Coloration patterns, a particular geographical distribution, or specific features distinguished one species from another.[71] It was not just Temminck who identified these diagnostic characters for defining taxa; in fact, this had become the standard procedure since Cuvier. But naturalists seemed unable to agree which characters were to be used for the definition of a specific genus or species.

---

69 On the influence of Jussieu on Cuvier see Stevens, *Development of Biological Systematics*, 65–72.

70 Temminck, "Introduction troisième partie," in *Manuel d'ornithologie*, 2nd ed., vol. 3, xv.

71 Temminck, "Discours sur l'ordre des pigeons," in *Pigeons et gallinacés*, vol. 1, 32–39; Temminck, "Introduction troisième partie," in *Manuel d'ornithologie*, 2nd ed., vol. 3.

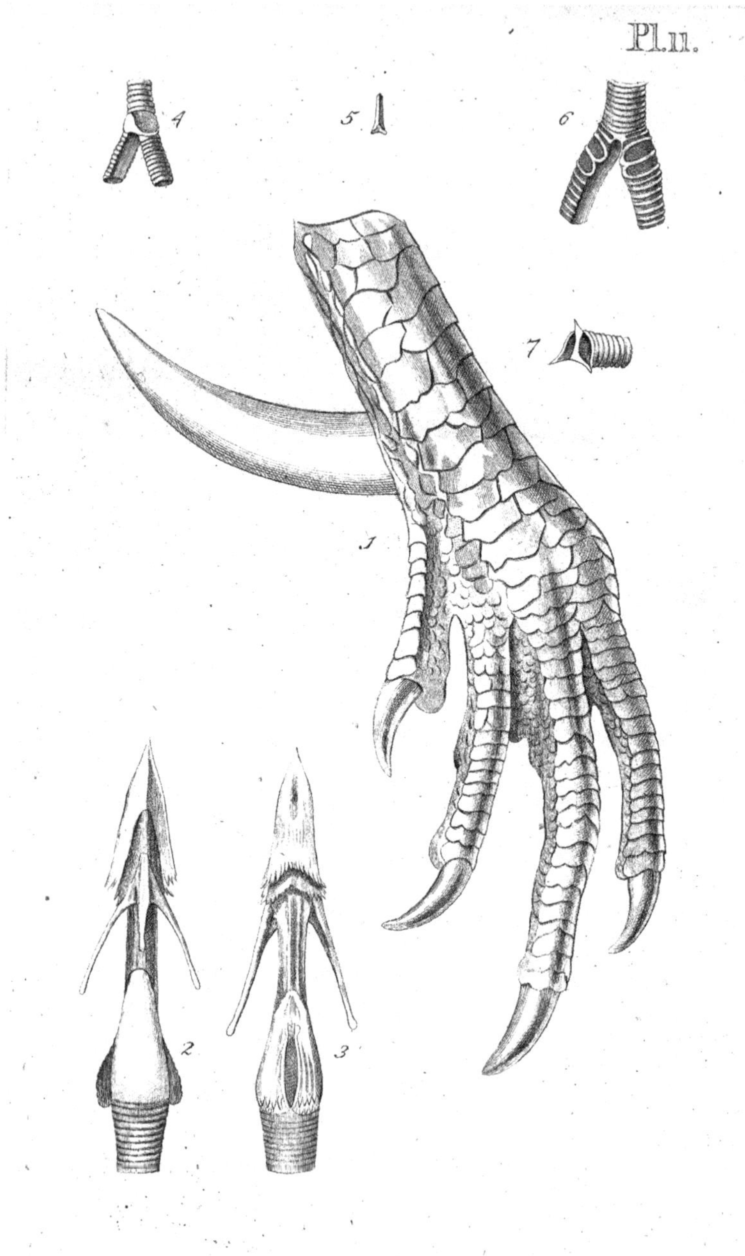

FIGURE 31 *Planche anatomique II*. Design proof for plate 2 in Coenraad Jacob Temminck, *Histoire naturelle générale des pigeons et des gallinacés*, vol 2 (1813)
NATURALIS BIODIVERSITY CENTER ARCHIVES, NL-LDNNAT_TEMMINCK_1_002

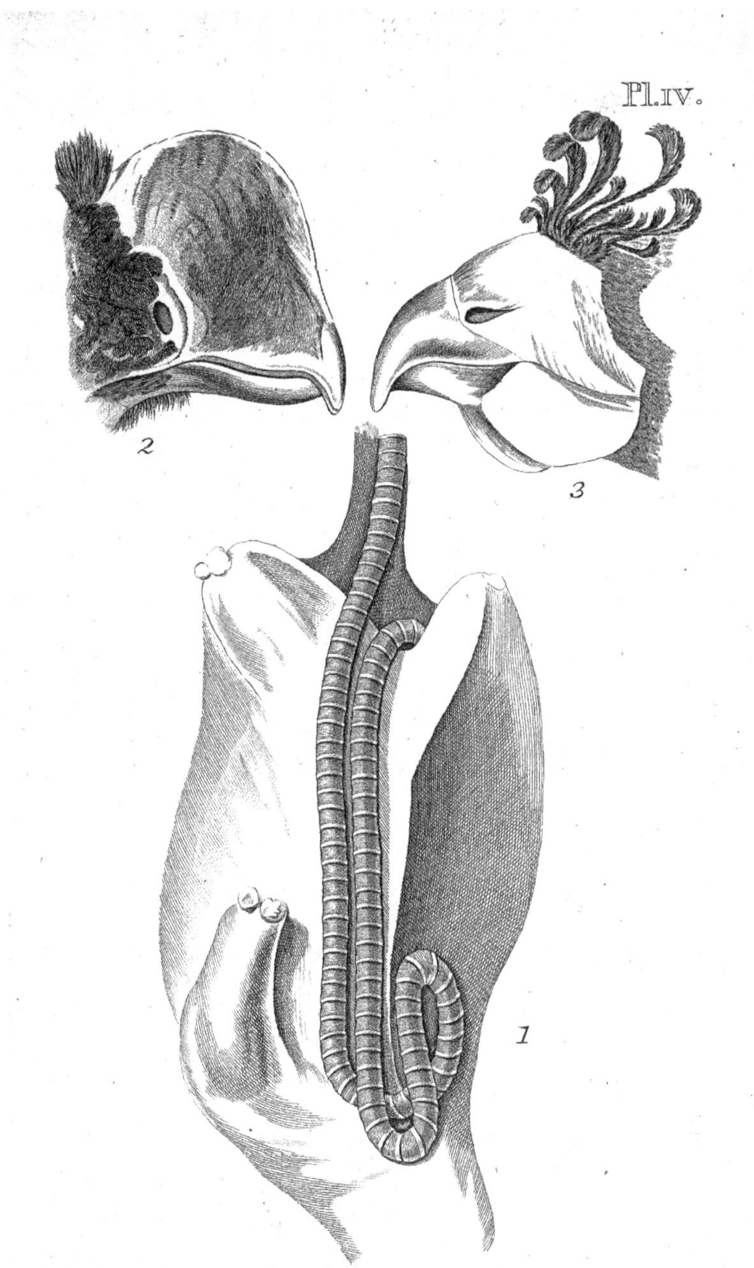

FIGURE 32 *Planche anatomique IV*. Design proof for plate 4 in Coenraad Jacob Temminck, *Histoire naturelle générale des pigeons et des gallinacés*, vol 3 (1815)
NATURALIS BIODIVERSITY CENTER ARCHIVES, NL-LDNNAT_TEMMINCK_1_004

Temminck's correspondent and friend Karl Illiger was one of the few naturalists of the period to discuss the species problem. He concluded, quite phlegmatically, that any classification of organisms as belonging to the same species was only tentative and temporary until their cross-fertility could be proved. In the meantime, he followed Cuvier's method of the correlation of parts to elucidate which were to be used for the establishment of species.[72] For Temminck, as for nearly all naturalists of the period, both genera and species required the selection and hierarchy of a few essential characteristics that captured their identity:

> As long as one knows most of the species that have been described and recalls them from memory, then when the description of a new species has to be undertaken, one immediately finds the main affinities or the characteristic differences. What matters is to have a perfectly satisfactory description of the type species; then those of the other species of the same genus are sketched with broad strokes: often, a single well-defined character is enough to ensure that the species cannot be confused, not even with those it resembles most at first glance.[73]

Apparently, experience and knowledge were the best tools, and there were no definite rules or guidelines that could help. The way naturalists delimited species involved comparing series of specimens and descriptions in the literature, and then selecting those characters that, for them, had diagnostic value for that species. Therefore, comparison and experience, however subjective, played a decisive role.

In 1844, Strickland noted with some dismay that ornithologists still had not figured out how to weigh zoological characters, which "must therefore be estimated by moral rather than by demonstrative evidence."[74] It seems that the general approach to defining genera was an intuitive one, at least for the time being. Strickland admitted as much when he wrote: "The precise rank in the scale of successive generalizations which shall be occupied by those groups we term *genera* is then a matter of *convenience*, and consequently of *opinion*." But he was optimistic: "More definite principles of classification may hereafter be

---

72 Illiger, "Einige Gedanken über die Begriffe," cited in Mayr, "Illiger and the Biological Species Concept," 172, 177.

73 Temminck, "Introduction troisième partie," in *Manuel d'ornithologie*, 2nd ed., vol. 3, xxv.

74 Strickland, "Report on the Progress," 217.

discovered, and meantime all that we can do is to arrange our systems according to sound reason and without theoretical prepossession."[75]

4 The Search for a Natural Classification System

Nineteenth-century discussions about the different classification methods were nothing new; naturalists had been concerned about these issues since the seventeenth century. For more than two centuries, the primary goal had remained to be the production of a natural system of classification. The term 'natural' referred to a system that reflected the true, real organization of all living forms and the laws of the natural world (as intended by God).[76] The highly esteemed botanist John Ray, for instance, favored a system that revealed the plants' natural affinities and relationships by using more characters than just the reproductive organs, that is, Linnaeus's system. The result was a more 'natural' system. Similarly, in the 1660s Francis Willughby – Ray's former pupil and friend – attempted to draw up what is now considered the first natural classification of birds, with a system based on anatomical features and characteristic traits.[77] Naturalists had become fixated on uncovering the natural system ever since. Two centuries later, however, there was still no consensus on what approach to classification would eventually lead to a natural system.

Naturalists were very aware of the differences between three types of systems: artificial systems (these were meant to help catalog diversity with no theoretical implications), classification systems constructed on partial knowledge (these were not comprehensive enough to reflect natural laws), and systems reflecting the divine order in the natural world (i.e. the system of nature itself). The search for the natural system was so vital for zoology that to label a classification system as artificial automatically implied that the system was inadequate or defective. In 1826, James Ebenezer Bicheno, a British colonial official with a keen interest in botany, published an article about the contrasts between artificial and natural systems and their functions:

> [T]he artificial and the natural systems aim at two very distinct objects, which are in some measure incompatible with each other. The one is to make us acquainted with individuals; and the other, founded upon acquaintance with individuals, to combine them according to their

75 Strickland, 218 (emphasis original).

76 Farber, *Emergence of Ornithology*, 80–91.

77 Birckhead et al., "Willughby's Ornithology."

> characters, so as to abridge the labour of reasoning, and to enable us to ascend from particular to general truths.[78]

If the goal was to analyze, Bicheno explained, divisions and subdivisions might be very convenient. Nevertheless, to truly understand nature, these subdivisions and the characters used to define them were, in fact, an obstacle. He pointed out the "confusion of analysis and synthesis" that was taking place in many classification systems, mainly in entomology, "a defect which may be attributed chiefly, I apprehend, to the attempt which both we and our continental neighbours have made to combine the natural with the artificial system. We have aimed at analysis and synthesis at the same time."[79] The main question was not whether there was an underlying order in nature, but rather, how could zoologists turn those artificial systems into natural systems. Was this to be achieved through the empirical, inductive study of nature? Or should naturalists rely on their theoretical insights and construct classification systems in a deductive way?

Temminck's approach was definitely inductive. Through careful study of the main characters of birds, including every variation, their behavior, and their geographical distribution, he would discover nature's laws. He insistently stated this in every work he published. He relied on Cuvier's methods of comparative anatomy to delineate the main groups in zoology, and on the careful study of collections to refine Cuvier's system. The accumulation of knowledge through patient research, building on the work of previous naturalists, was the path towards transforming the artificial systems into natural ones and slowly closing the gap between them. Temminck's confidence in reaching a natural classification may have been bolstered by the fact that he was mainly interested in birds and mammals, groups that had been well-studied since the time of Linnaeus and which were relatively small when compared with those of insects or plants.

Temminck believed in 1815 that ornithologists had reached a point when they could draw the rough outline of the organization of the class of birds:

> We hope that we will succeed, with time, in creating a better method of classification. The only way to achieve this goal is by a meticulous examination of nature; no genus, no subdivision, not even the acceptance of a new species can have a place in such a method, before a conscientious study of the living animals, or their undamaged remains, has been made

78 Bicheno, "Systems and Methods," 481.
79 Bicheno, 492.

> by trustworthy naturalists; we should no longer admit the multitude of animals described from observations of travelers and vague information, which collectors seem to have introduced in books only to enlarge the catalog of nomenclature.[80]

Empirical research was the foundation of the natural system. Introduce too many changes to the foundation, and the whole building would come crumbling down "like a house of cards."[81] Temminck rejected Vieillot's classification system of 1816 because, by changing the work of his predecessors, Vieillot discarded preceding advances by other ornithologists. He could not accept that Vieillot had chosen to create an entirely new system of classification: "Mr. Vieillot [...] has shattered all the systems that have been published previously as well as the modern ones."[82] By creating countless divisions and subdivisions, naturalists were constructing a system so complicated that it could never lead to an understanding of the true nature of the relations between groups. An artificial system should not be anything other than an auxiliary tool, a rough version of the final system. Neat and straightforward, it was meant to serve as a "universal central point of reference," the basis for all subsequent research.[83] For Temminck, knowledge emanated from the very act of classifying.

By contrast, other naturalists saw classification systems not as a foundation on which a natural system could be built, but as a practical tool intended to help order the natural forms. For Vigors, an understanding of nature did not depend upon classification: it was the classification systems that emanated from knowledge. As knowledge evolved, so did the classification systems, as they were secondary to science. Therefore, when Vigors embraced Quinarianism, he employed his quinarian approach to overhauling the classification of birds. His groups of birds reflected the extraordinary quinarian rule that all groups were linked to the number five.

William Sharp Macleay defined the quinarian principles for the first time in 1819, in his *Horae Entomologicae*, and William Swainson quickly adopted them. Vigors soon followed and constructed a classification of birds based on these principles. Quinarianism proposed that all taxa were divisible into five groups. If a group consisted of four or fewer taxa, it followed that there was a species or genus that was yet to be discovered. On occasion, quinarians willfully added new taxa to fill in a gap or merged groups to ensure that each arrangement

80 Temminck, "Avant-propos," in *Manuel d'ornithologie*, xv.
81 Temminck, "Introduction troisième partie," in *Manuel d'ornithologie*, 2nd ed., vol. 3, xiii.
82 Temminck, *Observations sur la classification*, 6.
83 Temminck, "Introduction troisième partie," in *Manuel d'ornithologie*, 2nd ed., vol. 3, xi.

contained taxa in multiples of five. Different arrangements were also possible, and Swainson, in particular, allowed for the possibility of groups of three and nine.[84] On the whole, their fellow naturalists met Vigors' and Swainson's claims that their classifications were 'natural' with understandable skepticism.

Methodology was as much in dispute as the genus concept. Swainson, in his *Preliminary Discourse on the Study of Natural History* of 1836, explained that the study of natural history followed two methods. The 'analytical method' was based on empirical research, and therefore inductive, while the 'synthetic method' relied on deduction, that is, the basic principles were assumed to be correct prior to any investigation. Bicheno would have agreed on this point. Only the first method would lead to the correct classification, which he thought was the quinarian system. According to Swainson Quinarianism had not been discovered earlier because naturalists had focused on details instead of searching for universal principles.[85] Like Vigors, he claimed that the circular principle of variation was the result of empirical research, and that it had emerged naturally from the zoologists' acquired knowledge. In other words, it was a natural system. This was surprising, to say the least, taking into account that Bicheno, like Temminck, favored the same inductive or analytical method but wholeheartedly rejected Quinarianism. In any event Swainson considered Temminck's classification in the *Manuel d'ornithologie* to be artificial, yet excellent:

> Of all those which have been framed *without* a reference to the general laws of the natural system, it is decidedly the best. This may appear unmerited praise, when we perceive that the very foundation, or, in other words, the primary divisions, are forced and unnatural. [...] These [Temminck's divisions], however, when viewed in reference to artificial arrangement, – and the author is evidently unacquainted with any other, – are very clear, and, consequently, excellent. The genera, it is true, are few, but they are defined with great care, and evidence of acquaintance with this class of zoology far superior to that possessed by any of the moderns.[86]

---

84 Vigors published his bird classification in a series of papers in the 1820s, including his "Observations on the Natural Affinities." Swainson discussed the quinarian classification extensively in the *Natural History and Classification of Birds*.

85 Swainson, *Preliminary Discourse*, 109.

86 Swainson, *Treatise on the Geography*, 184 (emphasis original).

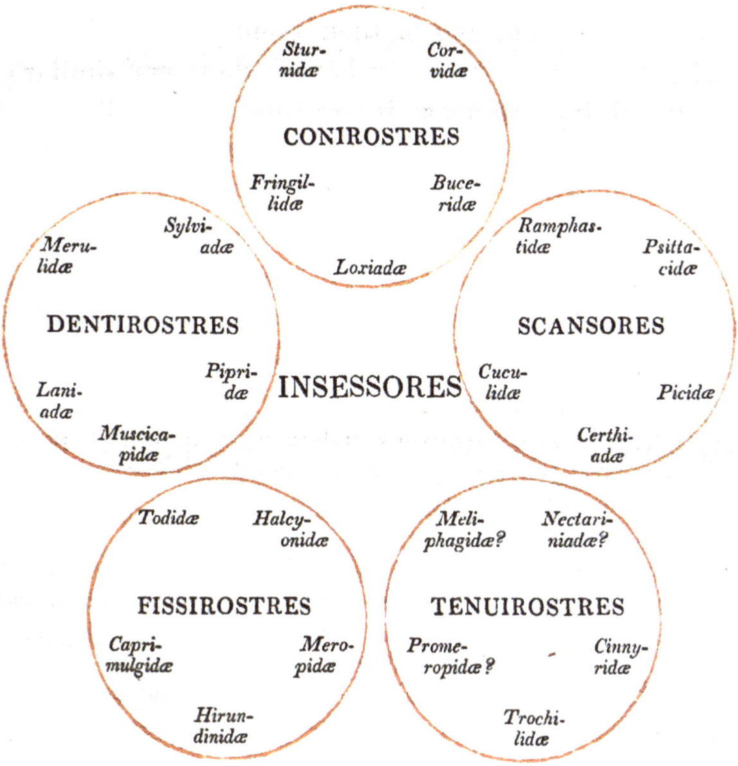

FIGURE 33 Diagram for the orders and families of birds, organized in circles of five according to the quinarian principles. In Nicholas A. Vigors, "Observations on the Natural Affinities that Connect the Orders and Families of Birds," *Transactions of the Linnean Society of London* 14 (1825): 468
IMAGE FROM THE BIODIVERSITY HERITAGE LIBRARY. DIGITIZED BY MISSOURI BOTANICAL GARDEN, PETER H. RAVEN LIBRARY

Swainson valued the clarity of Temminck's divisions. His quinarian contemporary Vigors clearly did not. He accused Temminck of working on an artificial system with large, unnatural genera to suit his ideas, bending nature to his views. Such practice was "subversive to the sound principles that regulate the science."[87]

Ironically, quinarians received the very same criticism for their systems of classification, i.e. that they were creating groups to suit their geometrical system. Nevertheless, Quinarianism enjoyed considerable popularity in Britain

87 Vigors, "Observations on the Nomenclature," 183.

for a few decades. Temminck complained in 1835: "One gets lost in circular affinities, in the most bizarre analogies; the trinary and quinary systems divide opinions and one arrives in the arena of hypotheses and philosophical ideas, which lead to endless and pointless debates; all this to explain this beautiful nature and the wonderful and grandiose order that governs it."[88] Similarly, Hugh Strickland, who had developed a complete theory on classification methods by then, noted: "the natural system is an accumulation of facts which are to be arrived at only by a slow inductive process, similar to that by which a country is geographically surveyed."[89] Therefore, he concluded, all systems based on *a priori* system, which attempt to classify forms in conformity with a preexisting theory, were erroneous. Quinarians claimed exactly the same, but did the opposite, that is, classified from the basis of a previously accepted thesis.

Strickland proposed the study of affinities, i.e. the similarities between species in those characters vital for their existence (the equivalent of the modern concept of homology in biology), for constructing the natural classification system. By measuring the morphological distance between species, Strickland built a system that could be depicted visually in a diagram, much like a geographical map, placing species into a two-dimensional space.[90]

In 1856 Alfred Russel Wallace joined the debate with an article cautiously entitled "Attempts to a Natural Arrangement of Birds." In it he rejected Quinarianism and adopted Strickland's map-making method: "When this has been done for all [families of birds], we may be able to arrange the whole group so as to present to the eye a view of the relations of the several parts, and then, and then only, shall we be able to determine whether any and what subdivisions can be established."[91] While Strickland's method was popular in Britain, Temminck made no mention either of Strickland's method or of his concept of affinity, although Temminck built his systems of classification upon very similar premises.

In the meantime, in Germany, a particular group of natural philosophers known as *Naturphilosophen* developed an entirely different approach to natural history. For them, classification systems reflected their belief system regarding nature, much as the Quinarian system did, but they had a more holistic theory. *Naturphilosophie*, which flourished from 1790 to about 1830, looked for relations between the natural objects and for forces driving natural processes,

88 Temminck, "Introduction troisième partie," in *Manuel d'ornithologie*, 2nd ed., vol. 3, xv.

89 Strickland, "True Method of Discovering the Natural System," 185.

90 See, for instance, Di Gregorio, "Strickland on Affinities and Analogies"; O'Hara, "Representations of the Natural System."

91 Wallace, "Attempts at a Natural Classification of Birds," 216.

applying *a priori* principles to their systems of classification.[92] *Naturphilosophen* like Lorenz Oken or Johann Jakob Kaup, like all other naturalists, searched for the ultimate natural system to organize living beings in all their diversity. Their starting point was the idea that knowledge derived not from experience, but from reason and divine inspiration. The naturalist was, to borrow a phrase from Breidbach and Ghiselin, "a kind of divine mind reader," whose classifications reflected his insight into God's plan as manifested in nature.[93] This led to apparently illogical and even somewhat esoteric classifications which, however, appear more comprehensible when viewed in the context of their conception.

Lorenz Oken devised a classification system for animals based upon the human five senses: touch, taste, sight, smell, and hearing. The underlying theory was that animal groups were physical representations of the five senses. The resulting groups were named accordingly as Dermatozoa (invertebrates), Glossozoa (fish), Ophthalmozoa (mammals), Rhinozoa (amphibians), and Otozoa (birds).[94] Oken developed his system further by including the sequence of development of the brain and other vital organs in animals. He published the result in 1833 in his *Allgemeine Naturgeschichte für alle Stände*, where he explained his theories in detail. The final arrangement is summarized by Breidbach and Ghiselin as follows:

A. Viscera animals
  I. Gut animals (infusorians, coelenterates etc.)
  II. Vessel animals (mollusks)
  III. Skin animals (worms and insects)
B. Flesh animals
  IV. Bone animals (fish)
  V. Muscle animals (amphibians including reptiles)
  VI. Nerve animals (birds)
C. Sense animals
  VII. Sense animals (mammals)[95]

Lower levels were similarly divided into three or five groups. Unsurprisingly, his classification was received with skepticism. Nevertheless, the main preoccupation of *Naturphilosophen*, quinarians, Temminck, Strickland, and every other naturalist remained the same: the search for a method that would bring order

---

92 Stresemann, *Ornithology*, chap. 10; Farber, *Emergence of Ornithology*, chap. 7.
93 Breidbach and Ghiselin, "Lorenz Oken and 'Naturphilosophie'," 220.
94 Oken, *Allgemeine Naturgeschichte* and *Grundriss der Naturphilosophie*.
95 Breidbach and Ghiselin, "Lorenz Oken and 'Naturphilosophie'," 228.

to this ever-growing number of species, which pointed towards an underlying natural order.

If anything, the lack of agreement on a philosophy of natural history had brought to light that the existing knowledge was not enough to fully comprehend the relations between taxa. In 1840 Strickland wrote: "we must first prove the existence of a real *natural system*, a subject which involves an enquiry into the designs of creative power, one of the most awful themes which the human intellect can attempt."[96] Following the failures of the classification systems based on natural philosophies like Quinarianism and *Naturphilosophie*, a general feeling of disappointment and a good dose of prudence may have prevented naturalists from developing a theoretical foundation for classification. For most naturalists, the only reasonable thing to do was to keep studying nature, accumulating facts and data, and improving their systems. The search for universal principles as promoted by Swainson in 1835 was abandoned towards mid-century, and most publications then focused on the description of new material and its arrangement within the best-known classification systems, building on those of Latham, Illiger, and Temminck.

Temminck's dismissal of theories was not unique. In the Netherlands, the empiricist tradition was particularly important in the nineteenth century. At the time Dutch scientists, including natural historians, were committed to applied and practical sciences, and few of them pursued the philosophical questions of their particular fields.[97] There was, however, a general feeling of hope in zoology that the laws of nature could, indeed, be discovered. Temminck pragmatically advocated patience and hard work: "In the present state of science, and surrounded as we are by this great wealth of newly discovered objects, it seems preferable to limit our research, for the time being, to the exact knowledge of the species, and to identify them carefully, in so far as this is possible."[98]

5 The 'Parliamentary Practice'

From the turn of the nineteenth century onward, natural historians spent much time and energy discussing nomenclature, the use of sections and genera, and the validity of their classification systems. These technical questions

96 Strickland, "Observations upon Affinities and Analogies," 220 (emphasis original).

97 Van Berkel, Van Helden, and Palm, *History of Science in the Netherlands*; Theunissen, *Nut en nog eens nut.*

98 Temminck, "Introduction troisième partie," in *Manuel d'ornithologie*, 2nd ed., vol. 3, xviii.

mirrored their views on methods and their reluctance to compromise. Their goal was the same; their methods, however, were not. Approaches and theories varied to such an extent that conflicts were unavoidable. To achieve any agreement was challenging, not least because there was no central authority to arbitrate the disputes. Even worse, each of them considered himself to be the authority on the matter. As Peter Stevens puts it: "One signal that all was not well was the unending proliferation of new systems and the failure to develop any way of evaluating characters and groups other than by trying to achieve a consensus with present and past masters of the art."[99]

In Britain, the polarization between reformers like Vigors and the more conservative naturalists like Strickland accentuated the situation. Newly created scientific journals allowed naturalists to publish their views – provided they were in line with those of the editors.[100] The 1837 issue of the *Magazine of Natural History* included eight articles exclusively devoted to discussing the issues of nomenclature and the status of new genera, including Hugh Strickland's first set of rules of nomenclature. A year later, the *Magazine* included a series of articles by Strickland and the Irish naturalist William Ogilby in a back-and-forth discussion of these rules. Their tone was harsh and personal, and in all probability, raised more than a few eyebrows.[101] Ogilby summarized the matter as follows:

> In short, these 'Codes for Nomenclature' put me in mind of nothing so much as Mrs. Malaprop's account of the "laws of CONFUSION, the Chinese Philosopher, 'Where you feel like a needle going astray, *With its one eye out*, through a bundle of hay'."[102]

Entertaining as this exchange may be, it also reflects the difficulties of regulating zoological nomenclature. The times when a Linnaeus or a Cuvier could impose his views for decades by using his social and scientific prestige were over.

---

99 Stevens, *Development of Biological Systematics*, 265.

100 McOuat, "Species, Rules and Meaning."

101 Ogilby, "Further Observations on 'Rules for Nomenclature'"; "Letter in Reference to Mr. Strickland's Observations"; "Observations on 'Rules for Nomenclature'"; Strickland, "A Few Words of Explanation in Reference to Mr. Ogilby's Letter"; "Remarks on Mr. Ogilby's 'Further Observations'"; "Reply to Mr. Ogilby's 'Observations'."

102 Ogilby, "Further Observations on 'Rules for Nomenclature'," 284 (italics original). Mrs. Malaprop is a character from Richard Sheridan's 1775 play *The Rivals*. The character became known by her verbal blunders, or malapropisms.

The process through which an agreed-upon code of nomenclature eventually emerged exemplifies this shift. Strickland's first set of rules for nomenclature, published in 1837, received almost exclusively negative reactions. The main reason for this was that the rules were laid down by one man and, therefore, had no social or political weight. On his own, Strickland did not carry enough scientific authority. In 1841 he suggested to the president of British Association for the Advancement of Science (BAAS), John Phillips, the appointment of a committee to work on the problem of zoological nomenclature, following "the parliamentary practice," by which the committee's proposal would be reviewed and approved as if it were a law.[103] This committee, consisting of Hugh Strickland, Charles Darwin, Sir Richard Owen, the Reverend Leonard Jenyns, and four other distinguished naturalists, presented a set of rules to the BAAS on June 27, 1842.[104] It was published in the following year by the BAAS and has been known since as the 'Strickland Code' or 'Strickland's Rules.'

The rules spread rapidly. They were published in several journals and translated into half a dozen languages. However, this does not mean that they were universally accepted. The rules were discussed, criticized, and altered. Finally, in 1895, the Strickland Code became the foundation for the modern International Code of Zoological Nomenclature.[105] Strickland's parliamentary modus operandi in zoology was an appropriate metaphor for the situation at the time. An individual man no longer had sufficient authority to determine rules of nomenclature or any other kind without reaching a consensus with his fellow naturalists. The Strickland Code was supported by peers and the BAAS, and soon re-published in the *Annals and Magazine of Natural History* and the *Philosophical Magazine*. The code was translated, circulated, discussed, and eventually accepted. In their report on the Rules of Nomenclature of 1842, the members of the committee depicted the situation accurately:

> By emanating from the British Association, it is hoped that the proposed rules will be invested with an authority which no individual zoologist, however eminent, could confer on them. The world of science is no longer a monarchy, obedient to the ordinances, however just, of an Aristotle or a Linnaeus. She has now assumed the form of a republic, and although

103 Strickland to J. Phillips, 20 September 1841; quoted in McOuat, "Species, Rules and Meaning," 506.

104 Strickland et al., "Report of a Committee."

105 McOuat, "Species, Rules and Meaning," 512; Melville, *Towards Stability*, 10. For more information on the gestation and evolution of the Strickland's code of zoological nomenclature, see also Knapp et al., "Stability or Stasis"; Rookmaaker, "Early Endeavours by Hugh Edwin Strickland."

> this revolution may have increased the vigour and zeal of her followers, yet it has destroyed much of her former order and regularity of government. The latter can only be restored by framing such laws as shall be based in reason and sanctioned by the approval of men of science; and it is to the preparation of these laws that the Zoological Section of the Association have been invited to give their aid.[106]

Natural history was not a monarchy or autocracy anymore; agreement had to be reached by consensus and, even more difficult, by making concessions. This was unfortunate for Temminck, to say the least.

Temminck was born in 1778, into a privileged patrician family in Amsterdam, which was a long line of wealthy and influential men and women with ties to the Dutch government, the VOC, and the WIC. Temminck never received a formal education in a school; he received private education at home. He had powerful friends in politics who helped him realize his dream of a national natural history museum and made him the head of the new institute. He had looked up to the Comte de Buffon and Georges Cuvier, from whom he had learned, among many other things, that an individual man could represent a scientific authority.[107] Temminck saw natural history as an oligarchy, the domain of a few competent naturalists whose achievements would be remembered forever:

> It is particularly the eloquent writings of Buffon that have given new appeal to this pleasant science, sounding a wake-up call around the world; and the order and the harmony in classification from the great Linnaeus, have equally helped increase the number of enthusiastic amateurs; illustrious scholars, who by following the writings of these renowned men, have earned the glory of seeing their names inscribed in the temple of memory.[108]

Temminck perceived himself as a member of this distinguished elite, partly because of his absolute confidence in his approach to systematics. He was particularly proud of the fact that he had based his classification systems on the examination of thousands and thousands of specimens from nearly all cabinets in Europe. This is not to say that Temminck was advocating objectivity in its modern sense; he was claiming authority precisely because of personal talents: his scientific status was the result of his experience and views. Daston

---

106 Strickland et al., "Report of a Committee," 107–8.
107 Temminck, "Avant-propos," in *Coup-d'oeil général*, vol. 1, ix.
108 Temminck, "Avant-propos," in *Manuel d'ornithologie*, vii.

and Galison have coined the term 'truth-to-nature' for the standard upheld by late-eighteenth- and early-nineteenth-century naturalists who, like Temminck, believed that hidden truths could only be uncovered by observation and by the particular skills of the experienced observer: "To see like a naturalist required more than just sharp senses: a capacious memory, the ability to analyze and synthetize impressions, as well as the patience and talent to extract the typical from the storehouse of natural particulars, were all key qualifications."[109] Temminck himself could have written this passage.

In the early 1830s Temminck ranked amongst the most celebrated ornithologists of his time. William Swainson declared that "of general systems [...], the most popular and esteemed are those of Illiger, Cuvier, Vieillot and Temminck."[110] He commended both the *Histoire naturelle des pigeons et gallinacés* and the *Manuel d'ornithologie* as excellent, exemplary works.[111] Neville Wood, in his annotated bibliography on ornithology, wrote about the *Manuel*: "Temminck is the first Ornithologist since the time of Linnæus, who promulgated a system worth attending to. In simplicity it almost rivals that of the illustrious Swede, and is infinitely its superior in exactitude and preciseness."[112] Wood's praise, however, should be taken with a pinch of salt. He was a controversial figure among ornithologists because of his caustic criticisms, excessive praise, and a tendency to change nearly every common (English) bird name known at the time.[113]

The *Manuel* rapidly became a standard work for other ornithologists and provided a basis for others to build upon. In recognition of his achievement, Temminck received his second honorary doctorate from the University of Jena in 1828 (the University of Groningen had awarded him his first honorary doctorate in 1819). Following its success, Temminck continued to work on his system and prepared a second edition of the *Manuel*, which he finally finished in 1840. Strickland wrote about the "long-expected supplements": "Although the author hesitates too much in adopting the generic groups of modern science, and does not sufficiently value the law of priority in nomenclature, yet the exactness of his descriptions and the general soundness of his criticisms will long render his work a valuable hand-book of European ornithology."[114]

Temminck appreciated some of the criticism he received, as long as he considered it constructive: "Concerning science, a work without criticism is

---

109 Daston and Galison, *Objectivity*, 58.

110 Swainson, *Natural History and Classification of Birds*, vol. 1, 200.

111 Swainson, 205–06, 208.

112 Wood, *Ornithologist's Text-Book*, 28–29.

113 Birkhead and Montgomerie, "Vile Passion for Altering Names."

114 Strickland, "Report on the Progress," 180.

like a meal without salt."[115] Indeed, Heinrich Boie's comments on the first edition of the *Manuel* had established contact between the two men and laid the basis for their friendship, as well as for Boie's career. The criticism of others, including Vieillot and Vigors, however, was harder to digest. Temminck considered theirs to be personal attacks on his judgment and authority. Remarkably, although Vigors criticized Temminck's genera, his fellow-quinarian Swainson praised them: "[Temminck's] merits in the arrangement of his generic groups, and the high finish he has bestowed upon them, have given to his system a prevalence and popularity above all others which have appeared since the days of Linnæus; next to whom, as an ornithologist, he assuredly ranks."[116]

The first few livraisons of his *Monographies de mammalogie* were also celebrated, mainly for their accuracy, clear descriptions, and nomenclature.[117] Initially at least, Temminck's meticulousness was much appreciated by his peers. In 1826 a review of his *Monographies* read:

> [W]hatever falls from the pen of so sedulous an enquirer, is well worthy of attention, founded as his facts universally are on the most patient and laborious investigation. That such has been the case on the present occasion, is proved, by the references to the various collections of Europe, nearly the whole of which have been visited in the progress of his work. Nor are the descriptions founded only on living or on set-up specimens; to trace them with more accuracy and to obtain a more ample view of their frequent variations, he has also recourse to the warehouses of furriers in all the principal commercial towns, without a continual examination of which, he repeats again and again, no certainty can exist with respect to the species of *Felis*.[118]

Another reviewer was equally impressed with the first three installments of the *Monographies*, and particularly with Temminck's treatment of the genus *Didelphis* (the large American opossum): "The synonymy of these species, which has hitherto been extremely confused, has been unravelled by M. Temminck with his usual sagacity and assiduity."[119]

After 1840 things changed drastically. The final installments of the *Nouveau recueil* were received with far less enthusiasm.[120] Swainson, who had praised

115 Temminck, "Introduction troisième partie," *Manuel d'ornithologie*, 2nd ed., vol. 3, lxii.
116 Swainson, *Treatise on the Geography*, 185.
117 "Analytical Notices of Books"; Drapiez, "Manuel d'ornithologie."
118 "Analytical Notices of Books," 533–34.
119 "Monographies de Mammalogie," in *The Zoological Journal*, 575.
120 See, for example "Bibliographical Notices," 267.

Temminck's earlier works, was not particularly impressed. He thought the descriptions meager, the illustrations stiff and unnatural, and deplored the lack of synonyms, specific characters, and descriptions of habitats. Overall, he wrote, "this work [is] far inferior to what it might have been, and what the scientific world expected, from the reputation of its authors."[121] The monographs on mammals received a similar response. Temminck was re-describing species already known, and this damaged his authority. Charles Bonaparte's criticism of Temminck's treatment of vesper bats reflects this change: "to correct the errors of eminent writers is the most effectual means to advance science, while the great veneration which is entertained for Temminck, and his *ex cathedrâ* tone, may in some cases prove fatal to truth."[122]

The fact is that Temminck's works on mammals were not lauded as much as his earlier works on birds. John Edward Gray (the older brother of George Robert Gray and the one who named Temminck's Tragopan) was particularly harsh in his criticism. He published a revision of the group of civet cats using specimens kept in the British Museum in 1864, and also summarized his ideas about Temminck in this rather condescending paragraph:

> M. Temminck was an eminent ornithologist, and has studied some groups of Mammalia, perhaps not with so much success. He was an amiable naturalist, but has carried his political anglophobia (so well seen in his 'Essay on the Dutch Colonial Possessions') into his zoological studies. This blinded him to the labours of the zoologists of this country, the richness of our collection, and thus rendered his observations in regard to their work not worthy of attention, as they otherwise might have been. It is to be observed that he never had a regular scientific training, never attempted to form scientific specific characters, and is rather to be regarded as a patron and amateur than as a scientific zoologist. He was the first in his country, as the late Earl of Derby was in this.[123]

However, if Temminck's lack of academic training did not prevent him from becoming "an eminent ornithologist," there must have been other reasons for the inferior quality of Temminck's studies of mammals. Whether Temminck's anglophobia – a condition he most certainly suffered from – caused him to

121 Swainson, *Natural History and Classification of Birds*, vol. 1, 222.

122 Bonaparte, "Observations of the State of Zoology," 22 (italics original).

123 Gray, "Revision of the Genera and Species of Viverrine Animals," 505. Gray was referring here to Edward Smith Stanley, thirteenth earl of Derby and Lord Stanley (1775–1851), politician, naturalist, president of the Linnean Society and Trustee of the British Museum.

ignore the work of British zoologists is difficult to prove, but it may very well have been the case.

Temminck seemed to neglect a large part of the British zoological works, perhaps in part because of his antipathy towards England. These feelings had been ignited by Raffles' role in the conquest of Java during the Napoleonic Wars and made worse afterward by Raffles' criticisms of Dutch rule in the 'East Indies' in his *History of Java*. When Temminck took up the task of re-writing the same history for his *Coup-d'oeil général sur les possessions néerlandaises dans l'Inde Archipélagique* (the book Gray referred to), he gave a very different account about nearly every single episode – before, during, and after the period of British rule.[124] His views on the administration of the British Museum – he was baffled by the composition of its Board of Trustees – and the fact that English naturalists published exclusively in English added fuel to the fire.[125]

Britain made a rather late entry into the field of systematics, which had been dominated by continental European naturalists until the 1830s. When British naturalists engaged with systematics, they changed the course of ornithology and, with it, Temminck's prestige. As the rest of Europe was recovering from the Napoleonic wars and other political and economic difficulties, natural history flourished in Britain. By the 1840s, the research at and collections in the British Museum of Natural History were challenging those at both the Muséum National d'Histoire Naturelle in Paris and 's Rijks Museum in Leiden.[126] As the British Museum expanded, English naturalists caught up with their French, German, and Dutch counterparts. By then Temminck was struggling for government funds and spent much time on administrative tasks as well as fighting the increasing tremor in his hands, so that he had practically abandoned ornithology. These were hard times, both for 's Rijks Museum and for Temminck's scientific prestige.

Temminck found himself in the middle of a transition within natural history, and as a consequence, his authority was widely recognized at first but challenged later. Temminck could not grasp this change. The confrontations with his peers exhausted and exasperated him. In 1846, at sixty-eight, he wrote to his colleague and friend Martin Heinrich Karl Lichtenstein about his disappointments:

> The new era is getting beyond me, and however much one reads one feels that it is impossible to make progress, and one seems to oneself quite

124 Temminck, *Coup-d'oeil général*, vol. 1, "Avant-propos" and chap. 1.
125 Gordin, *Scientific Babel*.
126 Limoges, "The Development of the Muséum"; McOuat, "Cataloguing Power."

FIGURE 34 *Portret van Coenraad Jacob Temminck, directeur van het Rijksmuseum van Natuurlijke Historie*. Lithograph by Johannes Christiaan d'Arnaud Gerkens, ca. 1850–1856
LEIDEN UNIVERSITY LIBRARIES, DIGITAL COLLECTIONS, COLLECTION J. T. BODEL NIJENHUIS, BN 1379

> stupid and ignorant in the new jargon of systematics and nomenclature, or at least too weak to swim against the stream, and through a sort of disgust frightened away from even attempting any sort of opposition. There is something revolutionary in every contemporary activity; we shall not live to see the settling of this ferment, but I cannot say that I cherish any great expectations of the final result, and I rejoice therefore that I have lived in a time when there were still scientific authorities and when everything proceeded more tidily.[127]

Indeed, the number of books, journals, societies, and collections that needed to be processed had increased rapidly since 1840. Temminck, it appears, could not keep up with this fast pace and was appalled by what he perceived as a lack of deference towards scientific authorities. He lost both his political influence in the Dutch government and his prestige in ornithology. Naturalists were now quick to dispute the great experts of their time, and they had a vehicle for their praise and criticism: the new journals and magazines on natural history. The time of the monarchs, as Strickland put it, was over.

127 Temminck to Hinrich Lichtenstein, 27 July 1846; quoted and translated in Stresemann, *Ornithology*, 150–51.

CHAPTER 7

Systematics and Natural History: 1800–1850

While naturalists were occupied pursuing the ever-elusive natural system and establishing the foundations of a meritocracy, systematics became an increasingly narrow field of study. It had its goals, place, methods, questions, and a growing community of practitioners. And with specialization, the boundaries between the different fields of natural history started to materialize. Natural history fragmented into incipient but recognizable subdisciplines, including comparative anatomy, systematics, geology, Humboldt's proto-ecology, physiology, phytogeography, mineralogy, and climatology. The mainly descriptive and ahistorical natural history of the eighteenth century evolved rapidly into a practically unrecognizable discipline by the 1850s. By then, 'natural history' and 'classification' were no longer synonymous. However, there was no consensus on what actually constituted natural history or which fields it included. Also, the disciplinary hierarchy was far from being settled, which had profound consequences for naturalists, their communication platforms, their institutions, and educational programs, and encouraged competition for both prestige and resources. The fact that classification had been naturalists' primary purpose for centuries did not ensure its continuation in the twentieth century. Furthermore, the position of systematics among all other fields in the natural sciences was as controversial as the standardization of classification itself.

1 Defining 'Natural History'

When Temminck published his first monograph on pigeons and gallinaceous birds in 1813, he justified the time and effort that had gone into his work with religious and utilitarian arguments. Temminck claimed that the study of natural history, or more precisely, the "study of the three realms of nature," not only brought joy and inner peace to its practitioner, but also brought the practitioner closer to the Almighty.[1] He also listed the benefits of natural history for man and society. For example, it allowed man to know himself and the workings of his internal organs, it made the domestication of various species of

1 Temminck, "Discours préliminaire," in *Pigeons et gallinacés*, vol. 1, 7–9.

DOI:10.1163/9789004441491_009

animals for profit and pleasure possible, it helped in finding fertile grounds for crops, and to circumnavigate the globe.[2]

These kinds of spiritual and utilitarian considerations were common in books of the late eighteenth and early nineteenth centuries, but Temminck would never again write an introduction like this. Just two years later, without including a word on the benefits of ornithology for the human spirit or society, Temminck presented his *Manuel* essentially as what it was: a theoretical, systematic work on birds, intended to increase the knowledge of the European avifauna. In the introduction, he explained that he aimed to correct errors and fill gaps in the existing works on the subject, introduce order to the nomenclatural chaos, and improve the description of genera; and then, in a very straightforward manner, he offered an index of the relevant genera. Temminck had dispensed with the philosophical preambles, but highlighted the utility of natural history for other reasons: "We seem to forget that the study of nature, apart from its intrinsic importance from the moral and scientific point of view, greatly influences civilization and industry, driving its springs, many of the arts and the propagation of enlightenment to all classes of society."[3]

What did Temminck mean by 'natural history'? In the *Histoire naturelle générale des pigeons et des gallinacés* he noted that natural history comprised anatomy, systematics, applied botany, physiology, geology, and geography, but not astronomy.[4] A quick and by no means exhaustive look at some other definitions from this period, i.e. from 1790 to 1840, reveals that natural history was not as well-defined a concept as its widespread use may suggest. Naturalists disagreed on what natural history was, which branches of the natural sciences belonged to it, and on whether or not it was part of their work to delve into philosophical considerations.

In 1817 Georges Cuvier expressed his concern about this lack of an agreed definition of natural history and felt the need to clarify: "[Since] few people have an accurate idea of natural history, it seems necessary to start our work by clearly defining the goals of this science and establishing the distinct limits between natural history and its related sciences."[5] He defined natural history, or "physique particulière" (particular physics), as the study of natural objects, including minerals and living beings, while the other natural sciences belonged to the "physique générale" (or general physics, including physics,

2 Temminck, 13–16.
3 Temminck, "Introduction troisième partie," in *Manuel d'ornithologie*, 2nd ed., vol. 3, xxxvi–xxxvii.
4 Temminck, "Discours préliminaire," in *Pigeons et gallinacés*, vol. 1, 7.
5 Cuvier, "Introduction," in *Règne animal*, vol. 1, 1.

optics, astronomy, and meteorology), and chemistry.[6] All these branches of the natural sciences pursued the same goal, that is, the discovery of the general laws governing nature, but each employed different methods. Cuvier's classification of the natural sciences appears almost verbatim in other textbooks from the beginning of the nineteenth century, like John Stark's *Elements of Natural History* or Jan van der Hoeven's *Handboek der dierkunde*.[7]

Jean-Baptiste de Lamarck had similar views on the scope of natural history. For Lamarck, natural history was the branch of the natural sciences that pursued "the general and detailed knowledge of our globe and all its products born or made on its surface and its interior."[8] He excluded physics, chemistry, and astronomy from natural history, which he divided into three large areas: zoology, botany, and mineralogy. For the first two, and thus the study of living beings, Lamarck introduced the term 'biology,' very much in the same sense in which we understand it today.[9] Zoology encompassed comparative anatomy, physiology, the description of life histories, geographical distribution, and classification.[10] Cuvier and Lamarck agreed, if on nothing else, that in their methodology, physics, optics, and chemistry were sciences based on experimentation and mathematical calculations. By contrast, natural history was based entirely on observation and comparison, while still making use of the principles governing the rest of the natural sciences.[11]

The entomologist William Sharp Macleay had a slightly different view on the matter. He questioned whether natural history was a science of observation alone, as it also depended on experience, and stated that chemistry could not be considered a science of experience alone, since it also required observation. He wondered whether mineralogy should not be part of chemistry rather than natural history. Most mineralogical discoveries depended on calculation, so that mineralogy was more closely related to chemistry than to natural history.[12]

In yet another interpretation of the disciplinary boundaries, the botanist Augustin P. de Candolle considered natural history part of the experimental sciences, like physics and chemistry, which were all intimately linked and

6 Cuvier, 2.
7 Stark, "Introduction," in *Elements of Natural History*, vol. 1, 2; Van der Hoeven, "Inleiding," in *Handboek der dierkunde*, vol. 1, 3.
8 Lamarck, "Sur l'histoire naturelle en général," 3–4.
9 Lamarck, *Hydrogéologie*, 188.
10 Lamarck, "Discours préliminaire," in *Philosophie zoologique*, 1–2.
11 Lamarck, "Sur l'histoire naturelle en général," 8–9; Cuvier, "Introduction," in *Règne animal*, vol. 1, 4, 7.
12 Macleay, *Horae Entomologicae*, vol. 1, 174–75.

depended on each other's progress.[13] For de Candolle, astronomy, geology, mineralogy, and meteorology were all branches of natural history that investigated inanimate matter, while zoology and botany were the branches studying living beings. He also explained that, due to the differences between animals and plants, anatomy played the most important role in zoology, while botany had to settle for the study of the species' external characters and geographical distribution. De Candolle divided botany into three main subdisciplines: "Glossologie," which dealt with the terminology of plant parts; "Taxonomie," the theory of classification; and "Phytographie," the "art of describing plants in the way most useful to the progress of science." "Botanique descriptive" and "Synonymie Botanique" were applied branches of phytography.[14] De Candolle also considered vegetal physiology and pathology to be branches of botany, as well as all applied disciplines like agriculture and botanical pharmacy. All these disciplines were, as Lamarck and Cuvier remarked, related to each other.[15]

The study of humans posed similar problems. Jan van der Hoeven published, among many other things, a series of articles on the natural history of man. In the first of these articles, he warned the reader: "*Anthropology* should not be confused, as sometimes happens, with the natural history of man." The former, he explained, was a wider field, and it included the latter, and the natural history of man was an "anthropological science," but was not anthropology itself.[16] According to Van der Hoeven the natural history of humankind dealt with two issues: first, with the differences between man and all other animals to determine the position of humankind among all living things, and second, with the differences "between different peoples, paying particular attention to the different tribes of man."[17] In order to address the first issue, comparative anatomy and physiology were essential, while the second required the study of ethnic groups.

The term 'naturalist' was as elusive as that of natural history. In 1821 Macleay expressed concern about the ease with which some collectors adopted the title of 'naturalist':

> Naturalists, it is said, may be classed like the objects of their study into genera and species; and in this classification places may be found for the comparative anatomist and physiologist down to the mere collector who

13 Candolle, "Introduction," in *Théorie élémentaire*, 2.
14 Candolle, 19.
15 Candolle, 8–9, 16–19.
16 Van der Hoeven, "Bijdragen tot de natuurlijke geschiedenis van den mensch," 87 (italics original).
17 Van der Hoeven, 88–89.

> hoards a shell or pebble, simply that he may be gratified in the possession of that which his neighbour wants. The variety of pursuits embraced in the comprehensive term Natural History, is without doubt multifarious; but it may be questioned whether the title of *naturalist* be merited by the cultivator of any one or two of these pursuits, or indeed by any person who deems even the most ignoble of them unworthy of his attention.[18]

Temminck definitely considered himself a naturalist. But of what kind? He had evolved from a "mere collector" to a systematist. Perhaps the difficulty of defining a 'naturalist' lay not so much in semantics but in the assumption that the role was the same for each practitioner of natural history. However, a definition did not need to be too complicated. For Temminck, a naturalist was an "observateur de la nature" (nature's observer) – a definition that was not very precise, but undeniably inclusive.[19]

Temminck was not as keen on definitions as some academics were. His books were not student textbooks. Since he wrote for fellow naturalists who worked on the classification of birds, he did not offer any classical definitions of the terms he employed, i.e. the kinds of definitions found in the writings of Cuvier or Van der Hoeven. For these professors, who held appointments in academic institutions, explications were mandatory. Temminck had no teaching obligations – in fact, he did not wish to have any – and the museum did not take part in the activities at Leiden University. Quite possibly, defining natural history or explaining classification methods at length seemed to him a superfluous exercise.

These few and somewhat random examples indicate that while most early-nineteenth-century naturalists agreed roughly on the question of what natural history was, the concept was not definite. In Stevens words: "There is no monolithic natural history in the nineteenth century."[20] For most naturalists, natural history was a branch of the natural sciences that included zoology, botany, and mineralogy.

In addition to the scope and arrangement of each branch of the natural sciences, there were three other points of discussion that proved to be much more significant for the practice of natural history: should natural history include philosophical or theoretical discourse? What was the status of anatomy and physiology in zoology? And lastly, what was the status of systematics in relation to the other branches of natural history?

---

18 Macleay, *Horae Entomologicae*, vol. 1, 327 (italics original).

19 Temminck, "Introduction troisième partie," *Manuel d'ornithologie*, 2nd ed., vol. 3, lxiv.

20 Stevens, *Development of Biological Systematics*, 218.

2 The Issue of Philosophical Arguments

The inclusion of philosophical considerations in natural history was, by far, the most difficult of the three questions. From Lamarck's and Geoffroy Saint-Hilaire's ideas on organic evolution to the *Naturphilosophen* and the Quinarians, there was a colorful array of theoretical proposals to choose from. Although locally popular at some point or another, none of them were to last, and none of them provided naturalists with a framework that was both predictive and generally applicable in both zoology and botany. As a result, systematists grew increasingly weary of *a priori* theories. Macleay admitted as much when presenting what would be known as the Quinarian System: "This part of my work [...] takes more the character of an hypothesis, and as such deserves more suspicion." He was quick to clarify that this was not a hypothesis in the 'Newtonian sense,' but one "entirely dependent upon observed facts which its object is to connect." He added, however, that there were just not enough observations available yet to move beyond the stage of a hypothesis.[21]

Cuvier's theory of the 'conditions for existence' and the principle of the correlation of parts might be the most striking exception to the systematists' general rejection of *a priori* theories. Interestingly, his theories never came to be perceived as philosophical, for several reasons. Like Macleay, Cuvier claimed that his findings were based on observation and comparison. Moreover, Cuvier's ideas provided systematists with a strong starting point for the classification of animals which, when applied to the higher taxonomic categories, gave classifications stability and predictability, and indicated that these categories reflected the organization of nature. Finally, his opposition to organic evolution, as proposed by Lamarck and Geoffroy Saint-Hilaire, reaffirmed Cuvier's image as someone opposed to speculation.[22] In general, while not entirely unchallenged, the validity of Cuvier's theories remained largely undisputed in zoology.

Lamarck considered the search for universal laws an integral part of natural history. So did Cuvier. But according to Lamarck, natural historians should also inquire into the causal principles behind forms and the relations between living beings and their environment. He believed that most naturalists neglected to do this:

---

21 Macleay, *Horae Entomologicae*, vol. 1, 167. Macleay quoted Newton: "Quicquid [enim] ex phaenomenis non deducitur, hypothesis vocanda est" (Whatever is not deduced from the phenomena, has to be called an hypothesis.) See Thomson and Blackburn, *Sir Isaac Newton's Principia*, 530.

22 Farber, *Emergence of Ornithology*, 81; Stevens, *Development of Biological Systematics*, 197–209.

> It would be a real hindrance for the natural sciences to persist in seeing only the observed objects, their form, their size, their external parts, even the smallest, their color, etc., and if those dedicated to such studies refrain from rising to higher considerations, such as the investigation of the nature of the objects they study, the causes of the modifications and variations to which these objects are subject, the relations of these same objects among themselves, and with all the others that we know, etc.[23]

Lamarck thus considered both zoology and botany to include two closely dependent fields: 'natural philosophy,' which was devoted to the development and discussion of the principles of biology, and 'observations,' i.e. the descriptive and classificatory part, which was based on observation and comparison.[24] Although some naturalists agreed with Lamarck on the need for a philosophical basis for natural history, his *Philosophie zoologique* and his transmutation theories were controversial and, ultimately, ridiculed. Cuvier himself considered Lamarck's incursions into philosophy whimsical at best: "A system established on such foundations may amuse the imagination of a poet; a metaphysician may derive from it an entirely new series of systems; but it cannot for a moment bear the examination of any one who has dissected a hand, a viscus, or even a feather."[25] The Cuvierian view endured.

Due to the severe criticisms of the theories proposed during this period (Lamarckism, numerical classifications, *Naturphilosophie*), systematists cast aside any theoretical approaches to classification and focused on a more empirical approach.[26] Amongst them was Temminck, who despised any philosophically tinted theories.[27] Jan van der Hoeven wholeheartedly agreed with Temminck on this point. He noticed that the human mind tended to look for order and resorted to hypotheses, too impatient to gather the necessary facts first. Science should be based on accurate and comprehensive knowledge, and then "the true theory will follow it, without anybody knowing how and when, and the solid building of science will be erected on the broad and firm basis of experience, to remain standing when the fairy castles of systems and opinions have evaporated and disappeared from memory."[28]

---

23 Lamarck, "Discours préliminaire," in *Philosophie zoologique*, 12–13.

24 Lamarck, "Sur l'histoire naturelle en général," 8–9.

25 Cuvier, "Biographical Memoir of M. de Lamarck," 15.

26 Farber, *Emergence of Ornithology*, 140–41; Mayr, *Growth of Biological Thought*, 203; Stevens, *Development of Biological Systematics*, 222.

27 Temminck, "Introduction troisième partie," in *Manuel d'ornithologie*, 2nd ed., vol. 3, xv, in response to Quinarianism.

28 Van der Hoeven, "Bijdragen tot de natuurlijke geschiedenis van den mensch," 91.

Karl Illiger thought along similar lines. He pointed out the necessity of enlarging the bulk of described species, which in his view was the only means available for confirming or rejecting the existing systems, "for we can never have enough observations if we want to use them as bases for entirely reliable conclusions."[29] Remarkably, the departing point of quinarians like Macleay was quite similar, or so Macleay claimed:

> In natural philosophy a system has usually indeed been considered as synonymous with an hypothesis; but the two ideas expressed by these words have of late been very properly distinguished by observing that though a mere fiction or hypothesis may explain phænomena, yet a system is a certainty that must be deduced by these.[30]

Macleay did not consider his own quinarian system a hypothesis, but the result of the discovery of the elusory natural system – after all, he noted, "the strength of this beautiful theory" was irrefutable because "different persons have respectively stumbled upon it in totally distinct departments of the creation."[31]

Most naturalists were not in the habit of explaining their views in extended chapters or articles.[32] Hugh Strickland is among the few exceptions, since he devoted essays to the species and genus concepts, helped define terms like 'affinity' and 'analogy,' and wrote on the methods to achieve a natural classification in ornithology.[33] Like Strickland, the botanist Joseph Hooker explained the classificatory principles of his works mostly in essays that preceded his botanical works.[34] He stressed the need of explaining the systematists' theoretical views on the origin, variation, and dispersion of species, as "the actual amount of knowledge we possess on such subjects is so very limited, that few experienced naturalists are inclined to pronounce positively upon them, whilst the majority offer no opinion at all."[35]

29 Illiger, "Einige Gedanken über die Begriffe," xxxiii; translated in Mayr, "Illiger and the Biological Species Concept," 173.

30 Macleay, *Horae Entomologicae*, vol. 1, x.

31 Macleay, "Remarks on the Identity of Certain General Laws," 63. For a discussion of numerical classification in botany, especially in Victorian Britain, see Endersby, "Classifying Sciences."

32 Mayr, "Illiger and the Biological Species Concept"; Farber, *Emergence of Ornithology*; Stevens, *Development of Biological Systematics*; Endersby, *Imperial Nature*.

33 Illiger, "Einige Gedanken über die Begriffe"; Strickland, "Observations upon Affinities and Analogies" and "True Method of Discovering the Natural System."

34 For example, Hooker, "Introductory Essay," in *Flora Novae-Zelandiae*, vol. 2; *Flora Indica*; *On the Flora of Australia*.

35 Hooker, "Introductory Essay," in *Flora Novae-Zelandiae*, vol. 2, i–ii.

Indeed, carefully articulated premises for classification, such as the ideas of the continuity or the fixity of species, were rare. The theories of naturalists from this period, with a few exceptions, have to be pieced together from their classification systems and the introductions to their works.[36] This failure to provide general explanations for natural diversity was intimately related to the general aversion towards any incursion into the realm of philosophy.

The line separating hypothetical conjectures from theoretical premises was rather indistinct during this period. Although Temminck emphatically defended the idea of an empirical, observation-based approach to systematics and rejected any *a priori* construction, he did embrace typological thinking as the basis for his genera and, by extension, as the foundation of his classificatory practice. His law on the distribution of animals around the globe was also born from the type concept. Temminck argued that his law emerged from an observation of the facts, but this was not strictly speaking the case. This was a hypothesis based on the idea of morphological types and extrapolation from a few observations. The Belgian botanist August Drapiez noted the strong philosophical background of Temminck's classification in the *Manuel d'ornithologie*: "The work we are reviewing here is, in this respect, beyond reproach, and the second edition, which closely followed the first, proves that to the present day, these studies have a truly philosophical purpose."[37]

Temminck had tried to marry the type concept to the idea of a continuous nature. He meticulously identified variations that deviated from the types within each group (whether it was an order, a genus, or a species) and then decided which type these variations resembled the most, classifying them accordingly. By 1840 Temminck had started doubting the possibility of delimiting genera and species in a clear-cut way, which led him to wonder if species as such even existed in nature, and he recognized the problems of definitions of genera based on a morphological type concept.[38] These problems had forced him to define the set of characters that should be used to delimit genera as precisely as possible; he also kept his genera as large as possible, i.e. the number of species they included. Nonetheless, he carried on, confident that the truth would reveal itself with time and patience, even when the sheer amount of forms and varieties being discovered each day undermined his theoretical framework.

---

36 See also Mayr, "Illiger and the Biological Species Concept," 167; Stevens, *Development of Biological Systematics*, 133.

37 Drapiez, "Manuel d'ornithologie," 276.

38 Temminck, "Introduction troisième partie," in *Manuel d'ornithologie*, 2nd ed., vol. 3, xix.

While Temminck was convinced of the possibility of achieving a natural classification by this method, others, such as Jussieu and Lamarck, concluded that continuity in nature implied necessarily and by definition that any classification was artificial.[39] Different starting points inevitably led to different definitions of what constituted a natural taxon, to a different methodology, and accordingly, to different classification systems. In any case, the idea of continuity was as much a hypothesis as were Temminck's morphological types.

In Britain, Hugh Strickland also openly declared his aversion towards philosophical theories in natural history. In 1844 he envisioned a time when "more definite principles of classification may hereafter be discovered, and meantime all that we can do is to arrange our systems according to sound reason and without theoretical prepossession."[40] Temminck would have emphatically agreed with Strickland's pragmatic approach. Yet, Strickland felt some sympathy for those taking the innovative road. He thought that the Quinarian system, however wrong, had contributed to the advancement of ornithology "in no small degree." Quinarians helped define the concepts of 'analogy' and 'affinity,' and had promoted fruitful discussions. He concluded: "Alchemy was the parent of chemistry, astrology of astronomy, and quinarianism has at least been one of the foster-parents of philosophical zoology."[41] Regarding Lorenz Oken and his animal classification based on the five senses of man, Strickland applauded his creativity: "This doctrine is far too fanciful to stand the test of common sense, but it is certainly very ingenious, and we may admit that *se non é vero é ben trovato*" (If it is not true it is well imagined).[42] However, Temminck could not bring himself to be amused by these ideas, while Hooker exclaimed: "I have been groaning over 'Rejuvenescence' que Diable! When is this German rubbish to end?"[43]

Strickland was ultimately, if knowingly, trapped in the same paradox as Temminck. While he claimed to avoid any metaphysical concepts, the foundation for his classification system was based on admittedly ill-defined concepts, including those of species and of the essential characters used to define species.

39 Lamarck, "Origine & histoire de l'établissement des genres," in *Encyclopédie méthodique: Botanique*, vol. 2, 630; Stevens, *Development of Biological Systematics*, 25–28.

40 Strickland, "Report on the Progress," 218.

41 Strickland, 172.

42 Strickland, 176 (italics and orthography original).

43 Hooker to Thomas Henry Huxley, 12 September 1854, in Huxley, *Life and Letters of Sir Joseph Dalton Hooker*, 425. Hooker had just read a translation of Alexander Braun's *Betrachtungen über die Erscheinung der Verjüngung in der Natur*, or the "Rejuvenescence of Plants." See also Endersby, *Imperial Nature*, 217.

In an article of 1841 confidently entitled "On the True Method of Discovering the Natural System in Zoology and Botany," Strickland remarked:

> It is probable that most naturalists at the present day have an instinctive belief in the existence of a natural system in Zoology and Botany, but there are very few who if questioned on the subject could give any clear explanation of the grounds of their belief, of the nature of that system, or of the mode by which a knowledge of it may be attained. The uncertainty which hangs over the subject is doubtless owing to the obscure and metaphysical nature of some of the principles involved, and still more to the vague conceptions and crude theories which have been promulgated on the subject.[44]

Immediately afterward he added: "The postulate with which I commence the inquiry is, to let it be granted that there are such things as *species*, distinct in their characters and permanent in their duration."[45] He then proceeded to define the natural system as an arrangement of species according to the resemblance of their essential characters. However, he did not define species or explain what made a character essential, except noting that the latter question was linked to an estimate to the physiological importance of the characters in a Cuvierian sense.

Even though zoological and botanical classification were necessarily based on a set of underlying, and often unquestioned, assumptions, most naturalists thought of natural philosophy and systematics as separate fields. Cuvier's views were universally accepted as empirically proven, with the exception, possibly, of catastrophism. With the arrival of Lyell's *Principles of Geology*, published in three volumes between 1830 and 1833, geology launched itself into the theoretical, speculative arena after decades of controversy, during which geologists had sought to avoid discussions by focusing mostly on stratigraphy. Lyell popularized gradualism, a theory opposed to Cuvier's catastrophism and that proposed that slow, gradual geological processes accounted for all the changes the earth had suffered up to the present moment, including the extinction of fossil forms. According to James Secord, the *Principles* was "the most thoroughgoing vindication of geology as a sophisticated and philosophical enquiry."[46]

---

44 Strickland, "True Method of Discovering the Natural System," 184.

45 Strickland, 184 (italics original).

46 Secord, *Visions of Science Books*, 144; for an account of these and other controversies in the history of nineteenth-century geology, see Hallam, *Great Geological Controversies*.

Botany was in a similar predicament as zoology and geology. Hooker fervently hoped that new ideas would arrive to help botany become a more accurate and methodical science. In his 1856 review of de Candolle's *Géographie botanique* he lamented "the vagueness of its [i.e. botanical geography] principles, the inexactness of its methods, the puzzling complexity of its phenomena, and the purely speculative character of those hypotheses upon which all inquirers base their efforts to explain its facts and develop its laws."[47] Until these shortcomings in botany were overcome, Hooker would mainly work on defining the principles for his classification system, focusing mostly on genera, orders, and classes.[48]

Classification was not atheoretical, but naturalists could not agree on which theoretical premises were universal and true. They were occupied with the search for a widely accepted philosophical foundation for classification for at least a century.[49] Temminck was convinced that by elaborating on Cuvier's methods and enlarging and refining classifications, one could finally build the natural system, the perfect classification with objective groupings and a stable nomenclature that reflected the true and divine order in nature. Yet he was constrained by his typological thinking and his use of characters to delimit any taxonomic group, as were all naturalists working on zoological classification.

All classifications were based on particular concepts (genus, species, essential characters, affinity, and the natural system, for instance), but not necessarily on a well-articulated explanatory theory about the overall structure of the natural world. Even those who avoided *a priori* hypotheses based their understanding of cause and effect, natural diversity, geographical patterns, and the origin of (and reason for) similarities between groups on pre-existing concepts. More than a century later, in 1966, the German writer and philosopher Carl Gustav Hempel accurately explained the relation between classification and hypothesis: "If a particular way of analyzing and classifying empirical findings is to lead to an explanation of the phenomena concerned, then it must be based on hypotheses about how those phenomena are connected; without such hypotheses, analysis and classification are blind."[50] For now, universally accepted theories remained elusive.

The general mood was one of expectation. Naturalists awaited the moment when they could take a step back from their classifications and see, almost visualize, a natural order – not unlike an artist stepping back from a large

47 Hooker, "Notices of Books: Géographie botanique raisonée," 249.
48 Endersby, *Imperial Nature*, 211–13.
49 Farber, *Emergence of Ornithology*, 113–14.
50 Hempel, *Philosophy of Natural Science*, 13.

fresco, or somebody finishing a large-scale jigsaw puzzle without the aid of an image showing the entire picture from the beginning. Around the 1840s, the general belief was that it was more information, not more hypotheses, that was needed. An understanding of the big picture would emerge from more specimens, more data on their geographical distribution, and more research on anatomy and physiology. No theory would instantaneously answer all questions about the general organization of the natural world. With their classification systems, Temminck, Strickland, and Hooker intuited an order that was already there, a structure slowly being uncovered, while Oken and Macleay put forward a system that, in their eyes, explained the natural order. However, in general, systematists were looking for explanations within their classificatory systems and the physical objects that constituted them. The rise and great popularity of experimental biological sciences during the last quarter of the nineteenth century might be partly due to this expectation.

3 The Status of Anatomy and Physiology

Anatomy and physiology had been closely linked to medicine for centuries and only after the turn of the nineteenth century applied to systematics, mainly thanks to Cuvier and Geoffroy Saint-Hilaire. In the first half of the nineteenth century, most naturalists, if not all, and especially zoologists agreed that both fields were part of natural history. After that, things quickly changed.

In 1800 Karl Illiger advocated the use of Cuvier's principles of comparative anatomy as the starting point for choosing characters to define species and genera.[51] Temminck thought that anatomy and physiology were "the fundamental sciences and starting point" for the construction of zoological classifications.[52] Likewise, the zoologist Henri Milne-Edwards, who worked at the Muséum and was as devoted to Cuvier's principles as Temminck, defined anatomy and physiology "as the foundation of Natural History."[53] And for the entomologist John Obadiah Westwood, comparative anatomy was so crucial that he defined entomology as "that branch of zoology which treats of the insect tribes, as restricted by the knowledge obtained by the elaborate researches of modern comparative anatomists."[54] After Cuvier, zoology was

---

51 Illiger, "Einige Gedanken über die Begriffe," xli.

52 Temminck, "Introduction troisième partie," in *Manuel d'ornithologie*, 2nd ed., vol. 3, xv–xvii.

53 Milne-Edwards, "Cours élémentaire de zoologie," in *Cours élémentaire d'histoire naturelle*, 11.

54 Westwood, "Introduction," in *Entomologist's Text Book*, 3. For a discussion of the transformation of entomology during this period, see Lindroth, "Systematics Specializes

based on anatomy and physiology, and complemented by systematics, the life histories, and the geographical distribution of animals. It seems surprising, then, that naturalists could not agree on the question of whether physiology and anatomy truly belonged within natural history. This was possibly due to the fact that these disciplines had been associated with medicine for much longer than with zoology.

The focus of many natural history journals of the time reflected the special status of anatomy and physiology. In 1824 a comprehensive journal for the natural sciences was launched in France, the *Annales des sciences naturelles: Comprenant la physiologie animale et végétale, l'anatomie comparée des deux règnes, la zoologie, la botanique, la minéralogie et la géologie*. The title reveals that the editors considered physiology and anatomy separate disciplines from zoology and botany.

The introduction to the first issue includes the editors' compelling explanation of the interplay between these disciplines as well as their wide choice of subjects. Zoologists were able to classify animals by comparing their internal organs, selecting the characters most relevant to their survival, and subsequently grouping them according to their similarities. Botanists, however, had been obliged to use external characters and Linnaeus' sexual system because of the obvious lack of such internal organs in plants. Therefore, the editors noted, zoological anatomy focused on function, while botanical anatomy studied form. Animal physiologists concentrated on pathology; vegetal physiologists basically worked on chemistry. Inevitably, the immense variability in animal body plans had led to a lack of standardization in zoology, and animal classification was in disarray. Ichthyologists, ornithologists, conchologists, and entomologists all used different systems and were oblivious to the classification principles of others. Within their constraints, botanists had achieved a uniform system of classification, but anatomists treated plants as if they had no organic functions, and chemists studied plants as if they had no organs. Consequently, botany and zoology had been drifting apart. There was, therefore, the need for a journal encompassing all disciplines.[55] In this journal, natural history as such had all but disappeared.

Meanwhile, in London, the editorial board of the *Zoological Journal* was debating whether or not to include particular articles in the first issue, which was due to appear in October 1824. They considered essays on comparative anatomy, on zoological classification, essays on particular groups, on animal chemistry, and on "palaeontography" as the most important subjects, in this

between Fabricius and Darwin," and Winsor, "The Development of Linnaean Insect Classification."

55 Audouin, Brongniart, and Dumas, "Introduction."

order. Comparative anatomy "claims a very high rank in Zoological researches. [...] It has, at length, in great measure rescued one branch of natural history from the confusion and absurdity in which, whilst the structure of the habitation only and not that of the inhabitant was considered, its arrangements were involved."[56] Nevertheless, the board chose an entirely different subject to open their first issue: "An Inquiry Respecting the True Nature of Instinct, and of the Mental Distinction between Brute Animals and Man," by John Oliver French.

The essay by French explored the relation between consciousness, rationality, morality, and intelligence. This is hardly a subject one might expect in a zoological journal now, but studies on instinct versus intelligence in animals had occupied naturalists for decades. Frédéric Cuvier had dedicated quite some time to study it, from the moral faculties of dingoes to the intelligence of orang-utans. He even designed an experiment using beavers as subjects.[57] The editors noted that, as French's essay delved into the development of intelligence in "the brute creation" (that is, animals other than humans), "it necessarily enters into the detail of facts purely Zoological and in the highest degree interesting – and if the main question be metaphysical, it is from natural history alone that its ingenious author derives his arguments in discussing it."[58] Philosophical and moral considerations were interesting to zoologists, as long as they included the study of all living beings.

Ten years later, in the Netherlands, Jan van der Hoeven and his friend, the doctor and botanist Willem Hendrik de Vriese, launched a new journal devoted to natural history, and they struggled with the same questions as the editors of the *Zoological Journal*. The journal was a continuation of the *Bijdragen tot de natuurkundige wetenschappen* (Contributions to the Natural Sciences), which had stopped publication two years previously, in 1832. The *Bijdragen* covered all branches of the natural sciences, from natural history to astronomy, hydrology, chemistry, and even engineering. The title of the new journal was rather succinct compared to its French counterpart. It was *Tijdschrift voor natuurlijke geschiedenis*, or Journal of Natural History, and in itself already indicated a substantial reduction of the number of subjects that were to be included in it.

Van der Hoeven and De Vriese described the scope of the *Tijdschrift* as follows: "natural history of the three kingdoms, plant and animal anatomy, and general physiology."[59] Articles on human pathological anatomy were also

56 Bell et al., "Introduction," iv.
57 Flourens, *Observations de Frédéric Cuvier*; Wallace, *Mountains and Marshes*, 104.
58 Bell et al., "Introduction," vi.
59 Van der Hoeven, "Voorberigt," ii.

FIGURE 35 *Portret van Jan van der Hoeven, hoogleraar te Leiden*. Engraving after a drawing by Johann Peter Berghaus, 1849
LEIDEN UNIVERSITY LIBRARIES, DIGITAL COLLECTIONS, COLLECTION J. T. BODEL NIJENHUIS, BN 661

welcome; after all, animal and human anatomy were closely related. Essays on applied and medicinal botany were equally considered interesting and therefore fell within the journal's scope. The editors concluded that they would not make a "fearful choice," as their goal was only to promote science, and all scientific research was, in one way or another, of practical interest.[60] However, after the first year, the editors considered it wise to expand the title of the journal slightly: the *Tijdschrift voor natuurlijke geschiedenis en physiologie* added 'physiology' to the original title and thereby divorced physiology from natural history. In a *Naschrift*, or postscript to the end of the first volume, Van der Hoeven and De Vriese declared that the original rationale had not changed, but the title had been extended "to avoid misunderstandings."[61] Unfortunately, the editors did not elaborate on the nature of these misunderstandings, but their amended title indicates their doubts on including physiology and anatomy in natural history. The last was, apparently, regarded as mostly descriptive and classificatory.

While Van der Hoeven and De Vriese split physiology from natural history, the German professors Carl Theodor von Siebold at Freiburg University, Ernst Heinrich Ehlers at the University of Erlangen, and Albert Kölliker at Würzburg went a step further. They founded the *Zeitschrift für wissenschaftliche Zoologie*, or Journal for Scientific Zoology, in 1848 (only four years after the *Tijdschrift voor natuurlijke geschiedenis en physiologie* had been founded). They introduced their new journal as follows:

> We desire to give our journal the most scientific character possible [...]. To this purpose we exclude all announcements of new genera and species that do not relate to this task, unless these offer us a more thorough-going insight into plant and animal structure, into the life history of animals and plants, or in the lawful organization of the organic realms.[62]

For them, there was a difference between "simple notes and natural history news" and "the truly scientific side of botany and zoology." Anatomy and physiology deserved, undeniably, the label 'scientific,' but classification and nomenclature somehow fell short. Morphological studies were no longer included in natural history, which was equated with classification, as Van der Hoeven and De Vriese had done. This was a significant shift. As Lynn Nyhart puts it: "When

60 Van der Hoeven, ii–iii.
61 Van der Hoeven and De Vriese, "Naschrift van het eerste deel."
62 Ehlers, Kölliker, and Siebold, "Prospectus," in *Zeitschrift für wissenschaftliche Zoologie* 1, no. 1; quoted and translated in Nyhart, "Natural History and the 'New' Biology," 429.

researchers choosing to separate themselves from natural history identified the part (taxonomy) with the whole (natural history), they made invisible much of the intellectual work that gave taxonomic work its life."[63]

This process started to gather momentum in the 1850s and would continue during the second half of the nineteenth century. It had a profound effect on the status of both the experimental sciences and systematics. By 1900 the notion that natural history was mainly occupied with classification and life histories was ingrained in the scientific and public minds, and 'biology' came to be used as the term that encompassed all disciplines dealing with living beings – just as natural history had been in the 1800s, although this also included geology and mineralogy.[64]

It is possible that the editors of the *Annales des sciences naturelles* had pointed to the source of the problem of defining 'natural history.' The different methods used in botanical and zoological studies, mainly in systematics, physiology, and anatomy, may have been responsible for the confusion about the exact definition of each of these disciplines. The boundaries of anthropology and even the very definition of what it meant to be human were also indistinct. Some defined natural history by the object of its research, others by its goals, and yet others by its methodologies. To complicate matters further, there was no consensus on whether theoretical premises should be accepted as a valid starting point. The experimental sciences were on the rise, and classical terms like natural history were being redefined. All these issues had far-reaching consequences, because the status of each discipline determined its place within the natural sciences, the authority of its practitioners and scientific institutions, and its public image, stability, and progress.

4 Systematics within Natural History

At the beginning of the nineteenth century, the field of botany included the practice of classification and applied botany, mainly in medicine, agriculture, and horticulture. As the century progressed, botany's prestige as a scientific endeavor and even its popularity diminished. Around 1830 systematic botany had lost much of its allure, and scientists were more excited about experimental

63 Nyhart, 429.

64 For literature on the decline of systematics from the end of the nineteenth century to today see, for example, Doolittle, "Practice of Classification and Theory of Evolution"; Endersby, "Descriptive and Prescriptive Taxonomies"; De Queiroz, "Systematics and the Darwinian Revolution"; Mayr, *Growth of Biological Thought*, chap. 6; Secord, "The Crisis of Nature"; Stevens, *Development of Biological Systematics*, chap. 9.

research than about descriptive work.[65] By contrast, zoology was an amalgam of several disciplines. Comparative anatomy, paleontology, theories of organic change, *Naturphilosophie*, pathological anatomy, physiology, systematics, and ethology all belonged to the sphere of zoology. Some had more defined goals and methods than others, and all contributed in some manner to zoological classification. There was then, as there has always been, a distinct hierarchy of fields within the natural sciences, with some carrying more prestige than others. This was not a rigid order, but a fluid one, subjected to change and, naturally, debate. The place of systematics in the natural sciences was discussed as much as practically any other aspect related to it. As the meaning of the term 'natural history' changed, the status of zoological classification changed with it.

Between 1800 and 1850, naturalists held two different and opposing attitudes towards systematics. Most saw it as a refinement of the groundwork of comparative anatomy and paleontology prepared by Cuvier and the steadfast path towards uncovering nature's – and God's – plan. Although based on anatomy, systematics was considered a discipline in and of itself. However, some naturalists saw classification and nomenclature as a practical tool, and certainly not as an independent discipline. This apparently straightforward dichotomy – between systematics as a discipline or a tool – depended on a single fundamental question: should natural philosophy be the basis for classification? This primary question influenced other aspects of systematics, like its goals, methods, practice, or the subject of its study.

Adherents of the first perspective, who considered systematics to be a true discipline of natural history, included, as one would expect, Cuvier himself, as well as Illiger, Temminck, and Strickland. In 1805 Cuvier made the distinction between anatomy and systematics (which he referred to as "natural history") quite clear when he wrote:

> The development of methods is the goal of natural history itself. Anatomy receives them, as it were, ready-made. It is from them that anatomy takes its first directions: but soon it illuminates them with its own light; anatomy is the best test for their truth; it is by applying a method of natural history to anatomy that we are able to judge whether or not it is in accordance with nature.[66]

By "methods" he was referring to the natural classification systems arrived at by following his principles of correlation and the conditions for existence.

65 Stevens, 209–15.

66 Cuvier, *Leçons d'anatomie comparée*, vol. 1, 64.

Anatomical laws and classification systems emerged independently, and each of them could be used to test the truthfulness of the other, that is, to see whether they reflected nature's organization. In practice, however, Cuvier focused mostly on comparative anatomy to define the higher ranks of animal taxa, without working out the details of the lower groups like genera and species.[67] As Macleay put it, apparently with a touch of disappointment: "no person of such transcendent talents and ingenuity ever made so little use of his observations towards a natural arrangement as M. Cuvier."[68] Cuvier himself admitted that his classification of birds in his *Tableau élémentaire de l'histoire naturelle des animaux* (1798) was the weakest of this work: he had kept the Linnaean genera and Buffon's subgenera.[69] Temminck set himself the task of finishing what Cuvier had left undone: applying his principles to classify the lower ranks of birds and mammals, i.e. the orders, genera, and species. In 1817 Cuvier incorporated the classification of birds made by Meyer, Wolf, Temminck, and Levaillant, among others, in his *Règne animal*, although he kept the higher taxa as published in the *Tableau élémentaire*.[70]

Naturalists such as Illiger, Temminck, Milne-Edwards, and Strickland thought that systematics should follow Cuvier's methodology. Comparative anatomy would guide them in their major groupings, while systematics would either prove or disprove the solidity of these higher taxa. Comparative anatomy was at the very base of systematics, but it was not synonymous with systematics. Temminck repeatedly remarked that while his classification methods were built upon the principles of comparative anatomy, osteology, and physiology, classification was an independent endeavor:

> Many anatomists find themselves hindered in their research because they have neglected to study the characters provided by zoology: they believed that comparative anatomy alone would guide them in their studies of the secondary levels [i.e. lower taxa]; and they have lost sight of this auxiliary science because of their underlying belief in the importance of anatomical research. Their research cannot be of use to us today, as it is often impossible to determine with exactitude on which species of animal they based their observations, while those that are still carrying

67 Farber, *Emergence of Ornithology*; Coleman, *Georges Cuvier, Zoologist*, 67.
68 Macleay, *Horae Entomologicae*, vol. 1, 326.
69 Cuvier, "Préface," in *Tableau élémentaire*, vi.
70 Cuvier, "Préface," in *Règne animal*, vol. 1, xxiii.

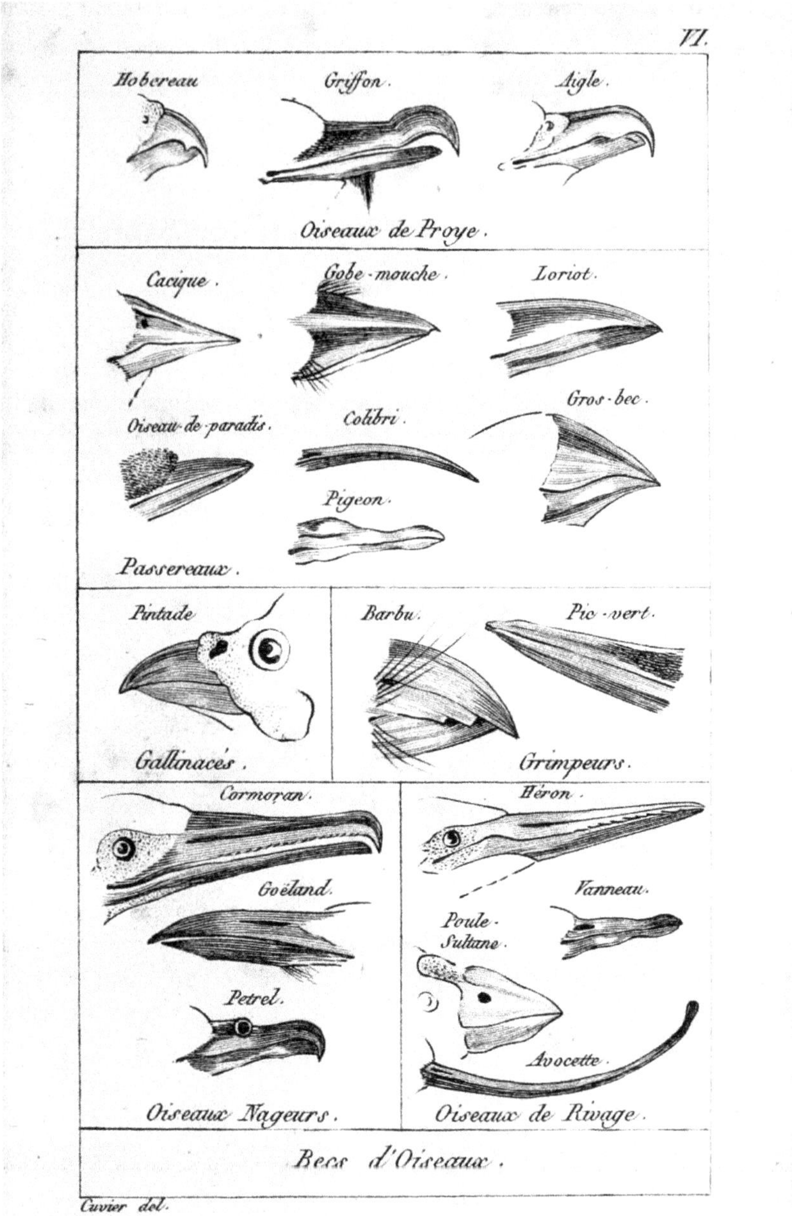

FIGURE 36 *Becs d'Oiseaux*. Engraving after a drawing by Georges Cuvier. In Georges Cuvier, *Tableau élémentaire de l'histoire naturelle des animaux* (1798): plate 6
IMAGE FROM THE BIODIVERSITY HERITAGE LIBRARY. DIGITIZED BY THE UNIVERSITY OF TORONTO, THOMAS FISHER RARE BOOK LIBRARY

> on without the help of any basic notions of zoology are not contributing to this field, which pursues the perfection of the natural system.[71]

Here, for some reason, Temminck used the term 'zoology' as a synonym for systematics. He included the comparison of external characters, the scrutiny of varieties (geographical as well as variations due to age and sex), life histories, and geographical distribution within it, which were all indispensable for defining the lower taxa, especially genera and species.

In a section titled *Desiderata of Ornithology* from his 1844 "Report on the State of Ornithology," Strickland revealed similar thoughts. According to him, comparative anatomy provided naturalists with essential precepts, but they still needed ecological, behavioral, and geographical information for the purposes of classification. As important as it was, anatomy alone was insufficient for achieving the main goal of systematics: the uncovering of the natural system. Nonetheless, Strickland wished for a better understanding of bird anatomy and suggested that someone – a student, he specified – should draw up a classification of birds based on their anatomical characters alone. If that Cuvierian classification agreed with one based on external characters, he concluded, "its reality and truth" would be proven, just as Cuvier had done almost forty years previously.[72]

Lamarck, as usual, had other ideas. He stressed the concepts of transmutation and continuity, and inevitably saw systematics differently than Temminck. With continuity in the foreground, classifications were necessarily artificial constructs – catalogs to record the described species and the gaps in the existing knowledge. None of the groups above the level of species (genera, families, orders, and classes) existed in nature; they were mere instruments to facilitate research.[73] For Lamarck, classification was a tool. His focus was on theories explaining the diversity of living beings, cause and effect, and relationships. In all probability, Lamarck did not think that the classifying effort in itself merited a high status within natural history.

On the other side of the channel, the quinarians also hoped that classifying living beings would lead them to fill the gaps they now saw in their model of nature. But these were not the same gaps that Lamarck perceived. The quinarians' goal was to complete their perfect mathematical circles as the representation of order in nature. Lamarck, by contrast, was classifying a natural world

71 Temminck, "Introduction troisième partie," in *Manuel d'ornithologie*, 2nd ed., vol. 3, xv–xvii.

72 Strickland, "Report on the Progress," 217, 219.

73 Lamarck, *Philosophie zoologique*, 33–34.

that he believed to be continuous, and the groupings he made were therefore never meant to be a reflection of nature's structure: there was no structure other than a linear succession of beings that evolved up a ladder of complexity. In any case, for both Lamarck and the quinarians, with their approach to classification set up *a priori*, systematics was a means to an end, and never a scientific discipline without a philosophical component. Macleay had observed that classification had become many naturalists' ultimate goal:

> The almost exclusive attention which has of late years been unfortunately lavished on *Nomenclature* and *Systematic Arrangement* – on the means in short, and not on the end of the science – has with ignorant persons diminished the importance of the study of Natural History itself. Let us hope that the slur will be soon entirely obliterated by those naturalists who have already shown that they are not to be deterred from the investigation of affinities by great names, because, forsooth, these may have preceded them in the annals of science.[74]

He insisted that "nomenclature is not a department of natural history, but only a convenient instrument whereby an acquaintance with it may the more easily be cultivated."[75] While equating nomenclature with classification went too far for Vigors, he fully agreed with Macleay.[76]

Philosophy bestowed a higher status to any scientific field. In Victorian Britain the tradition of gentlemanly science lingered on, and as Jim Endersby notes, "only a gentleman could be philosophical." Therefore, in Britain, the status of classification was necessarily related to gentility: "although the mathematical sciences were the most prestigious and best-rewarded, numbers alone did not bring status […]. It was more important that a discipline be considered 'philosophical.'"[77]

In the Netherlands, Jan van der Hoeven took on the task of defining natural history and its subjects for a textbook on zoology. He considered systematics (for which he used the term 'taxonomia') to be a *kunst*, a skill or art, not a truly scientific practice. Van der Hoeven discussed the importance of achieving a natural classification system using Cuvier's definitions of "méthode" and "analyse," but in his view, classification was not a discipline as such. However, it was indispensable for natural history. Without it, one could not

74 Macleay, *Horae Entomologicae*, vol. 1, 10, fn. b (italics original).
75 Macleay, 24.
76 Vigors, "Observations on the Natural Affinities," 180.
77 Endersby, "Classifying Sciences," 64.

learn about nature, understand others' writings, or even communicate with other naturalists.[78]

In France, Milne-Edwards shared this line of thought with Van der Hoeven. To understand animals and plants, one had to study the relations between living beings as well, describe the peculiarities they offer, which could be used as characters, learn about their geographical distribution and then name and classify them.[79] However, Milne-Edwards saw classification as a method, a tool that helped zoology (which he broadly defined as "the Natural History of Animals") move forward and "unburden one's memory."[80] Concerning its importance in natural history, he wrote:

> The introduction of natural methods for the classification of living beings is one of the greatest services rendered to natural history: it has changed the image of this science and has made compelling that part of botany and zoology that had been up to now the most arid.[81]

The relevance of classification, for him, was closely linked to the "natural methods." The best systems were those based on anatomy and physiology (as opposed to artificial systems), but nonetheless, classification was a tool, not a scientific discipline.

There was yet another field of study contending for leadership in natural history, and it proposed an entirely new approach to the study of nature. Its leading ambassador was Alexander von Humboldt. He saw nature as an organic structure, as one system of interconnected living things which were subject to historical processes. His was what we would now call a multidisciplinary approach: the discovery of the laws of nature required the fieldwork that consisted of observing, measuring, and mapping phenomena.[82] Humboldt recognized the importance of systematics as the basis for descriptive natural history (i.e. classification), suggesting that he did not regard it as a discipline within natural history, but as an instrument. It was an essential tool, but a tool nonetheless. For instance, descriptive botany was indispensable for the advancement of medicine, culture, art, and even philosophy, but it was

78 Van der Hoeven, "Inleiding," in *Handboek der dierkunde*, 28.
79 Milne-Edwards, *Cours élémentaire d'histoire naturelle*, 281–91.
80 Milne-Edwards, "Cours élémentaire de zoologie: Notions préliminaires," in *Cours élémentaire d'histoire naturelle*, 11.
81 Milne-Edwards, 288.
82 See also Outram, "New Spaces in Natural History"; Walls, *The Passage to Cosmos*.

equally important to establish the geography of plants, which he understood as a field in itself.[83]

In the Netherlands no-one embraced Humboldt's approach more closely than Caspar Reinwardt. His inaugural lecture of 1823, when he was a newly appointed professor at Leiden University, was a tribute to fieldwork in the Humboldtian sense. According to Reinwardt, all things in nature existed only in interaction and cooperation with each other, and none of its elements were isolated from the whole.[84] His discourse stressed the importance of measuring, observing, and analyzing cause and effect. Museum researchers and curators were necessarily limited in their knowledge and perceptions because the collection specimens were studied out of their natural context.[85] Rather unkindly he compared the naturalist who never traveled with stagnant water: "In the same way as filth accumulates in stagnant water, one can notice a certain impurity, lowness, and narrowness of judgment in those who are studying Nature but have never gone beyond the boundaries of their native country or of any particular place," and these shortcomings eventually resulted in "innumerable errors and trivialities" in their understanding of nature's workings and structure.[86] The study of objects outside their natural environment had become "a common shortcoming, spreading through the entire field of Natural History."[87] Providentially, Temminck was in Paris at the time of Reinwardt's inaugural lecture.

Even more radical than Reinwardt was the Leiden professor Sebald Brugmans, who had had a strong contempt, even aversion for systematics. Although he had obtained his doctoral degree from the University of Groningen by studying and classifying local minerals, he slowly turned against systematics in favor of applied research. In a lecture at Leiden University of 1787, he advocated more botanical research, and while recognizing the advances made by systematics in botany, he claimed to be much less interested in plant names and classifications and much more in the role plants played in nature, their function, and their usefulness.[88] Brugmans promoted the careful study of the local flora, telling his colleagues that he did not mean "a catalog of herbs or a mere index, nor a simple Flora, as it is called."[89] He regularly warned his students against the futility of trying to pigeonhole nature, which was so rich, complex, and

83 Humboldt and Bonpland, *Essai sur la géographie des plantes*, 13.
84 Weber, "Hybrid Ambitions," 29.
85 Reinwardt, *Redevoering van C. G. C. Reinwardt*, 22–23.
86 Reinwardt, 23–24.
87 Reinwardt, 29.
88 Brugmans, *De accuratiori plantarum indigenarum notitia maxime commendanda*, 6; De Jonge, "Sebald Justinus Brugmans," 26.
89 Brugmans, 10.

FIGURE 37 *Portret van Caspar G. C. Reinwardt, hoogleraar te Leiden.* Engraving after a drawing by August Erxleben [no date]
LEIDEN UNIVERSITY LIBRARIES, DIGITAL COLLECTIONS, COLLECTION J. T. BODEL NIJENHUIS, BN 1165

indivisible, in useless classification systems. His medical and anatomical collections reflected his ideas in the same manner as the systematic collections in Groningen reflected Van Swinderen's views. For Brugmans comparative anatomy and pathology were decidedly superior to systematics. As his biographer dramatically phrased it: "the zeal for system is the grave of truth!" (*systematis ardor veritatis sepulcrum!*).[90]

90 Capadose, *Lofrede op Sebaldus Justinus Brugmans*, 165. See also De Jonge, *Het vergeten fenomeen*, 19.

Despite this disregard for systematics, Brugmans could not dispose of it completely, if only because his lessons at the university required a structure. The plants needed to be named in one way or another. Reluctantly, he used Linnaeus' *Systema vegetabilium* and *Philosophia botanica* from 1774, a long-outdated system.[91] Incidentally, his dislike for classifications did not seem to trouble his relationship with Temminck. In 1813, after receiving a copy of Temminck's *Histoire naturelle des pigeons*, Brugmans wrote to him: "All ornithologists are in your debt for this beautiful Monograph, and all will eagerly await its continuation."[92] Temminck had always been welcome to work on Brugmans' collection whenever he needed to collect anatomical information from Brugmans' skillfully made zoological preparations.

Overall, the search for the general order in nature constituted the central preoccupation of natural history during this period, and one might expect that this pursuit would define systematics as a discipline. This was not the case. Naturalists decided upon the scientific status of systematics not by its main goal (the natural system) but by its methods, i.e. whether they were empirical or philosophical. If one believed that a natural system was to be attained by descriptive and classificatory zoology, within agreed rules and by Cuvier's anatomical principles, then systematics was indeed a scientific discipline. But if *a priori* hypotheses guided the pursuit of the natural system, then systematics was not a true discipline but a tool for testing the validity of one's natural philosophy. A general model of the organization of nature was put forward first and then tested by specialist examination of one or a few groups. This also involved hypothesizing about the relations between species and higher taxa, and about cause and effect. None of this is necessary for an empirical process, where the classification method borrows its philosophy and the definitions of concepts from Cuvier's comparative anatomy. It is because of this additional intellectual exercise that natural philosophers could not consider bare systematics a discipline. Among them were Lamarck, Macleay, Swainson, Oken, and Van der Hoeven. The primary requisite for raising systematics to the status of a scientific discipline was, therefore, not the search for the natural system, but the rejection of a pre-existing theory as a method to uncover it.

The status of any particular classification system (not of systematics itself) depended on whether or not it approached that ideal, the ever-evasive natural system. A natural classification, as defined by Temminck and Strickland (i.e. based on anatomical and physiological characters complemented with

91 De Jonge, "Sebald Justinus Brugmans," 35.

92 Brugmans to Temminck, 24 December 1813, Naturalis Biodiversity Center Archives, Archive C. J. Temminck, NAT_ARC_TEM_00254.

external characters, life histories, and geographical distribution) was regarded as the highest achievement. A natural system uncovered by similar methods but based on an *a priori* model of nature – the quinarian system, for example – would also rank very highly in the eyes of naturalists. However, the path towards the natural system was long and winding, and even if classifications still did not reveal the expected natural order, there was hope that, with more information and new specimens, these classifications would become natural. These in-between classifications were seen by naturalists as temporary tools, a work in progress pursued in anticipation of more knowledge and material that was yet to come.[93] If the system was good enough, it would hold true against new facts and become a natural one.

By contrast, the construction of artificial systems was very different altogether. Often based on one or a few characters, such classifications were lesser, although convenient, products. A commonly used analogy during this period for the artificial system had been coined by Georges Cuvier: an artificial classification was like a 'dictionary' in which words were arranged alphabetically and not according to their meanings.[94] An artificial system was useful for the quick identification of species within the system, but it provided no insights into the relations between the organisms. Therefore, the relevance of a particular classification system depended on whether it served practical purposes, like a dictionary, or led to uncovering the structure of nature. However, as methods in classification and nomenclature were constantly revised, the relative importance of classification systems was always in flux.

In sum, there were very few agreements on very few matters. From the type and genus concepts to explanations for geographical distribution patterns, and on to appropriate nomenclature and classification practices, the need for standardization was blatantly evident to all. Yet solutions did not emerge until the triumvirate of France, Germany, and the Netherlands was forced to accommodate British naturalists and their 'parliamentary practice.' To the end of the nineteenth century, the frequent discussions and occasional insults continued, mostly because of a lack of agreement on how the perfect natural classification system could be achieved.

The issue of the status of systematics at the time was equally a rather complicated matter. Natural history had fragmented into several disciplines, each with its own sets of goals and practices, and systematics was losing the hegemony it had held during the eighteenth century – which is not to say it had

---

93 Temminck, "Introduction troisième partie," in *Manuel d'ornithologie*, 2nd ed., vol. 3, xi; Strickland, "Report on the Progress," 177.

94 Cuvier, "Introduction," in *Règne animal*, vol. 1, 9–10.

sunk to the lowest status. Since the Parisian professors Cuvier, Lamarck, and Geoffroy Saint-Hilaire had applied themselves to the field, zoological systematics had been flourishing. Museums, monographs, and journals multiplied, and systematists became a paid group of professionals. The discovery of the natural system seemed to be imminent yet frustratingly elusive. The tensions between field and museum naturalists – and between amateurs and professionals – also contributed to changing the discipline's position. By the middle of the nineteenth century systematics had become a more democratic field, and consequently the platforms used by systematists to communicate also changed. Disagreements about natural philosophy and the general reluctance to tackle the question of universal laws would ultimately accelerate the decline of the status of systematics.

The breaking up of natural history into subdisciplines brought various changes with it. With the emergence of a physical natural history in Humboldt's spirit, the predominance of anatomy and physiology, and the popularity of the relatively new experimental sciences, systematics had to compete for visibility, resources, and status.[95] The main question was, it seems, whether all these disciplines were complementary to each other, or subordinate. The hierarchy was far from established towards 1840, but by the end of the century systematics had fallen towards the bottom of the list. This had a profound impact on systematics. For any discipline its status determines where, when, and by whom it is practiced, how it is communicated, the channels of communication it uses, how it contributes to scientific debates, the available resources, and whether it is being taught and practiced at universities. The more prestige a field has, the more attention and resources it receives.[96] Knowledge and skills can be passed on not only through publications and specialized journals, but also via university curricula that mold the next generation of practitioners. The consequences of the low status of systematics around the 1850s are still felt today.

95 Farber, *Finding Order in Nature* and *Emergence of Ornithology*; Nyhart, "Natural History and the 'New' Biology"; Stevens, *Development of Biological Systematics*.

96 The consequences of the low status of systematics in the post-Darwinian era are discussed by Endersby in "Descriptive and Prescriptive Taxonomies." See also Curry and Secord, "Natural History and its Histories."

CONCLUSION

The Emergence of Systematics

When Louis Napoleon was still king of the Netherlands, Temminck was denied admittance to the Koninklijk Instituut van Wetenschappen. The reason for Temminck's exclusion was apparently his lack of formal academic training. Indeed, Temminck was tutored at home and had never set foot in a university. Nevertheless, by 1836 he was finally admitted to the First Class of the Koninklijk Instituut. By then Temminck was boasting two honorary doctorates (one from the University of Groningen and the other from the University of Jena), as well as a number of royal distinctions and awards. He was also the notable director of 's Rijks Museum van Natuurlijke Historie, whose collections ranked amongst the best in Europe. He had published several monographs that had given him international prestige. Twenty years later, however, his fame was diminishing, and his authority dimming.

In the past, Temminck's career and its ups and downs were considered from different angles, with each biographer focusing on a different phase in his life, which has resulted in an incomplete and seemingly contradictory narrative. However, Temminck's career is best understood in the context of Dutch politics, the development of systematics, and in its entirety. It mirrors three main events that shaped natural history between 1800 and 1850: the European colonial expansion, the rise of a scientific meritocracy, and the emergence of systematics as a discipline. From this perspective, it becomes clear that Temminck's career, priorities, and status depended directly on the path of natural history.

When Temminck was in his thirties, he earned his place on the international stage with his early monographs on birds. They were received with praise, mainly for straightening out the chaotic nomenclature of ornithology and for his clear, unambiguous definitions of genera. Temminck's main contribution to ornithology may have been the resulting increase in the accuracy, clarity, and method of zoological classification. The process of working in his father's cabinet had confronted him with a body of literature that was incomplete, heterogeneous, and sometimes unreliable. To solve this, he had proceeded to match the original descriptions of Linnaeus, Brisson, Levaillant, Illiger, Meyer, Latham, and other naturalists with the specimens they had used, which were dispersed all over Europe. Traveling for months from cabinet to cabinet, he built an extensive network of correspondents – something he would never have been able to do had he not been born into a patrician family. Years of

 | DOI:10.1163/9789004441491_010

effort paid off, and he produced several indexes of species with complete synonymies and references to existing literature, which were generally welcome and exceedingly useful for other naturalists. This recognition allowed him to take up remunerated posts in national collections, which paved the path for his scientific career.

Temminck's rising prestige owed as much to his monographs as to the political situation in the Netherlands. The young nation needed to compete with the other European powers, and its king understood that. When Temminck put forward a plan for scientific development – a national museum of natural history – the government (which, notably, employed many of his friends) quickly subscribed to it. Naturally, the nation's colonial ambitions also played a role in the foundation of the museum. Natural history (in the broadest sense) was, after all, the science aiding the utilization of useful plants, spices, fertile soils, exotic animals, and valuable minerals. Government-sponsored expeditions set sail to study and collect the products of nature for the benefit of both the museum and the national economy. During the 1830s the devastating colonial wars, a rebellious Belgium, and the ensuing economic crisis put an end to these expeditions and Temminck's golden years. From then on, he had to devote much of his time to keeping the museum running as both he and the institute faced decline. He succeeded well not only in keeping the museum in working order, but also in expanding it thanks to his admirable networking skills and relentless determination. Specimens continued to enrich the collections for decades to come, while an army of competent naturalists curated and researched them. The Leiden museum was, and still is, a testament to his vision. Temminck's scientific prestige, however, suffered several blows. Systematics as he knew it – and practiced – changed faster than he was able to adapt.

Despite the initial success of Temminck's early scientific publications, his work fell short on several issues – or at least, in the eyes of his peers. Temminck's worst offense, they thought, was his repeated neglect of the elegant principle of priority. Not all naturalists were pleased with this practice. Frédéric Cuvier and George Robert Gray, for instance, allowed no exceptions to the rule of priority, while others, such as William Swainson or Thomas Horsfield, thought it appropriate to change incorrect epithets – especially if these were not their own. Any change questioned one's credibility and authority, and the discussions of nomenclature were fierce. The fact that even such a seemingly simple rule as the principle of priority stirred such passions reveals just how complicated the issue of standardization was, and how seldom the naturalists agreed with each other. Even so, during the first half of the nineteenth century, several deliberate efforts were made to standardize nomenclature, which were, however, unsuccessful. Finally, in Britain, naturalists found a way to solve their differences: they came together as a community when the BAAS institutionalized the code of nomenclature that had been proposed by Hugh Strickland, Charles Darwin, Richard Owen, and five others.

The selection of scientific names was not only a matter of convention; each name was linked to a description, and groupings based on inaccurate descriptions made subsequent classifications confusing. Names were important because they transmitted ideas: they reflected relations between varieties, species, and genera – categories that, in turn, were as much in dispute as scientific names. Any challenge to a naturalist's proposed arrangement touched upon much deeper questions, namely, on the naturalist's ideas on the organization of nature. Classifications were, after all, expected to mirror the true natural order. At that point, this order was still nothing more than an intuitive idea, but naturalists were confident that laws similar to those for, say, the planetary harmony would eventually be forthcoming. Discovering nature's plan had been the main goal of classification for centuries, and by the 1850s there was a general expectation that it would eventually emerge from natural classification systems. The question of how this was to be achieved, however, would keep naturalists occupied for decades to come.

A particularly difficult issue they faced was the circumscription of 'natural' groups. In general, they embraced the principle of continuity, which made delimiting species something of a paradox. While Cuvier's four *embranchements* were generally acknowledged as four separate groupings that applied to the highest ranks in zoological classification, species and varieties of related organisms formed a continuum that was increasingly difficult to divide into natural groups. Genera proved to be extremely challenging. Even if naturalists admitted that species were arbitrary groupings, genera seemed to many to be natural, separate entities. Temminck believed that a genus comprised a group of species that belonged together as manifestations of an ideal 'type,' a basic form of organization. His type carried with it the key to unlocking the organization of nature: it stood at the foundation of Temminck's law on the geographical distribution of animals, which stated that similar types live along the same latitude on the globe. Also, types were for Temminck fixed categories, as immutable as the species were. Everything in Temminck's work revolved around the concepts of type and genus – fixed, eternal, and definite. Different concepts would necessarily have led him to other views about nature.

Temminck's type concept as an organizing category was not necessarily at odds with his idea of continuity in nature: without groupings, nature would become shapeless, but Temminck believed nature to be structured. He found continuity in variations of species as well as in species that resembled each other but belonged to different genera. But ideal genera were discrete entities, and recognizable as such: the bears, the monkeys, the fruit bats, the ibises, the leopards. Similarly, his belief in separate acts of divine creation allowed Temminck to explain the geographical patterns of distribution, but it did not imply that he saw nature as a composite of discrete groups. What Temminck's writings suggest is that he believed not in a linear chain of beings but a web, a reticulum of related groupings. His classificatory work and his law of the patterns of mammal distribution manifest that, despite the emphasis he put on defining groups, Temminck's focus was on relationships, similarities, and the overall organization of the fabric of nature, with an emphasis on continuity above division. This also explains Temminck's reluctance to split big genera into smaller groups and the fact that his interpretation of the distribution of mammals around the globe was the opposite of Buffon's, Humboldt's, and de Candolle's: different forms occupied different regions which were separated by insurmountable barriers. They focused on faunas made up of different species where Temminck saw faunas comprising similar genera.

But what is a 'genus'? In order to deal with this question, Temminck advocated the use of a few, carefully selected characters for all genera of a particular class. He spent years perfecting this selection, considered which to use for genera and which for species, and tested it throughout all his groupings. He followed Cuvier's principles of the correlation of parts and the subordination of characters. In this way, he succeeded in defining genera consistently for all groups of birds while using a limited set of characteristics. Each genus so defined was an unambiguous entity, but necessarily a large one (in that it contained many species). Temminck then resorted to 'sections' to divide large genera into smaller groups, especially when he was not sure he had found definite characters that would allow him to make new genera. Others, like Nicholas Aylward Vigors, preferred to split genera into smaller units and found the use of sections confusing and unnecessary. They based their decisions upon a different understanding of what a genus was. Significantly, by 1844 Hugh Strickland still lamented that the general approach to defining genera was an intuitive one, a matter of convenience more than a natural principle, and Alfred Russel Wallace concluded that the type was neither fixed nor eternal. Interestingly, in the end, the search for perfectly defined genera caused Temminck to doubt the existence of types and even of that of species.

Like nomenclature, the breaking up of large Linnaean genera became a battleground because of the underlying belief that systems should reflect the divine plan behind all these wonderful organisms. For these systems the concept of 'genus,' not 'species,' was essential. The definition of genus hinged on the fact whether or not it was a natural entity, which was inextricably connected to the type concept. And the concept of type, in turn, depended on each naturalist's idea of the divine blueprint for creation. These concepts – the type and the genus – were crucial in nineteenth-century systematics as they carried the basic premises for constructing the natural system. If naturalists did not succeed in defining them, the natural system would never materialize. The whole enterprise of systematics between 1800 and 1850 depended on the genus concept.

It is important to stress that the intended improvement of the foundations laid by Cuvier does not imply that zoological systematics up to 1850 was simply producing an increasingly complex version of Cuvier's system, like some kind of elaborate and fastidious book-keeping. The main issues in systematics were the integration of a myriad of diverse systems of classification, as well as the desire to arrive at universally accepted rules of nomenclature, to find a way of harmonizing geographical patterns of distribution and classifications, to define a method of setting clear boundaries around genera, and to find a satisfactory procedure for weighing the characters used in classification. Standardization was crucial, and although it had not yet been achieved by the mid-nineteenth century, it was slowly emerging as zoological systematics developed. Such efforts to set guidelines and methods for classification were concentrated in the hands of an expanding but definite group of naturalists who worked in specific settings.

In the Netherlands, comparative anatomy was practiced and taught at universities, but systematics in the natural history museum. This separation excluded systematics from the universities and put it at the center of the activity at 's Rijks Museum, notably for two hundred years. In other countries, however, the situation was very different. At the Muséum National d'Histoire Naturelle, anatomy and systematics coexisted in the same institution, and sometimes in the same person. Additionally, French professors also held posts at academic institutions, while in Victorian Britain, gentlemen of independent means – naturalists not employed by museums or universities – enjoyed a better reputation in the scientific community. In any case, whether in museums, academia, or private cabinets, classificatory endeavors were exclusively collection-based.

The practice of systematics itself, which consisted of identifying, naming, describing, and recognizing relations, was a comparative exercise, and comparison required extensive collections and libraries – none of which are available while doing fieldwork in the colonies. If extensive collections were indispensable to a panoramic view of natural diversity, Humboldtian naturalists necessarily needed the experience and the measurements from fieldwork to understand the interactions among organisms and their environment. In other words, naturalists were claiming different spaces for different practices.

In general, naturalists were aware of these separate approaches, and they also recognized that they had different goals. Agreeing on the most desirable and important goal at a specific time was another matter altogether.

At the same time, a process of community building accompanied this localization of systematics. Naturalists dealing with zoological classification were aggregating into a collective with specific dynamics and platforms. By 1840, the era of Linnaeus, Buffon, and Cuvier was giving way to a period following an altogether different model of organization. The transition was set in motion by the dramatic increase in the number of naturalists engaging with classification, especially in Britain. Societies, magazines, and clubs quickly came into being and became places for naturalists to debate, criticize, and review each other's works. At this point, no individual could attempt to set rules on nomenclature or any other matter, however well-received his work was and regardless of his status in an institution.

Community building necessarily implied peer review. Any proposed name, system of classification, or general principle was open to discussion. Consensus was necessary, if elusive, for the much-needed process of standardization. Temminck, Vieillot, Vigors, Swainson, and Strickland all had to present their best arguments to convince each other of the truth of their systems. They required robust reasoning, extensive knowledge of their fields, and a keen mind to respond to their reviewers. This situation, chaotic as it might seem, was the fertile ground from which systematics evolved.

The emergence of a meritocracy played a decisive role in pushing systematics forward. It came of age by the mid-nineteenth century, partly due to the mentioned development within the social infrastructure of natural history. Communication promoted convention. Discussions were settled by argument, and the process of the democratization of natural history cast aside those who claimed authority based on past successes, social status, or, in Temminck's case, both. As Britain took the lead in natural history, the naturalists' community changed, and with it necessarily the ways of communication and language, English becoming increasingly important.

After 1835 Temminck was noticeably absent from this new community of naturalists. He continued to publish extensive monographs that took him decades to finish, partly because of his fixation on detail and partly because of the constraints of time caused by his management of the museum. He did write a handful of short articles, only four of which dealt with theoretical issues. It took him an inordinate amount of time to formally describe a species that was new to him. Birds and mammals had enjoyed enormous popularity for centuries, and an extensive community of eager and more productive naturalists dedicated time to describing them – the sooner, the better. It is hardly surprising, then, that Temminck's writings were outdated even before they were published. He published them nonetheless, inevitably introducing species that had been already described by others. This slowly eroded his authority.

Temminck's lack of urgency in publishing might seem paradoxical, especially since he was very much aware of the need to put order into the complex business of naming species, but it makes sense in the context of his self-perception. He saw himself as part of an elite group of naturalists who were authorities on the study and understanding of nature, including Cuvier, Illiger, Meyer, or Latham. Additionally, Temminck felt a certain disdain for British naturalists or, at least, for British natural history. He routinely ignored their work, even when Britain began to take the lead in natural history. This transition affected him, as well as Van Marum, Reinwardt, Strickland, and others, but he was not able to adapt. The complexities of running the museum in the middle of the Dutch economic and political crisis, the concerns around the Natuurkundige Commissie, and the tremor that had developed in his hands made his situation even more difficult. Work piled up, and resources dwindled. The old autocracy to which Temminck had belonged for decades was now gone.

While Temminck's influence faded after the 1830s, he had played a central role in the early decades of the century in the standardization of names and methods. He had promoted the establishment of a museum independent from the university, which was dedicated exclusively to the practice of systematics. Temminck's legacy was mainly established by his resolving the synonymy of late-eighteenth- and early-nineteenth-century ornithology, the foundation of 's Rijks Museum van Natuurlijke Historie, and its monumental collection of specimens, which was worthy of international recognition. His work raised the scientific quality of zoological classification and its status for nearly three decades. Also, although presumably unwittingly, Temminck ignited discussions that led his contemporaries to refine the definitions, methods, and practices in systematics.

What Temminck witnessed during his career was the emergence of systematics as a discipline, one that had been traditionally part of natural history and even been synonymous with it. The driving forces behind this process were standardization, increased rigor, specialization, and the rise of a meritocracy. By the middle of the century, systematics had advanced a set of goals, methods, terms, and concepts unique to its subject matter; and there was a developing self-regulated social structure that, although struggling to establish order, relied on an incipient peer-evaluation system. The discipline had its own journals and societies, and a space in which to practice: the natural history collections. Naturalists seemed to be conscious of the fact that, with their efforts of forming structure and standardized methods, they were creating a new discipline.

With the breaking up of natural history into several separate domains, a disciplinary hierarchy emerged. The status of systematics depended directly on the question whether or not it relied on an *a priori* hypothesis to achieve its primary goal: uncovering the order of nature. If systematics itself – the very act of describing and classifying – eventually lead to explanations about order and causality, it would be acknowledged as a true discipline. Naturalists who believed this to be the case defended systematics as independent of natural philosophy and free of any theoretical assumptions. Temminck, Illiger, and Strickland, for example, subscribed to this idea. By contrast, others regarded systematics as a tool, i.e. an aid to help prove – or disprove – theories about the organization of nature. Lamarck, the *Naturphilosophen*, and quinarians like Swainson and Vigors considered the practice of classification a useful instrument. They saw classifications and explanatory arguments as separate entities. Therefore, for them, systematics ranked very low among the disciplines – regardless of whether or not they thought their classifications to be natural or artificial. The status of systematics within natural history depended on whether it could be an explanatory discipline in itself, rather than a merely descriptive one. Temminck possibly even formulated his law on the distribution of animals in an attempt to raise the status of systematics.

The challenge of harmonizing classification systems with natural philosophy was exceedingly difficult. As systematics proved unable to offer explanations of cause and effect, its status plummeted and never fully recovered. In fact, for the past two centuries systematics has been trying to gain prestige by incorporating into its domain the most pressing conceptual questions of the time. By the mid-nineteenth century systematics pursued the discovery of order in nature. A century later it attempted to redefine itself as 'evolutionary systematics,' a discipline designed to answer evolutionary questions such as the processes behind speciation and population genetics.

Curiously, systematics struggles now, as it did then, to maintain a position in the higher ranks of the disciplinary order. Given that evolutionary processes render all taxonomic categories artificial and human-made, the concepts of species, genus, and type are now necessarily agreed-upon constructs. This does not mean that all taxa are characterized uniformly in each branch of zoology, but they are not arbitrary. Systematics has fine-tuned its methods, and international codes and agreements regulate its practices. Central databases with numerous contributors facilitate access to the wealth of information now available, and there are reference collections and data ready to be used from all corners of the world. There is even such a thing as 'cybertaxonomy.' All accepted taxonomic publications are peer-reviewed to avoid what is known as 'taxonomic vandalism,' the malpractice of describing new species by plagiarizing the work of others or without sufficient research to support the description. Often these faulty descriptions appear in obscure magazines and websites that have not undergone the process of peer review, much to the exasperation of the experts that have to straighten the subsequent muddle – as was frequently the case two centuries years ago. But unlike their peers in Temminck's time, experts are now assisted by the International Commission for Zoological Nomenclature, the organ that regulates the Code of Nomenclature (a direct descendant from Strickland's Code) and the nomenclatural practice, and arbitrates for the community. Nevertheless, despite the modern scientific methodology of systematics, its practitioners are still haunted by images of dusty museum drawers and eccentric European men wearing wide-brimmed hats and carrying nets for butterfly hunting.

The main issue today seems to be, just as in the nineteenth century, that scientific inquiry is more enthusiastically applauded if it is concerned with causal explanations of natural phenomena. Two hundred years after Strickland and Temminck decided to avoid any incursions into the realm of hypothesis, systematists today once again find themselves defending the prestige of their field. What seems to be overlooked, however, is that systematics and its results underpin practically all other zoological and applied research, as well as conservation efforts.

Any discipline's status constitutes much more than personal and professional prestige. It directly affects the attention and resources a discipline receives and the number of students it can attract. Today museums face financial cuts, taxonomy is seldom taught at the universities, and the discipline is suffering from the so-called 'taxonomic impediment': there is too much taxonomic information for the number of available experts to be able to process it. Similarly, the low status of systematics has resulted in a paucity of historiography: the existing literature is mostly episodic. The conceptual history of systematics is still fragmentary. There is more to learn about the philosophical premises of classification, what classifications represented, and why. A more detailed history of the foundations and aims of zoological classification, embedded in the pertinent social and cultural contexts, would clarify how and why systematics developed, and why its status within the natural sciences has been a matter of discussion for more than a century.

Appendix

Quotations

Introduction

9 Temminck's werken over de vogelen leerden eene rangschikking die den grootsten bijval vond. Men stond verbaasd over de naauwkeurige beschrijvingen, de naauwgezette synonymen en de menigvuldige opgaven omtrent de levenswijs en de zeden der vogelen, die alles overtroffen welke vroeger geleverd waren. (Witkamp, "Vier-en-twintig voorgangers," 199.)

10 [D]en man, die zich tot een systematischen zoöloog van den hoogsten stempel zoude ontwikkelen. (Vrolik, "Levensbericht van Coenraad Jacob Temminck," 66.)

16 Temminck's betekenis voor de wetenschap, speciaal voor de ornithologie en mammalogie, is groot. Hij publiceerde veel grondslaggevende werken en deed dit tot op hoge leeftijd. (Holthuis, *Rijksmuseum*, 24.)

Chapter 1. From Catalogs to Monographs

1 Men wil u niet, mijn waarde Temminck, omdat gij geen geleerde zijt. (Susanna, "Levensschets van Temminck," 74.)

8 Je me rappelle avoir assisté dans mon enfance à un diner chez M. Ameshoff, qui, pour faire étalage de la magnificence de sa ménagerie, faisoit servir sur sa table, non seulement des Pauxis des Hoccos et différentes espèces de Faisans exotiques; mais aussi les Sarcelles à évantails de la Chine et les Canards de la Louisiane, se trouvoient sur sa table, lors de ce festin digne des tems d'Héliogabale. (Temminck, "Histoire naturelle générale des gallinacés," in *Pigeons et gallinacés*, vol. 2, 458, fn. e.)

13 De Administratieve Municipaliteit van Amsterdam […] gelast zynde, om alle nuttelooze of onwaardige plaatselyke Ambtenaaren en Bedienden, van hunne posten te ontzetten, en dezelven met waardige en kundige Vaderlanders te voorzien. (Administrative Municipality of Amsterdam to Coenraad Jacob Temminck, 6 April 1798, Naturalis Biodiversity Center Archives, Archive C. J. Temminck.)

19 [A]l vroeg was ik op en herdacht met vreugde het voorrecht dat ik weder genoot deze dag te beleven, zijnde onze trouwdag. (Van Lynden-de Bruïne, *In vogelvlucht*, 26.)

23 Hij was zeer sober in zijn levenswijze, ongemeen net en naauwkeurig op zijn kleeding en omgeving. (Snellen van Vollenhoven, "Coenraad Jacob Temminck," 68.)

26 [J]e ne pouvois me lasser d'admirer celui de M. Temminck, Trésorier de la Compagnie des Indes, & la brillante Collection qu'il renferme. J'y remarquai

une foule d'objets précieux que je n'avois jamais vus en France. Tout m'y parut extrêmement rare, & de la conservation la plus pure. (Levaillant, *Voyage de M. Le Vaillant*, 1.)

31 M. Temminck, quoique jeune encore, est un homme à distinger de tous ces curieux qui, n'aimant que les espèces brillantes et à beau plumage, entassent à grands frais collections sur collections, sans aucun goût pour l'étude de la science. (Levaillant, *Oiseaux d'Afrique*, vol. 5, 61.)

32 Je dois à Le Vaillant et à ses écrits mes premières pensées et mon premier élan en histoire naturelle; ses ouvrages et ses conseils me servirent de guides dans l'étude pratique de cette science, et je me plais à reconnaître qu'il posa, à son retour d'Afrique, les premières bases d'une collection qui, passant depuis du père au fils, fait aujourd'hui le plus bel ornement du Muséum des Pays-Bas. (Temminck and Meiffren Laugier de Chartrouse, *Nouveau recueil*, vol. 5, livraison 64, plate 380, text, [1–2].)

37 Etes vous toujours content de notre fino T.-, pour moi je ne le suis guére? J'ai reçu une lettre de Geneve de Mr. Gosse que n'est pas fort content non plus aussi que Mr. Boissiere [?], les echanges qu'il ont fait avec lui. Il parroit qu'il a attrappe tout le monde. Vous savez sans doute qu'il est president d'une Academie, directeur d'une Ecole de peinture, Chambellan du Roy et enfin Chevallier de Union. Dieu veuille que tous ses honneurs le changent un peu. (Levaillant to Prêtre, 15 January 1809, Staatsbibliothek zu Berlin (SBB), Manuscripts Department, Slg. Darmstaedter Afrika 1780: Levaillant, François.)

40 Allé vous maintenant etre 3-mois ou 6 et meme une année sans me repondre touchant mon argent il faut convenir que vous etes un diable d'homme. (Levaillant to Temminck, 27 July 1812, Naturalis Biodiversity Center Archives, Archives C. J. Temminck.)

42 Serait-ce pour s'être prononcé avec trop de franchise sur des écrits de ce genre, et pour avoir émis une critique sévère, peut-être souvent trop amère, que ce savant ornithologiste s'est vu réduit dans ses vieux jours au plus strict nécessaire, sans aucune espèce d'encouragement et sans cueillir le plus léger fruit de ses travaux, au milieu des récompenses, des grâces et des titres donnés à ses compatriotes, et accordés dans tous les temps au vrai mérite et aux hommes célèbres dont la France s'honore. (Temminck and Meiffren Laugier de Chartrouse, *Nouveau recueil*, vol. 3, livraison 4, plate 23, text, [4].)

56 Il aura pour titre Histoire générale et complette [sic] des Gallinacés [...] ou les Amateurs trouveront representés environ quatre-vingt individus nouveaux, de la plus–part desquels aucun Auteur n'a fait mention. (Temminck, "Preface," in *Catalogue systématique*, vi.)

63 Tous les moyens mis en œuvre pour en appeler contre un acte si arbitraire, furent sans effet, et ma voix ne put alors s'élever contre l'intrigue soutenue par

des protecteurs puissants; les journalistes refusèrent de placer mes reclamations dans leur feuilles; même celle en réponse à l'article que le nouvel auteur avait fait publier par ces journeaux, me fut interdite. (Temminck, *Pigeons et gallinacés*, vol. 3, 643.)

64 [L]es désagréments que j'ai recueillis pour fruit de mon désintéressement envers les Editeurs doivent être mis au rang des évenemens familiers de nous jours. (Temminck to Hoffmannsegg, 29 July 1812, in Stresemann, "Briefwechsel von Temminck mit Hoffmannsegg und Illiger," 258.)

65 Vous en devés être mécontent jusqu'à la faire réimprimer. D'ailleurs j'en félicite nous autres pauvres Ornithologistes, à qui un ouvrage somtueux est défendu, que Vos remarques sans doutes précieuses sur une classe intéressante nous seront par cela accessibles. (Illiger to Temminck, 14 September 1812, in Stresemann, "Briefwechsel von Temminck mit Hoffmannsegg und Illiger," 260.)

74 A cette fin l'ouvrage portatif que je leur offre, donne une description concise et exacte, non-seulement de chaque espèce, mais aussi de ces variétés, tant de sexe que d'âge, ou simplement de celles qui sont accidentelles. (Temminck, "Avant-propos," in *Manuel d'ornithologie*, viii.)

78 Comment oser se flatter de voir jamais en seul homme embrasser l'étude de l'ensemble de la création? Pour décrire avec précision, et d'après de la nature, une seule classe du règne animal, la vie de l'homme le plus actif suffirait à peine. Le meilleur moyen que les naturalistes puissent choisir pour faire connaître une serie de résultats des faits présentés en masse, paraît être celui de publications par *Monographies*. (Temminck, "Discours préliminaire," in *Monographies de mammalogie*, vol. 1, viii–ix.)

Chapter 2. From Collector to Director

1 [J]'en dois l'hommage à la réunion des savants et des protecteurs zélés, dont les travaux tendent au noble but d'éclairer les sciences par les lumières de leur esprit, de les protéger et de les encourager. (Temminck, "Dédication," in *Manuel d'ornithologie*.)

20 [O]ndanks eene Franse post door mij uit pecuniare noodzakelijkheid is aangenoomen, deszelfs waarneeming mij niet heeft doen ophouden Nederlander te zijn[.] (Temminck to Falck, February 1813, Naturalis Biodiversity Center Archives, Archive C. J. Temminck.)

31 In dit alles zie ik gaarne alzo, wanneer men van U en elders toevoer van Naturaliën ontvangt, en dan te meerdere drang kan aangevoerd worden, om het locaal te veranderen, en om des te meer, alzo wij ondervonden hebben dat de bedomptheid en uitslag van de zo dikke Muren gebrek aan Lugt en Licht, de thans exteerende objecten aanmerkelijk doen deterioreren, vooral de Insecten, bij bij de welke men wel gedurig dien de te stooken, om de vogtige lugt te verwrijven,

en om niet door den tijd Campignons op de tonnen te plukken. (Temminck to Reinwardt, 25 November 1816, NL-HaNA, BiZa 1813–1848, 2.04.01, inv.nr. 4032B.)

32 Voor dit gebrek, behoeft men geen de minste vrees te hebben bij eene expositie conform aan die van het Museum van Parijs, van Weenen en van Londen, en zo iets diergelijks moet ook zo ik vertrouw door den tijd in U plan vallen. (Temminck to Reinwardt, 25 November 1816, NL-HaNA, BiZa 1813–1848, 2.04.01, inv.nr. 4032B.)

34 Ik moet hier nog bijvoegen, dat hoezeer ik Reinwardt bijzonder Hoogacht, en aan Zijn Ed. door oprechte en hartelijke vriendschap op het naauwst verbonden ben; ik nimmer van Zijn Ed. de nietige en Subalterne Commissie van een Kabinet Opzichter of van een Conservateur, op mij zoude genomen hebben, hiertoe is de tijd mij te kostbaar. Ik dagt veel meer, gedurende zijne afwezigheid nuttig te kunnen zijn, en zo wel voor zijn belangens zorgende, iets wezentlijks voor de Studien der Wetenschappen en voor het bevorderen van dien te kunnen uitrigten. [...] De Goede Man, is mij geheel statig komen zeggen, dat zijne ouderwetse wijze van denken en inzien der zaken, hem dwongen om zich tegens mijn gevoelens en daden te moeten verzetten[.] (Temminck to the minister of public education, national industry, and colonies, 23 March 1819, NL-HaNA, BiZa 1813–1848, 2.04.01, inv.nr. 4073.)

39 De arme Reindert deze zomer gehuwd zijnde, en zijn vrouw moetende kramen, is op het verdienen van iets meeder dan Drie Guldens bedagt[.] (Temminck to Reinwardt, 25 November 1816, NL-HaNA, BiZa 1813–1848, 2.04.01, inv.nr. 4032B.)

41 [W]at mij ook aangaat, valt enigzints moeijelijk voor den Heer van Marum, alzo ik zederd nu ruim ses maanden zijn Ed. slegts tweemalen heb gezien, en gesproken; en ik het raadzaamer heb gevonden het Veld van Eer te verlaten[.] (Temminck to Reinwardt, 25 November 1816, NL-HaNA, BiZa 1813–1848, 2.04.01, inv.nr. 4032B.)

50 [C]hargé d'organiser les cabinets pour les Universités de Gand, Liège et Louvain, j'ai été forcé de donner à cette partie un tems assez considérable, ce qui même a été cause que je n'ai pu travailler pour moi; aujourd'hui que l'organisation de cette partie touche à sa fin, je reprends avec empressement le fil de mes précédentes relations. (Temminck to Naumann, 5 March 1817, in Stresemann and Thomsen, "Ornithologen-Briefe aus den Jahren 1816 bis 1820," 111.)

51 Je n'ai point des étudians à mes ordres comme les très illustres professeurs; Généralissime et à la fois soldat de ma collection, tout doit passer par mes mains. (Temminck to Lichtenstein, 18 September 1818, in Stresemann, "Aus C. J. Temmincks Briefen an H. Lichtenstein," 255.)

52 De natuurlijke historie der dieren en delfstoffen, welke, voor zoo verre de dierkunde betreft, met *anatomie comparata* zal moeten vereenigd worden. ("Besluit regelende het hooger onderwijs," in *Staatsblad der Verëenigde Nederlanden*, no. 14, art. 63, 41.)

59 [I]k laat de bezigtigers van het Museum ook altijd zoo geleiden, dat zij naar deze Systhematische orde de verzameling zien. (Van Swinderen, *Index rerum naturalium*, xi.)

63 Ik heb mij in alles het nut der studerende jeugd tot hoofddoel gesteld, mij dus alle mogelijke moeite gegeven, om slechts zoo vele genera bij elkander te krijgen, als ik, in het korte tijdperk van zes jaren (want voor zes jaren was hier nog niets) kon magtig worden, om den Studenten een Systematisch overzigt te kunnen geven over de geheele wetenschap, mijns inziens het eerste doel der Akademische studiën. (Van Swinderen, *Index rerum naturalium*, xiii.)

70 [J]e répugne un peu, à vous parler franchement, à risquer mes oiseaux uniques à un long voyage, dans lequel ils pouraient s'abîmer totalement, ou bien perdre de leur fraicheur et de leur pureté. (Temminck to Naumann, 14 October 1816, in Stresemann and Thomsen, "Ornithologen-Briefe aus den Jahren 1816 bis 1820," 106.)

76 Car une collection, comme la votre, distinguée par le nombre d'espèces et la beauté d'exemplaires, le rend extrèmement difficile, de trouver quelque chose, qui soit digne d'y entrer. (Illiger to Temminck, 14 September 1812, in Stresemann, "Briefwechsel von Temminck mit Hoffmannsegg und Illiger," 259.)

80 Il est entendu que dans l'avertissement à la nouvelle édition de mon Manuel, votre nom sera cité comme de droit et vous paraitrez comme ayant cooperé à former l'ouvrage le plus parfait qui existera de l'histoire naturelle des oiseaux d'Europe[.] (Temminck to Naumann, 23 July 1816, in Stresemann and Thomsen, "Ornithologen-Briefe aus den Jahren 1816 bis 1820," 101.)

85 M. le professeur Lichtenstein occupe un rang distingué dans la sphère des sciences sous le triple titre de savant, de naturaliste et de voyageur. (Temminck and Meiffren Laugier de Chartrouse, *Nouveau recueil*, vol. 5, livraison 60, plate 361, text, [1].)

88 Le jeune Kuhl de Hanau, attaché depuis peu a l'une de nos universités m'accompagnera et me sécondera aussi dans cette année par ses traveaux; c'est un jeun home dont je fais beaucoup de cas et que je veux diriger pour l'envoyer quelque jour faire des voyages dans les pays peu connu[s] de nos colonies dans l'Inde. (Temminck to Naumann, 5 March 1817, in Stresemann and Thomsen, "Ornithologen-Briefe aus den Jahren 1816 bis 1820," 111.)

90 Het beeld was welgemaakt, dog vond het een aakelig gezigt deze gestalte te zien, want hoewel ik meer mommus en in gebalsemde liggaam gezien had, nimmer was dit mij voorgekoomen een mensch dat opgezet was en gelijk de vogels vertoond wierd. (Van Lynden-de Bruïne, *In vogelvlucht*, 163.)

94 Il m'a été communiqué par M. Dupont l'aîné, qui l'avait reçu de M. Bonelli, pour savoir de moi si je le regardais comme une espèce nouvelle, et c'est depuis ma décision qu'il a été envoyé à M. Temminck. Ne croyez pas que je mets une grande importance à faire connaître le premier une espèce nouvelle; mais je dois éviter

de passer pour un auteur qui s'approprie les faits des autres en changeant les noms, moyen employé très-souvent par certains savans. (Vieillot to Lesson, June 28, 1828, in Lesson, "Réclamation de M. Vieillot," 144.)

95 La figure lithographiée et très en diminutif d'un sujet bien plus jeune que celui donneé pl. 288 de ce recueil, vient de paraître dans un nouvel ouvrage sur les oiseaux de proie d'Europe, sous le nom d'Aigle intermédiaire; l'auteur n'a sans doute pas su que nous avions fait mention de cet oiseau sous le nom indiqué ci-dessus. (Temminck and Meiffren Laugier de Chartrouse, *Nouveau recueil*, vol. 1, livraison 49, plate 288, text, [1], fn. 1.)

106 [J]'ai taché de rendre nos traveaux en quelque sorte cosmopolites et les noms adoptés et sanctionnés dans d'autres musées ont en partie fait place aux étiquettes existantes, tous les noms de Illiger ont été adoptés et tous les noms de Vieillot rayés[.] (Temminck to Lichtenstein, 20 May 1820, in Stresemann, "Origines des 'Planches coloriées'," 39.)

108 Le Souverain forme t'il le plan d'établir secondé par de tems propices un Museum voué aux sciences naturelles; et dans un pareil cas serait il porté à m'en confier la direction. (Temminck to Falck, 18 April 1814, Naturalis Biodiversity Center Archives, Archive C. J. Temminck.)

109 Les titres de notre respectable ami R. à l'employ de directeur, suffisent pour me faire renuncier à mon projet. J'ai communiqué à cet ami tous les details de ce plan, ainsi que vôtre lettre, afin de me mettre à l'abri du soupçon d'aviser par lui-même dans sa carrière. (Temminck to Falck, 18 April 1814, Naturalis Biodiversity Center Archives, Archive C. J. Temminck.)

112 Men had dan teffens een Nationaal Museum gesticht, als een kern aangelegt, om door zijne doubletten en door wel aangewende zorgen der directeuren, te kunnen strekken tot vermeerdering der Universiteits Cabinetten, en teffens ingerigt, om tot een voldoende vraagbaak te dienen, voor dezulken, welke zig in de onderscheide vakken der Natuurkunde door vergelijkingen de gewenschte volledige kennis trachten eigen te maken. (Temminck to Reinwardt, 25 November 1816, NL-HaNA, BiZa 1813–1848, 2.04.01, inv.nr. 4032B.)

116 Een Museum te schichten, overeenkomstig met de waardij onzer verzameling, overeenkomstig den luister onzer Academie en de staat der Wetenschappen. (Temminck to the commissioner-general of education, arts, and science, 4 December 1816, NL-HaNA, BiZa 1813–1848, 2.04.01, inv.nr. 4032B.)

122 [H]oe of sulks ook het plan moge zijn, zo is het nogthans billijk dat Reinwardt bij zijne terugkomst niet als ondergeschikt of als eventueele plaatsvervanger worde teruggeschoven[.] (Temminck to Falck, 17 July 1819, NL-HaNA, Collectie 048 Falck, 2.21.006.48, inv.nr. 85.)

123 [W]ordt er niets van een nationaal museum, of blijft de inrigting voortdurende gebrekkig, dan zal men mij niet ten kwade kunnen duiden, dat ik mijne

begeerte en lust tot uitbreiding mijner natuurkundige kennis in een ander land voortzette[.] (Temminck to Falck, 17 July 1819, NL-HaNA, Collectie 048 Falck, 2.21.006.48, inv.nr. 85.)

129 Ik beware hem [deze brief] echter als de kiem bevattende van het thans gerealiseerde denkbeeld van een groot Museum der Natuurlijke Historie. (Colenbrander, *Gedenkschriften van Anton Reinhardt Falck*, 411.)

131 Het komt mij wijders voor, dat een Directeur der Botanische Tuinen, belast met het Professoraat van de Chimie, Botanie, Mineralogie, Vergelijkende Anatomie en alle Vakken der Zoologie, oneindig veel te doen heeft. (Temminck to Falck, 29 October 1819, NL-HaNA, ARA, 2.14.03, inv.nr. 5.)

Chapter 3. National Museum, National Expeditions

3 Geene, het zij pathologische hetzij phijsiologische praeperaten van de bijzondere organen der dieren, behooren tot den Omvang van dit Kabinet. ("Extract uit het Register der Handelingen en Resolutien van den Minister, voor het Publieke Onderwijs, de Nationale Nijverheid en de Koloniën, No. 30 Den 31 December 1820," in Gijzen, "'s Rijks Museum," 17.)

11 De graad, die ik bij mijne Promotie verkregen heb, en de rang die ik in andere opzigten in de Maatschappij bekleede, laten mij geen vrijheid om eene subalterne post te ambieeren. (Van Hoorn to Temminck, 24 August 1821, Naturalis Biodiversity Center Archives, Archive C. J. Temminck.)

17 Dat de Hr. Temm. Directeur van het Museum is en ik dus buiten den oorlog met Motten en Spekkevers kan blijven is mij waarlijk aangenaam. (Reinwardt to Van Marum, 1 January 1822, NHA, Archive Martinus van Marum 529-20b.)

21 De Directeur van 's Rijks Museum van Natuurlijke Historie te Leijden, adverteert bij deze, dat, hoezeer de galerijen nog niet zijn voltooid, en al de voorhanden zijnde voorwerpen niet kunnen geplaatst worden, er echter, aan diegenen, welke verlangen mogten hetzelve *in dien staat* te bezigtigen, daartoe gelegenheid zal worden verschaft, door het afgeven van toegangsbilletten, welke te bekomen zijn op de Papengracht[.] (Temminck, "Van de Directeur van 's Rijks Museum," in *'s Gravenhaagsche Courant*, May 8, 1822.)

36 Men kan het dier daardoor beter in deszelfs natuurlijke gedaante herstellen; iedere uitwendig zigtbare spier en beengedeelte scherper en zigbaarder doen voorkomen, en daareenboven, behoeft men naar deze wijze de huid niet meer met gaten te doorboren; daar dezelve thans, om den vooraf, wel en naauwkeurig gemaakten, vorm, als een gewaad wordt vast gespannen. De meerdere moeite, welke deze bewerking veroorzaakt, wordt rijkelijk opgewogen door den voortreffelijken uitslag. (Temminck, Report to the Ministry of Home Affairs on the Year 1828, 10 May 1829, Naturalis Biodiversity Center Archives, Jaarverslagen.)

42 [M]oet men veronderstellen dat de natuurkundige wetenschappen in het Brittische Koningrijk met de meeste luister vrugten moeten vortbrengen. Wel verre van eene zo gunstig tijd-perk te nuttte te maken, waar in zij op de zeen den skepter zwaagde; hebben zij niets wezentlijks uitgerigt. Ouden verlamde instellingen, beuzelagtige gebruiken onnutte personen en niets wezentlijk uitrigtende genootschappen met oude privilegien omhuld kwamen telkens in den weg, en staan nog tegenwoordig in houding, om alle goede instellingen te verijdelen. Niets wezentlijks kan in Engeland voor de Natuurkundige Wetenschappen ondernomen worden en tot stand komen, zo lang een Aartsbisschop van Canterbury de Lord Maior van London en de Speaker of the house of commons de opper rangen uitmaken van het Brittische Museum; en de president der Royal Society het regt kan hebben een chimist tot Directeur van de Zoologie te doen benoemen. (Temminck, Report to the Ministry of Home Affairs on the Year 1824, 3 October 1824, Naturalis Biodiversity Center Archives, Jaarverslagen.)

44 Om welke redenen laat ons het gouvernement altijd met Januari, Februari en maart zonder een penning disponibel, heeft men de gedachte dat wij als de woeschte dieren de winterslaap houden of van onze opgegaarde voorraad leeven als de hamsters. Geen voorschoten en geen betalingen op de gehorige tijd, eene fraaye wijs om huishouding te maaken! Nu, een ieder schikke zig zo hij kan[.] (Temminck to Thijssen, 28 February 1824, Naturalis Biodiversity Center Archives, Archive C. J. Temminck.)

50 Zoo noemde hy b.v. het Museum van Nat. Hist., 'het apenhuis', het Herbarium den 'Hooizolder'. (Schlegel, "Levensschets van Hermann Schlegel," 69.)

52 Het Rijks Museum van Natuurlijke Historie blijft nog steeds zijnen teregt verkregen roem op eene lusterrijke wijze handhaven en die roem wordt betrekkelijk, in geen geringe mate verhoogd, door het in 't nog loopende verval, waaronder soortgelijke Inrigtingen in andere, zelfs groote Rijken, lijdende zijn. Zoo lang de Hemel mij leven en krachten schenkt, hoop ik het voorwerp, waaraan ik meer dan dertig jaren al mijn tijd en vlijt, met onverpoosde inspanning heb besteed, voor zulk een droevigen toestand te behoeden, en ik zal daarin slagen, zoo lang ik mijne krachten mag geschraagd zien door de bescherming en ondersteuning der Regering, zoo als mij die, ik erken het dankbaar, tot den huidigen dag in ten deele geweest en waaraan het mij onder het krachtige bestuur van Uwe Exc. gewis ook is 't vervolg niet zal ontbreken. (Temminck, Report to the Ministry of Home Affairs on the Years 1847–1850, 26 November 1851, Naturalis Biodiversity Center Archives, Jaarverslagen.)

55 Vous dire que je m'occupe des Gallinacés et des Oiseaux d'eau, c'est vous dire combien je pense à vous, car si vous ête le Papa (non Pape) de tous les oiseaux, vous ête le Génie des poules et des Goelands[.] (Bonaparte to Temminck,

1 January 1851, Temminck Family Archives, A. M. van Lynden-de Bruïne private collection, no. 46.)

56 Et vivent les Gallinacés, les pigeons et les oiseaux d'eau! Ils reportent le vieillard au temps de sa jeunesse, et me font rajeunir d'un quart de siècle. (Temminck to Bonaparte, 8 January 1851, Temminck Family Archives, A. M. van Lynden-de Bruïne private collection, no. 47.)

57 Un vétéran en déclin de sa carrière, mais qui se berce encore de l'espoir de mourir les armes à la main. (Temminck to Wied-Neuwied, March 1854. Temminck Family Archives, A. M. van Lynden-de Bruïne private collection, no. 48.)

69 In aanmerking nemende de aanstaande terugkomst van den heer Reinwardt en willende meer en meer bevorderen de kennis van de natuurlijke gesteldheid van de voortbrengsels onzer bezittingen in de Oost-Indiën. [...] De Heeren H. Kuhl [...] en J. C. van Hasselt worden genoemd om, onder de orders van onzen Goverņeur Generaal, de Nederlandse bezittingen in Oost Indië te bereizen, met het bepaalde doel, om de de wetenschappelijke kenis van de voortbrengselen der Natuur, in die landen, uit te breiden. (Royal Decree no. 10, 2 May 1820, NL-HaNA, Rijksmuseum van Natuurlijke Historie te Leiden (RMNH), nummer toegang 3.12.17, inv.nr. 1.)

83 Deze zijn Kruidkundige nasporing zullen zich echter geenzins alleen mogen bepalen tot eene beschrijven van gedaante, nog tot de bepaling en onderscheiding van de geslachten en soorten der gewassen, maar zullen – aangezien door de Zending van den Heer Korthals naar Nederlands Indie, bepaaldelijk eene, voor de geweste, voor Nederland en voor Wederzijdsche betrekkingen onmiddelijk nuttige strekingen bedoeld wordt – eene meer algemeene, nuttige en vructhbare Kennis der gewassen tot oogmerk hebben. (Van Ewijck to Temminck, 10 November 1830, NL-HaNA, RMNH, 3.12.17, inv.nr. 4.)

84 Waarom heeft men aan K. [Korthals] niet vergund, naar zijne instructie te werken? Waarom zijn zijne tot Economie of Landbouw betrekkelijke, ingediende stukken afgewezen of onbeantwoord gebleven? Waarom de opzettelijk last, om zich niet dan met verzamelen bezig te houden? Zullen ook voor het Indisch Gouvernement moeten dienstbaar zijn? (Temminck's minutes [1836], Naturalis Biodiversity Center Archives, Archive C. J. Temminck, NAT_ARC_TEM_00453.)

91 Het is voor de wetenschap immer nuttig; meer dan een Exemplaar der voorwerpen van Natuurlijke Historie in bezit te hebben; zijnde overigens alle wel bewaarde voorwerpen derzelve, voor de verzamelingen van het Museum van gelijk aanbelang, zonder dat de mindere of meerdere schoonheid van kleur of vorm iets toe- of afdoet. (Temminck, *Voorschrift*, 28.)

98 Uw Ed. zult my wel ten goede gelieven te houden, dat ik my de vryheid neeme, te vragen, welke Prijs UE. Inclineeren mocht te betalen voor ogelen, viervoetige Dieren, insekten, Kapellen, visschen en slangen, misschien ook Bloemzaden en

drooge Planten. (Brehm to Temminck, 15 October 1821, Naturalis Biodiversity Center Archives, Archive C. J. Temminck, NAT_ARC_TEM_00179.)

101 Zo heb ik in afwachting van antwoord de voorwerpen gekocht, en ben reeds in het bezit van dezelven, maar spreekt aan niemand dan aan Boié iets van dit gebeurde. (Temminck to Thijssen, 13 May 1822, Naturalis Biodiversity Center Archives, Archive C. J. Temminck.)

Chapter 4. A Place for Systematics

1 Geene beschrijving der ontdekte of overgezonden voorwerpen, noch der waargenomen natuurverschijnselen, zal in een vreemde taal mogen geschieden of aan Buitenlandsche Genootschappen en Geleerden mogen toegezonden worden, dan met toestemming van onzen minister voornoemd, en na de vroegere bekendmaking door een der geleerde Genootschappen of tijdschriften van dit Rijk. (Royal Decree no. 251, 29 April 1820, art. 3, in Veth, "Overzicht van hetgeen," 22.)

32 In 1846 ontmoetten wij hem op een pad, waar wij hem niet verwachten konden. (Susanna, "Levensschets van Temminck," 65.)

36 On a souvent blâmé la nation hollandaise ainsi que son gouvernement, de leur *prétendue* réserve en tout ce qui concerne nos possessions d'outre-mer. J'avais à coeur de prouver que cette imputation n'avait pas le moindre fondement[.] (Temminck, "Avant-propos," in *Coup-d'oeil général*, vol. 1, xvi.)

37 L'utilité de ces voyages de découvertes est de nos jours généralement reconnue; nous pouvons les recommander sous le double point de vue des progrès qu'ils font faire aux sciences, et de l'influence qu'ils peuvent avoir sur les destinées futures de nos lontaines possessions; même les prôneurs d'une stricte économie, et qui ne consultent le plus souvent que le côte positif des dépenses, seront dans la nécessité de convenir que celles, d'ailleurs dans tous les temps très-modiques que les gouvernements consacrent aux progrès des sciences, offrent immanquablement un but réel, immédiat et d'utilité publique. (Temminck, *Coup-d'oeil général*, vol. 2, 424–25.)

39 De verrijkingen der Academiën, en van dat ondankbare *Brussel*, van die muitende steden van *België*, wier nog geblevene gedenkzuilen, in spijt van het bijgeloovigst Wandalismus, dienst zullen bewijzen aan de beschaving van een dom en dweepziek volk; eindelijk de werken van natuurlijke historie, waardoor de kunsten werden aangekweekt, doch die oproerige handen niet hebben geschroomd onder geregtelijke verzekering te leggen. (Temminck, "Algemeen overzigt van de orde der Cheiroptera," 316, fn.)

45 Leur contenu [du Manuel et mon système] est basé sur l'examen le plus sévère de la nature, sans aucune espèce de compilation; toutes les espèces ont été vues et souvent comparées entre elles dans tous les cabinets d'Europe; voilà peut-être

les seuls mérites de mon ouvrage, et la seule différence qui le distinguera de ceux publiés par des naturalistes sédentaires et de bibliothéque. (Temminck, "Introduction," *Manuel d'ornithologie*, 2nd ed., vol. 1, xxvii–xxviii.)

60 De natuur toont ons de volkomende éénheid en de onbegrijpelijkste rijkdom van verscheidenheden, zijn kent geen stelsel, geen verdeelingen, geene klassen noch soorten: neen! Het [is] thans buiten bedenking dat de vergelijkende ontleedkunde de eenige vaste grondslag moet zijn zoo veel van Systematische als Philosophische beschouwing van het Dierenrijk. (De Jonge, *Het vergeten fenomeen*, 19.)

64 Hij [Van Marum] nam in overweging, om bij de verzameling van opgezette dieren eene andere daarbij eigenaardig behoorende, meer wetenschappelijke verzameling van vergelijkende ontleedkunde aan te leggen, waarvoor hij in de Ontleedkundige School van Prof. Camper grooten lust en eenige hebbelijkheid in het vervaardigen van praeparaten verkregen had. ("Van Marum's betrekking tot de Hollandsche Maatschappij der Wetenschappen, etc.," NHA, Archive Martinus van Marum 529-9.)

76 Er gaat naauwelijks een dag voorbij, op welken geen nieuw Genus gevormd of geene nieuwe Species bestemd wordt, en nog verzamelen wij zeer veel, hetwelk voor 's hands niet verder in aanmerking wordt genomen, en hetwelk wij eerst in Europa denken te onderzoeken. (Kuhl and Van Hasselt to Temminck, 18 July 1821; published in Kuhl and Van Hasselt, "Pjihorjavor," 99.)

77 [E]ene Verzameling van Systhematische gelegde platen uit de beste schrijvers der Natuurlijke historie gecollecteerd, zijnde bijna een volkomen *Systhema Natura* in platen[.] (Will of Johan Conrad van Hasselt, 2 October 1823, Naturalis Biodiversity Center Archives, Archive Natuurkundige Commissie, MMNAT01_AF_NNM001001046_008.)

79 Goede aantekeningen maken, alles verzamelen en voor het verzamelde de meeste zorg dragen, zoo dikwerf het mogelijk is bezendingen naar het Moederland doen, ziedaar de hoofdinstructie voor een natuuronderzoeker in de tropische gewesten. (Temminck to the minister of public education, national industry, and the colonies, 21 March 1822; quoted in Veth, "Overzicht van hetgeen," 29.)

80 [Kuhl] had veel meer nut in Europa kunnen stichten, maar een veld van dien aard, hoe uitgebreid ook in het oog van anderen, vond hij te eng. [...] Ik wil niet ontveinzen en de uitkomst doet het blijken, dat Kuhl juist de man niet was om tot uitbreiding van een museum en tot verzamelen te gebruiken. (Temminck to the minister of public education, national industry, and the colonies, 21 March 1822; quoted in Veth, "Overzicht van hetgeen," 29.)

90 Il est nécessaire de faire observer à ceux qui compareront notre planche avec celle publiée par M. de Humboldt, que le dessin fait par ce savant et retouché par Barrabant represente la tête et le cou dessiné d'après un individu vivant;

ces mêmes parties, dans la planche ci-jointe, ont été dessinées sur un empaillé. Avant de la publier j'ai soumis ma gravure à M. de Humboldt; ce savant trouve que l'occiput est trop relevé, le dos trop bossu et les rides transversales du cou ne sont point indiquées; tous ces défauts sont des suites d'une préparation vicieuse de la peau dans l'individu monté. (Temminck and Meiffren Laugier de Chartrouse, "Catharte Condor," in *Nouveau recueil*, vol. 1, livraison 23, plate 133, text, [4], fn. 1.)

92 Y-a-t-il bien certainement un oiseau qui se trouve partout, des oiseaux identiques entre les tropiques dans les 2 continents p.e. en Afrique et en Amérique? Y-a-t-il des mammifères et oiseaux identiques dans les zones tempérées boréales et australes. Ignorant que je suis, je voudrais être exact dans les peu de pages que je dois consacrer à la Géographie des animaux. Les sources dans lesquelles je devrais puiser, ne me sont pas très abordables. Je désire un petit nombre d'exemples piquans [sic] relatif aux questions que je viens de toucher. Ce que Vous pouvez donner si facilement, c'est de l'or en barre. (Humboldt to Temminck, 16 August 1846, Naturalis Biodiversity Center Archives, NAT_ARC_TEM_01617.)

Chapter 5. Patterns, Laws, and Types

4 Les productions zoölogiques du Japon […] nous fournissent des preuves nouvelles sur la répartition geógraphique des êtres: elles servent à constater cette loi d'organisation animale, […] qu'il y a rapport d'organisation, de formes extérieures et de mœurs, entre le plus grand nombre des animaux qui habitent des latitudes correspondantes, quelque éloignées que puissent être entre elles les contrées où ils vivent et se propagent en liberté, sans que l'étendue plus ou moins vaste des mers qui les sépare, ait en cela, la moindre influence, comme sans égard à la distance énorme qui éloigne ces contrées, les unes des autres, par une suite non interrompue de terres. (Temminck, "Aperçu général," in *Fauna japonica*, vol. 5, 7.)

5 Cette loi est applicable, sous certains rapports, aux deux hémisphères, mais elle l'est plus strictement pour l'hémisphère septentrional, où existent des espèces plus exactement semblables entre elles, tandis que dans l'hémisphère méridional, les groupes et les espèces y sont repartis sur un modèle plus diversifié, quoique offrant, le plus souvent, une grande analogie dans les animaux qui vivent sous ces latitudes parallèles. (Temminck, "Aperçu général," in *Fauna japonica*, vol. 5, 7.)

7 Onder de van Japan bekomene zoogdieren heeft men nog die soorten niet aangetroffen, wier opmerkelijke vorm werkelijk van de bekende grondvormen (*types*) verschillen, zoo als die in Australie gevonden worden, behalve eenigen, die slechts weinig van onze bekende geslachts-afdeelingen verwijderd zijn; kunnende al de overigen in de systematische afdeelingen der europesche en

aziatische zoogdieren gerangschikt worden. (Temminck, "Kennis en verbreiding der zoogdieren van Japan," 287.)

11 Wanneer men al deze geringe, plaatselijke onderscheidingen in aanmerking wilde nemen, om haar te doen strekken tot vermeerdering van het getal soorten, zoude zulks, vooral wanneer zij ter naauwernood zijn op te merken, het zekerste middel zijn, om den voortgang der studie gedurende een' langen tijd te belemmeren [...]; van die vormen, welke door het gezigt naauwelijks te onderkennen zijn; van eene kleingeestige, tot strepen herleide uitmeting; met één woord, van het invoeren van karaktertrekken, welke met het scalpel in de hand en door een met de loup gewapend oog, moeten worden opgespoord; terwijl het geheel der vormen van deze naamsoorten onmiskenbaar en op den eersten blik, ten duidelijkste al de voornaamste karaktertrekken erkennen doet, welke aan den typus van de eenige, ware soort eigen zijn. (Temminck, "Kennis en verbreiding der zoogdieren van Japan," 292–93.)

12 Ce nouveau genre vient remplir une lacune très importante dans la série animale; il prouve, jusqu'à l'évidence, que plus le cercle de nos connaissances vient à dépasser les bornes jadis tracées, et que les découvertes nouvelles d'animaux dont nous n'avions jusqu'ici aucune notion se multiplient, plus aussi nous pouvons nourrir l'espoir de connaître, sous peu, tous les chaînons de cette chaîne non interrompue qui lie étroitement entre eux les groupes différents d'animaux, tant ceux anciens ou des siècles passés. (Temminck, "Description d'un nouveau genre de mammifères (Urotrichus)," 1.)

13 La vue d'une multitude d'individus de la même espèce, mais originaires de contrées différentes, nous indique clairement qu'il est souvent bien difficile de parvenir à déterminer strictement la limite entre les espèces analogues, et de dire positivement qu'il existe une séparation spécifique ou une démarcation bien tracée pour toutes les espèces que nous séparons comme telles. En résumé, l'existence des *Genres*, en donnant à cette dénomination l'explication conventionelle, me paraît un paradoxe: je suis même loin d'établir celle des *espèces* en axiôme. (Temminck, "Introduction à cette troisième partie," in *Manuel d'ornithologie*, 2nd ed., vol. 3, xix.)

16 [O]n doit définir l'espèce, la réunion des individus descendus l'un de l'autre ou de parens communs, et de ceux qui leur ressemblent autant qu'ils se ressemblent entre eux[.] (Cuvier, *Règne animal*, vol. 1, 19.)

22 [L]es espèces puissent avoir subi des altérations locales et se montrent sous de légères variétés; ce qui n'empêche pas que les dissemblances, par lesquelles ces variétés sont caractérisées, ne soient pas assez tranchées, pour qu'on puisse les isoler des types et les classer comme espèces réellement différentes. (Temminck, "Aperçu général," in *Fauna japonica*, vol. 5, 8.)

25 Il y a dans la nature un prototype général dans chaque espèce, sur lequel chaque individu est modelé, mais qui semble, en se réalisant, s'altérer ou se perfectionner par les circonstances; en sorte que, relativement à de certaines qualités, il y a une variation bizarre en apparence dans la succession des individus, et en même temps une constance qui paroît admirable dans l'espèce entière. (Buffon, "Quadrupèdes," in *Histoire naturelle*, 61–62.)

26 Le premier animal, le premier cheval, par exemple, a été le modèle extérieur et le moule intérieur sur lequel tous les chevaux qui sont nés, tous ceux qui existent et tous ceux qui naîtront, on été formés[.] (Buffon, "Quadrupèdes," in *Histoire naturelle*, 62.)

28 Quelle étrange erreur n'est pas celle, d'indiquer le Bizet ou Colombe Sauvage comme le père commun de toutes les différentes espèces de Pigeons, répandues sur la surface de la terre; voyons quelles sont les raisons qui ont pu porter M. de Buffon à statuer une loi que la nature désavoue presque à chaque individu? Je vais en faire connoître la cause; après quoi il me sera facile de prouver que la supposition de M. de Buffon est dénuée de toute vraisemblance. Le manque d'observations, faites sur la nature, est la principale source d'une erreur qui proflue d'une trop grande confiance dans son génie créateur; cette sécurité fatale pour la science a souvent perpétué les écarts d'un grand homme, dont on se plaît a juste titre, à respecter le mértie et à honorer les talents. (Temminck, "Discours sur l'ordre des pigeons," in *Pigeons et gallinacés*, vol. 1, 16.)

34 Ainsi les caractères les plus superficiels sont les plus variables; la couleur tient beaucoup à la lumière; l'épaisseur du poil à la chaleur; la grandeur à l'abondance de la nourriture [...] Ainsi, quoique le loup et le renard habitent depuis la zone torride jusqu'à la zone glaciale, à peine éprouvent-ils, dans cet immense intervalle, d'autre variété qu'un peu plus ou un peu moins de beauté dans leur fourrure. (Cuvier, *Discours sur les révolutions*, 59–60.)

49 Les trois grandes îles des régions équatoriales qui forment, avec quelques autres de troisième et de quatrième rang, le groupe désigné sur nos cartes géographiques sous le nom d'archipel de la Sonde, offrent, par leur position et par leur sol fertile couvert d'une végétation vigoureuse, le point le plus favorable à l'étude des animaux de la zone tropicale de l'ancien continent; par leur grand nombre réuni dans ces îles, la comparaison devient plus facile; tandis que là, mieux que partout ailleurs, on peut obtenir des données certaines sur la répartition géographique des espèces. (Temminck, "Faune des Iles de la Sonde et de l'empire du Japon," in *Fauna japonica*, vol. 1, iv–v.)

50 Les *Phalangers* et les *Couscous*, animaux à bourse [...] sont, pour l'ancien continent, ce que sont les *Sarigues* et les *Didelphes* dans les contrées parallèles du Nouveau Monde. (Temminck, *Coup-d'oeil général*, vol. 3, 113, 16.)

53 [L]a science qui conduit l'homme à lire dans les pages divines du livre de la nature, qui lui fait connoitre la partie animée et inanimée de la création qui élève insensiblement son ame exaltée à adôrer cet Être tout-puissant dont l'atôme rappèle l'existance[.] (Temminck, "Discours préliminaire," in *Pigeons et gallinacés*, vol. 1, 6.)

54 Il est donc plus probable que les élémens qui ont exercé leur influence sur la création ont reproduit les mêmes types à l'extrémité orientale de l'Asie et dans l'Amérique du nord, comme sous le climat de l'Europe[.] (Temminck, "Aperçu général," in *Fauna japonica*, vol. 5, 8.)

56 La partie septentrionale nourrit toutes ces espèces, dont les representants ou les types identiques existent dans les autres îles de l'archipel: de l'autre côte de ces montagnes, ou à leur versant méridional, se montrent les premiers répresentants d'une Faune qui semble différente, ou au moins transitoire entre les espèces propres aux Moluques et les groupes australiens, qui semblent faire partie d'une création toute particulière. (Temminck, *Coup-d'oeil général*, vol. 3, 168.)

57 Les *Phalangers* et les *Couscous*, animaux à bourse, qu'on ne voit nullepart dans les limites des îles de la Soende, fournissent ici [Celèbes] leurs premiers répresentants[.] (Temminck, *Coup-d'oeil général*, vol. 3, 111–13.)

70 La végétation sur son sommet [Gunung Bongsoe] porte tous les caràcteres des plantes alpestres d'Europe[.] (Temminck, *Coup-d'oeil général*, vol. 2, 82.)

78 Il serait également intéressant, d'offrir à côté de ce tableau esquissé à grands traits, celui des différences très notables qu'on observe dans la distribution géographique des animaux, selon les degrés de longitude où les espèces se trouvent reparties sur notre sphéroïde terrestre. Cette matière riche en faits et en conséquences, toute intéressante qu'elle nous semble, nous entraînerait trop loin du but spécial de ce mémoire, c'est-à-dire, de la description des mammifères du Japon. (Temminck, "Aperçu général," in *Fauna japonica*, vol. 5, 8–9.)

Chapter 6. Systematics Wars

8 Je n'ai épargné ni travaux ni moyens pour mettre l'*Index général* au niveau des connaissances actuelles en ornithologie, c'est-à-dire, que je l'ai épuré, autant que possible, des emplois doubles, triples et souvent quadruples dont Gmelin, Latham ainsi que plusieurs ouvrages, plus récens encore, sont encombrés. (Temminck, "Introduction," in *Manuel d'ornithologie*, 2nd ed., vol. 1, xviii–xix.)

11 Nomen genericum dignum; alio, licet aptiore, permutare non licet. [...] Abstinendum ab hac innovatione quæ numquam cessaret, quin indies aptiora detegerentur ad infinitum. (Linnaeus, *Philosophia botanica*, 196.)

12 Il n'est guere permis de soupçonner que notre savant a cru de bonne foi qu'on passerait son pillage sous silence, et que sa nouvelle classification et sa

nomenclature seraient citées comme les fruits de ses propres découvertes, et des conceptions de son génie. (Temminck, *Observations sur la classification*, 55–56.)

13 [U]n pillage un peu trop effronté, consistant à s'appropier les recherches et la nommenclature publiés par des auteurs contemporains, qui ne se trouvant point cités, voient les fruits de leurs conceptions altérés dans le sens des mots ou par une ortographie différente. (Temminck, *Observations sur la classification*, 7.)

16 [T]elle est la compilation que l'auteur donne pour le fruit de ses travaux, et comme un modèle que tous les ornithologistes présens et à venir doivent adopter. N'est-ce pas rappeler la fable du Geai? (Vieillot, "Ornithologie," 97–98.)

17 Rassurez-vous, trop scrupuleux Vieillot! Je ne rappellerai point à mon tour d'autres allégories du bon La Fontaine … (Temminck, "Introduction," in *Manuel d'ornithologie*, 2nd ed., vol. 1, xv.)

18 Sans doute déjà quelques-uns de mes lecteurs ont rendu justice à la conduite de M. Vieillot, et peut-être on me blâmera de répondre à des puérilités; mais attaqué, comme je le suis, par un censeur qui vise à la célébrité, non moins par les prétentions littéraires que par les travaux scientifiques, je montrerai à mon tour de la suffisance en empruntant d'autres armes que les siennes. (Temminck, "Introduction," in *Manuel d'ornithologie*, 2nd ed., vol. 1, xi–xii.)

20 Suivant ma manière de voir, il est préférable de conserver à une espèce telle ancienne dénomination, qui la fait reconnaître de tout le monde, (la composition de ce nom fut-elle même barbare au point de ne dériver ni de racine Greque ni de la langue Latine), plutôt que d'en substituer une autre à la place, dont la composition mieux choisie et plus grammaticale, seroit susceptible d'occasionner la plus legère méprise; car rien n'est plus funeste au développement de l'étude des sciences naturelles, et particulièrement de celle qui comprend l'histoire des oiseaux, que ces différentes opinions sur la dénomination des genres et des espèces[.] (Temminck, "Avant-propos," in *Manuel d'ornithologie*, xi.)

22 [I]l ne convient plus de suivre cette manière excessivement vicieuse […] d'emprunter pour dénomination des animaux, les noms de pays, d'îles ou de contrées, puisqu'il est rare qu'une espèce se trouve confinée dans les limites de nos divisions géographiques. (Temminck, *Monographies de mammalogie,* vol. 1, 132.)

25 On a critiqué, avec une apparence de raison, les noms de pays donnés aux objets que peuvent exister naturellement dans des contrées fort éloignées l'une de l'autre; mais si l'on considère, qui ces noms ne sont imposés qu'à des êtres qui sont tout-à fait nouveaux, et dont on ignore l'existence partout ailleurs que dans le pays qui les fait connaître, on leur trouvera l'avantage de conserver la trace de l'origine de ces êtres pour nous; et comme tout le monde sait, comme il est facile de convenir que ces désignations n'ont rien d'absolu, on reconnaîtra que de tous les noms qui ont une signification, ce sont ceux des pays qui présentent

peut-être les moins d'inconvénients. (Geoffroy Saint-Hilaire and Cuvier, "Ours de Sibérie," in *Histoire naturelle des mammifères*, livraison 42.)

35 Si nous avons différé à entretenir nos lecteurs de cet important recueil, nous n'en sommes que plus fondés dans l'éloge mérité que nous en devons faire; car la sixième livraison qui vient de paraître, égale au moins en beauté et en perfection, si elle ne les surpasse pas, celles qui l'ont précédée. C'est, au reste, ce que l'on devait attendre de la réputation des auteurs, et des droits que depuis long-temps ils se sont acquis à la reconnaissance des savans. ("Nouveau recueil," in *Annales générales*, vol. 7, 361.)

36 Cette belle entreprise se poursuit avec un zèle admirable, et ses auteurs reçoivent de toutes parts les témoignages de satisfaction qu'ils étaient en droit d'attendre des nombreux partisans de l'histoire des oiseaux. ("Nouveau recueil," in *Annales générales*, vol. 8, 395.)

43 [L]e système de ce savant [Latham] étant, de toutes les méthodes qui existent, la plus complète et la moins encombrée de citations à double emploi, fruits de la misérable compilation du professeur Gmelin, qui a eu le talent de former, de la 13e. édition de Linné, le livre le plus indigeste qui existe: aussi tous ceux qui s'obstinent encore à le suivre servilement, ne peuvent manquer de tomber dans les erreurs les plus grossières. (Temminck, "Avant-propos," in *Manuel d'ornithologie*, xxxii.)

46 [D]ivisée sans necessité absolue en tribus et familles, faisant partie d'ordres qui portent des caractères trop vagues et donnent trop de latitude; subdivisés en des genres multipliés sans nécessité absolue; ces genres (alors que l'on vouldrait suivre le plan tracé par l'auteur) trop peu nombreux en proportion des êtres connus, qui n'ont point fait partie de ses recherches, puis encore une subdivision en sections si nombreuses, qu'il ne faudra qu'un pas de plus pour que les espèces disparassent et qu'ainsi faisant l'ornithologie comprendra uniquement des genres et des sections. (Temminck, *Observations sur la classification*, 6–7.)

47 Mais quel sera le sort de ces *espèces anomales* en si grand nombre dans la classe des oiseaux? La minutie ne pourra pas toujours obvier au doute; dans ce cas, l'espèce est destinée à être ballottée sans cesse de genre en genre; elle prêtera bien plus facilement matière à l'erreur, et servira à fournir des emplois bien plus multipliés; les naturalistes auront moins de chances à pouvoir s'entendre [...] Dans le doute, pour se tirer d'embarras, on fait un *nouveau genre*. Ce mode de procéder est à l'ordre du jour, et chatouille complaisamment l'amour-propre[.] (Temminck, "Introduction troisième partie," in *Manuel d'ornithologie*, 2nd ed., vol. 3, xxviii–xxix.)

52 [L]a recherche [est] plus facile dans un *species* de cent cinquante à deux cents espèces, coupé par *sections* sans appellation distincte, facultatives à adopter, à modifier ou à supprimer totalement, sans que l'ordre systématique en souffre

atteinte, et sans entraver, par une série de dénominations nouvelles, les rapports établis entre les naturalistes, d'un bout du monde à l'autre. (Temminck, "Introduction troisième partie," in *Manuel d'ornithologie*, 2nd ed., vol. 3, xxx.)

53 La coupe par sections, empruntée d'une forme rigoureuse, me paraît préférable aux coupes plus sévères de genres: la première peut dépendre de la manière de voir, et est comparable, dans la société, aux règles du ménage; l'autre doit servir de règle universelle. (Temminck, "Introduction troisième partie," in *Manuel d'ornithologie*, 2nd ed., vol. 3, xxxi.)

54 L'amour-propre flatté, le droit, qu'assez généralement on se fait, de passer l'éponge sur le nom de ses devanciers, et de dire, au moyen d'un MIHI ou d'un NOBIS en lettres majuscules: *A moi la cargaison!* fait très-souvent donner la préférence à la coupe générique. (Temminck, "Introduction troisième partie," in *Manuel d'ornithologie*, 2nd ed., vol. 3, xxxii.)

56 C'est vainement qu'on inventerait encore double et triple de noms nouveaux pour former des groupes strictement méthodiques. Les mœurs de tous ces oiseaux étant, à quelques légères nuances près, absolument les mêmes, on n'a pu avoir recours à ce moyen pour sous-diviser ce grand genre. J'ai mis tous mes soins à comparer plus de cent espèces étrangères, avec nos espèces indigènes; le résultat de cet examen m'a confirmé dans l'opinion qu'il existe un passage graduel, sans démarcation aucune, d'une espèce à l'autre[.] (Temminck, *Manuel d'ornithologie*, 2nd ed., vol. 1, 342–43.)

64 Une bonne diagnose n'est pas si facile à former qu'on le pense. Les naturalistes auteurs ne savent pas toujours donner en peu de mots une définition nette des espèces. Une longue description hérissée de minuties est précisément celle qui embarrasse le plus. (Temminck, "Introduction troisième partie," in *Manuel d'ornithologie*, 2nd ed., vol. 3, xxiv–xxv.)

66 Lorsqu'on se propose d'établir une nouvelle coupe méthodique dans un genre déjà adopté, il est essentiel de connaître, par l'examen de la nature, toutes, ou du moins le plus grand nombre des espèces connues, au milieu desquelles ont été reparties celles qu'on veut isoler; il faut nécessairement avoir sous les yeux la sérié des êtres qui présentent plus ou moins d'affinité avec ceux qu'on se propose de séparer génériquement; il est encore urgent de savoir si le genre de vie, la nourriture, et tout ce qui tient aux mœurs et a la demeure de ces espèces, offrent des différences marquées avec celles reconnues à leurs anciens congénères. Il faut savoir distinguer de simples variétés de formes accessoires, dans une série naturelle d'espèces, de ces caractères d'un ordre supérieur, qui sont les indices d'une organisation et de mœurs disparates: le coup d'œil exercé du naturaliste praticien parvient d'ordinaire plus sûrement à déterminer avec justesse que ne saurait le faire celui dont les études ont principalement été vouées à la théorie

des sciences naturelles. (Temminck, "Introduction troisième partie," in *Manuel d'ornithologie*, 2nd ed., vol. 3, x.)

68 Je dois préalablement faire observer que Mr. Vieillot ne suit point une marche égale dans les caractères essentiels qu'il donne à ses genres; il fait usage de ces marques distinctives avec une espèce d'indifférence, les empruntant tantôt d'une partie tantôt d'une autre; ce procedé est vicieux au plus haut degré, il fait qu'on ne trouve point un nombre approchant égal d'objets de comparaison d'un genre avec l'autre. (Temminck, *Observations sur la classification*, 41.)

70 Il est certain que l'anatomie comparée et l'étude de l'ostéologie des animaux doivent être considérées comme sciences fondamentales et comme le point de départ du zoologiste qui veut travailler à élever l'édifice méthodique; mais ce n'est point exclusivement selon les lumières dont ces deux études lui fournissent les moyens, qu'il doit établir les groupes de second ordre. (Temminck, "Introduction troisième partie," in *Manuel d'ornithologie*, 2nd ed., vol. 3, xv.)

73 Lors-qu'on connaît le plus grand nombre des espèces décrites, et qu'on les rappelle à la mémoire, alors qu'il s'agit de calquer la description d'une espèce nouvelle, on trouvera, du premier coup d'œil, les principaux rapports ou les dissemblances caractéristiques. Il s'agit de ne rien laisser à désirer dans la description de l'espèce type; puis on ébauche à grands traits celle des autres espèces du même genre: un seul caractère bien défini suffit alors, le plus souvent, pour que l'espèce ne puisse pas être confondue, même avec celle qui offre au premier coup d'œil le plus d'identité. (Temminck, "Introduction troisième partie," in *Manuel d'ornithologie*, 2nd ed., vol. 3, xxv.)

80 [I]l est à espérer, qu'on parviendra avec le tems à créer une méthode plus parfaite. Le seul moyen pour atteindre à ce but est l'examen minutieux de la Nature; aucun genre, aucune sous-division, pas même l'admission d'une espèce, ne doit avoir lieu dans une semblable méthode, avant que préalablement les animaux vivants, ou bien leurs dépouilles non mutilées, ayent été soigneusement examinés par des naturalistes dignes de confiance; on n'admettra plus, sur les seuls renseignements des voyageurs, et sur une indication vague, une multitude d'animaux, que les compilateurs semblent avoir introduits dans les livres, dans le seul but d'augmenter le catalogue de nomenclature. (Temminck, "Avant-propos," in *Manuel d'ornithologie*, xv.)

82 Mr. Vieillot [...] bouleverse tous les systèmes qui ont paru précédemment et de nous jours[.] (Temminck, *Observations sur la classification*, 6.)

88 [O]n se perd dans les affinités circulaires, dans les analogies les plus bizarres; des nombres trinaires et quinaires partagent l'opinion, et on arrive dans l'arène des hypothèses et des idées philosophiques, qui mènent à des contestations sans fin comme sans but; le tout, pour expliquer cette belle nature et l'ordre merveilleux

et grandiose qui y règne. (Temminck, "Introduction troisième partie," in *Manuel d'ornithologie*, 2nd ed., vol. 3, xv.)

98 Dans l'état présent de la science, et entourés comme nous le sommes de cette grande affluence d'objets nouvellement découverts, il paraît préférable de borner, pour le moment, nos recherches à la connaissance exacte des espèces et à la détermination, autant que possible, rigoureuse de celles-ci, laissant aux méthodistes, prévenus en faveur de leur échafaudage compliqué, hérissé de termes techniques, le soin de répartir dans un système méthodique toutes ces nuances subtiles qu'ils veulent caractériser rigoureusement par des démarcations génériques. (Temminck, "Introduction troisième partie," in *Manuel d'ornithologie*, 2nd ed., vol. 3, xviii.)

108 Ce sont particulièrement les écrits éloquents de Buffon, qui, en sonnant l'éveille aux bouts de l'univers, ont ajouté de nouveaux charmes à cette étude aimable; l'ordre et l'harmonie, que la classification doit au grand Linné, n'ont pas moins contribué à augmenter le nombre de ses amateurs zèlés; d'illustres savants, en prenant pour guides les écrits de ses hommes célèbres, se sont acquis la gloire de voir leurs noms inscrits au temple de mémoire. (Temminck, "Avant-propos," in *Manuel d'ornithologie*, vii.)

115 En matière de sciences, un ouvrage sans critique est comme un mets sans sel. (Temminck, "Introduction troisième partie," *Manuel d'ornithologie*, 2nd ed., vol. 3, lxii.)

Chapter 7. Classification and Natural History: 1800–1850

3 On semble perdre de vue que l'étude de la nature, abstraction faite de tout ce qu'elle a d'important, envisagée sous le point de vue moral et scientifique, influe encore puissamment sur la civilisation, sur l'industrie dont elle fait mouvoir les ressorts, sur plusieurs branches des arts et sur la propagation des lumières dans toutes les classes de la société[.] (Temminck, "Introduction troisième partie," in *Manuel d'ornithologie*, 2nd ed., vol. 3, xxxvi–xxxvii.)

5 Peu de personnes se faisant une idée juste de l'histoire naturelle, il nous a paru nécessaire de commencer notre ouvrage, en définissant bien l'objet que cette science se propose, et en établissant des limites rigoureuses entre elle et les sciences qui l'avoisinent. (Cuvier, "Introduction," in *Règne animal*, vol. 1, 1.)

8 [L']Histoire Naturelle [...] a pour objet la connoissance générale et particulière de notre globe, et de toutes les productions qui naissent ou se forment à sa surface et dans son intérieur. (Lamarck, "Sur l'histoire naturelle en général," 3–4.)

14 Elle [Botanique proprement dite] se compose, 1.° de la *Glossologie*, c'est-à-dire, de la connaissance des termes par lesquels on désigne les organes des plantes et leurs diverses modifications; 2.° de *laTaxonomie*, ou de la Théorie des classifications

appliquée au règne végétal; 3.° de la *Phytographie*, ou de l'art de décrire les plantes de la manière la plus utile aux progrès de la science: d'où résultent, comme applications, la *Botanique descriptive*, c'est- à-dire, la description de toutes les espèces dont le règne végétal se composé, et la *Synonymie Botanique*, ou la connaissance des noms divers sous lesquels les plantes ont été désignées. (Candolle, "Introduction," in *Théorie élémentaire*, 19.)

16 [O]ok moet men, – hoewel zulks somtijds geschiedt, *Anthropologie* met natuurlijke geschiedenis van den mensch niet verwarren. De eerste is eene wetenschap van ruimeren omvang en sluit in hare uitgestrekte grenzen de laatste in; deze is eene anthorpologische wetenschap, maar zij is niet de *Anthropologie* zelve. (Van der Hoeven, "Bijdragen tot de natuurlijke geschiedenis van den mensch," 87.)

17 De twee hoofddeelen dezer natuurlijke geschiedenis [...] zijn: vooréérst, het onderscheid tusschen den mensch en de overige dieren, waardoor het standpunt van ons geslacht in de rij der op onze aarde levende wezens bepaald wordt; ten andere, het verschil tusschen menschen en menschen, onderling met elkander vergeleken, waarbij hoofdzakelijk op de verschillende menschenstammen gelet wordt. (Van der Hoeven, "Bijdragen tot de natuurlijke geschiedenis van den mensch," 88–89.)

23 Ce seroit une cause réelle de retard pour les sciences naturelles, si l'on s'obstinoit à ne voir dans les objets observés, que leur forme, leur dimension, leurs parties externes même les plus petites, leur couleur, etc.; et si ceux qui se livrent à une pareille étude dédaignoient de s'élever à des considérations supérieures; comme de chercher quelle est la nature des objets dont ils s'occupent, quelles sont les causes de les modifications ou des variations auxquelles ces objets sont tous assujettis, quels sont les rapports de ces mêmes objets entr'eux, et avec tous les autres que l'on connoît, etc., etc. (Lamarck, "Discours préliminaire," in *Philosophie zoologique*, 12–13.)

28 De ware theorie zal dan van zelve volgen, zonder dat men weet hoe en wanneer, en het hechte gebouw der wetenschap zal worden opgetrokken op den breeden en vasten grondslag der ervaring, om te blijven staan, wanneer de tooverkasteelen der stelsels en meeningen reeds in damp vervlogen en uit de herinnering verdwenen zijn. (Van der Hoeven, "Bijdragen tot de natuurlijke geschiedenis van den mensch," 91.)

37 L'ouvrage dont nous rendons compte, est, sous ce rapport, exempt de tout reproche, et la seconde édition qui a suivi de près la première, prouve qu'à l'époque actuelle, les études ont un but vraiment philosophique. (Drapiez, "Manuel d'ornithologie," 276.)

59 De wetenschappen, die er in behandeld zullen worden, zijn de volgende: natuurlijke geschiedenis der drie rijken, ontleedkunde der planten en dieren en algemeene physiologie. (Van der Hoeven, "Voorberigt," ii.)

60 Wij willen ons echter geene andere bedoeling voorstellen, dan de bevordering der wetenschap zelve en zullen ons door geene angstige keus laten leiden, om telkens eenige bepaalden invloed op het gewone leven, op de geneeskunst of op eenige toegepaste kennis te verlangen[.] (Van der Hoeven, "Voorberigt," ii–iii.)

66 [L]a formation des méthodes est l'objet de l'histoire naturelle proprement dite; l'anatomie les reçoit, pour ainsi dire, toutes faites. C'est d'elles qu'elle prend ses premières directions: mais elle ne tarde pas à leur rendre la lumière qu'elle en a reçue d'abord; elle est même la plus forte épreuve de leur bonté; et c'est en appliquant une méthode d'histoire naturelle à l'anatomie comparée, qu'on est bientôt en état de reconnoître si elle s'écarte ou non de la marche de la nature. (Cuvier, *Leçons d'anatomie comparée*, vol. 1, 64.)

71 Le plus grand nombre des anatomistes s'est trouvé embarrassé dans les recherches, pour avoir négligé l'étude des caractères que la zoologie fournit: ils ont cru que l'anatomie comparée pouvait à elle seule servir de guide dans toutes les observations de second ordre; et ils ont perdu de vue cette autre science auxiliaire, suivant leurs idées trop au-dessous de l'importance attachée aux recherches anatomiques. Leurs recherches ne sont guère en état de nous servir aujour-d'hui, vu que le plus souvent il est impossible de pouvoir déterminer avec précision sur quelles espèces d'animaux reposent leurs observations; tandis que ceux qui les poursuivent encore sans l'aide des premières notions en zoologie, ne répandent point de lumière très-utile dans l'étude qui a pour but le perfectionnement du système de la nature. (Temminck, "Introduction troisième partie," in *Manuel d'ornithologie*, 2nd ed., vol. 3, xv–xvii.)

81 L'introduction des méthodes naturelles pour la classification des êtres vivants est l'un des services les plus grands que l'on ait rendus à l'histoire naturelle: elle a changé la face de cette science, et a donné un puissant intérêt à la partie de la botanique et de la zoologie qui jusqu'alors avait été la plus aride[.] (Milne-Edwards, *Cours élémentaire d'histoire naturelle*, 288.)

86 Daareentegen, gelijk in stilstaande wateren zich gemeenlijk eenige vuilheid vertoont, zoo ook valt bij hen, die nimmer buiten de grenzen van hun geboorteland of eenig bijzonder oord gegaan zijn, maar, binnen dezelve besloten, zich tot de studie der Natuur begeven hebben, zekere onzuiverheid, laagheid en bekrompenheid van oordeel op te merken, waardoor zij, als zij somwijlen van hetzelfde een vrijer gebruik trachtten te maken, tot eene ontelbare menigte van dwalingen en beuzelarijen in de gansche uitlegging der Natuur vervallen zijn. (Reinwardt, *Redevoering van C. G. C. Reinwardt*, 23–24.)

87 Ofschoon de waarheid hiervan buiten alle bedenking is, is dit nogtans een gemeen gebrek, zich door de gansche Natuurlijke Historie verspreidende[.] (Reinwardt, *Redevoering van C. G. C. Reinwardt*, 29.)

89 Accuratiorem plantarum indigenarum notitiam dum commendo, non catalogu herbarum aut Indicem nudum volo, non simplicem ut ajunt, Floram. (Brugmans, *De accuratiori plantarum indigenarum notitia maxime commendanda*, 10.)

92 Alle Ornithologen hebben UwEd. waarlijk voor deze schone Monographie veel verplichting, en alle zullen ongemeen na de voortzetting daarvan verlangen. (Brugmans to Temminck, 24 December 1813, Naturalis Biodiversity Center Archives, Archive C. J. Temminck, NAT_ARC_TEM_00254.)

Bibliography of Coenraad Jacob Temminck's Works

The entries in this bibliography are arranged chronologically.

Temminck, Coenraad Jacob. *Catalogue systématique du cabinet d'ornithologie et de la collection de quadrumanes avec une courte description des oiseaux non-décrits suivi: d'une nôte d'oiseaux doubles et de quelques autres objets d'histoire naturelle offerts en échange*. Amsterdam: C. Sepp Jansz., 1807.

Temminck, Coenraad Jacob, and Pauline Knip-de Courcelles. *Histoire naturelle générale des pigeons*. Paris: Garnery, 1808.

Temminck, Coenraad Jacob. *Histoire naturelle générale des pigeons et des gallinacés*. 3 vols. Amsterdam: J. C. Sepp & fils; Paris: G. Dufour, 1813–1815. doi:10.5962/bhl.title.64844.

Temminck, Coenraad Jacob. *Manuel d'ornithologie, ou Tableau systématique des oiseaux qui se trouvent en Europe*. Amsterdam: J. C. Sepp & fils; Paris: G. Dufour, 1815.

Temminck, Coenraad Jacob. *Observations sur la classification méthodique des oiseaux, et remarques sur l'Analyse d'une nouvelle ornithologie élémentaire par L. P. Vieillot, auteur de divers ouvrages d'ornithologie, et un des collaborateurs du Nouveau dictionnaire d'histoire naturelle*. Amsterdam and Paris: Gabriel Dufour, 1817.

Temminck, Coenraad Jacob. "Sur le genre Bradype et description d'une espèce encore peu connue." *Annales générales des sciences physiques* 6 (1820): 204–18.

Temminck, Coenraad Jacob. "Sur le genre Hyène et description d'une espèce nouvelle découverte en Afrique." *Annales générales des sciences physiques* 3 (1820): 46–57.

Temminck, Coenraad Jacob, and Guillaume Michel Jerôme Meiffren Laugier de Chartrouse. *Nouveau recueil de planches coloriées d'oiseaux: Pour servir de suite et de complément aux planches enluminées de Buffon*. 5 vols., 102 livr., 600 pls. Paris: F. G. Levrault, 1820–1838. doi:10.5962/bhl.title.51468.

Temminck, Coenraad Jacob. *Manuel d'ornithologie, ou Tableau systématique des oiseaux qui se trouvent en Europe; précédé d'une analyse du système général d'ornithologie, et suivi d'une table alphabétique des espèces.* 2nd ed. 4 vols. Paris: Gabriel Dufour (vols. 1 and 2, 1820), Edmond d'Ocagne (vol. 3, 1835), H. Cousin (vol. 4, 1840), 1820–1840. doi:10.5962/bhl.title.130481.

Temminck, Coenraad Jacob. "Sur l'utilité des voyages des decouvertes dans les colonies du royaume des Pays-Bas." *Annales générales des sciences physiques* 7, no. 20 (1821): 172–78.

Temminck, Coenraad Jacob. "Account of Some New Species of Birds of the Genera Psittacus and Columba, in the Museum of the Linnaean Society [1819]." *Transactions of the Linnean Society of London* 13 (1822): 107–30.

Temminck, Coenraad Jacob. *Voorschrift, hoedanig te handelen met voorwerpen van natuurlijke historie, ten einde dezelve behoorlijk te verzenden en voor bederf te bewaren: Ten gebruike van het 's Rijks Museum van Natuurlijke Historie te Leyden.* Leiden: [printed for the author], 1825.

Temminck, Coenraad Jacob. *Monographies de mammalogie, ou description de quelques genres de mammifères, dont les espèces ont été observées dans les différens musées de l'Europe.* 2 vols. Paris and Amsterdam: G. Dufour et Edmond d'Ocagne (vol. 1, 1827); Paris: G. Dufour et Edmond d'Ocagne; Leiden: C. C. van der Hoek (vol. 2, 1835–1841), 1827–1841.

Temminck, Coenraad Jacob. "Blik op de dierlijke bewoners van de Sunda-eilanden en van de overige Nederlandsche bezittingen in Indië." *Bijdragen tot de natuurkundige wetenschappen* 3 (1828): 62–78.

Temminck, Coenraad Jacob. "Notice sur quelques espèces nouvelles d'oiseaux des genres Psittacus et Columba dans le Muséum de la Société Linnéenne (Lin. transact. XIII P.1. p. 107)." *Isis von Oken* 22 (1829): 1194–209.

Temminck, Coenraad Jacob. "Algemeen overzigt van de orde der Cheiroptera, en monographie van de geslachten Harpyia en Cephalotes." *Bijdragen tot de natuurkundige wetenschappen* 6 (1831): 301–34.

Temminck, Coenraad Jacob. "Monographie over een nieuw geslacht van knaagdier, onder den naam van Nyctocleptes." *Bijdragen tot de natuurkundige wetenschappen* 7 (1832): 1–8.

Temminck, Coenraad Jacob. "Over het nieuw geslacht Nyctophilus (Nachtvlieger), Leach." *Bijdragen tot de natuurkundige wetenschappen* 7 (1832): 143–45.

Temminck, Coenraad Jacob. "Over een geslacht der vleugelhandige zoogdieren, Bladneus genaamd." *Tijdschrift voor atuurlijke geschiedenis en physiologie* 1, no. 1 (1834): 1–16.

Temminck, Coenraad Jacob. "Coup-d'oeil sur la faune des Iles de la Sonde et de l'empire du Japon: Discours préliminaire destiné a servir d'introduction a la faune du Japon."

In *Fauna japonica, sive, Descriptio animalium, quae in itinere per Japoniam, jussu et auspiciis, superiorum, qui summum in India Batava imperium tenent, suscepto, annis 1823–1830* … edited by Philipp Franz von Siebold. Vol. 1, i–xxx. Leiden: J. G. La Lau, 1836.

Temminck, Coenraad Jacob. "Ueber das Geschlecht der Blattnasen (Rhinolophus. Geoffr.) von C. J. Temminck aus der *Tijdschrift voor natuurlijke geschiedenis* von van der Hoeven u. Vriese l. 1. übersetzt vom Herausgeber." *Archiv für Naturgeschichte* 2, no. 1 (1836): 81–100.

Temminck, Coenraad Jacob. "Over de geslachten Taphozous, Emballonura, Urocryptus en Diclidurus." *Tijdschrift voor natuurlijke geschiedenis en physiologie* 5, no. 1 (1838): 1–34.

Temminck, Coenraad Jacob. "Over de kennis en de verbreiding der zoogdieren van Japan." *Tijdschrift voor natuurlijke geschiedenis en physiologie* 5, no. 4 (1838): 273–93.

Temminck, Coenraad Jacob. ed. *Verhandelingen over de natuurlijke geschiedenis der Nederlandsche overzeesche bezittingen*. 3 vols. Leiden: S. en J. Luchtmans en C. C. van der Hoek, 1839–1842. doi:10.5962/bhl.title.114730.

Temminck, Coenraad Jacob. "Over eenige geslachten van zoogdieren, een deel der fauna der Japan uitmakende." *Verslagen en mededeelingen uitgegeven door de vier klassen van het Koninklijk Nederlandsche Instituut van Wetenschappen, Letterkunde en Schoone Kunsten* (1841): 208–16.

Temminck, Coenraad Jacob, and Hermann Schlegel. "Mammalia." In *Fauna japonica, sive, Descriptio animalium, quae in itinere per Japoniam, jussu et auspiciis, superiorum, qui summum in India Batava imperium tenent, suscepto, annis 1823–1830* … edited by Philipp Franz von Siebold. Vol. 5. Leiden: A. Arnz et socios, 1842.

Temminck, Coenraad Jacob. "Aperçu général et spécifique sur les mammifères qui habitent le Japon et les îles qui en dépendent." In *Fauna japonica, sive, Descriptio animalium, quae in itinere per Japoniam, jussu et auspiciis, superiorum, qui summum in India Batava imperium tenent, suscepto, annis 1823–1830* … edited by Philipp Franz von Siebold. Vol. 5, 1–9. Leiden: A. Arnz et socios, 1842.

Temminck, Coenraad Jacob. "Description d'un nouveau genre de mammifères (Urotrichus)." *Magasin de zoologie, d'anatomie comparée et de palaeontologie* 4 (1842): 1–5.

Temminck, Coenraad Jacob. *Coup-d'oeil général sur les possessions néerlandaises dans l'Inde Archipélagique*. 3 vols. Leiden: A. Arnz & comp, 1846–1849.

Temminck, Coenraad Jacob, and Jean-Charles Werner. *Les oiseaux d'Europe décrits par C. J. Temminck; Atlas de 530 planches dessinées par J.-C. Werner*. Paris: Baillière, 1848. doi:10.5962/bhl.title.104704.

Temminck, Coenraad Jacob. *Esquisses zoologiques sur la côte de Guiné. Pt. 1: Les mammifères*. Leiden: E. J. Brill, 1853. doi:10.5962/bhl.title.14828.

Temminck, Coenraad Jacob. “Over eenige nieuwe soorten van eekhoorns van den Indischen Archipel.” *Natuurkundig tijdschrift voor Nederlandsch Indië* 4 (1854): 343–48.

Temminck, Coenraad Jacob. “Zoologische schetsen van eenige soorten van hoenderachtige vogelen.” *Bijdragen tot de dierkunde* 6, no. 1 (1854): 49–50.

Bibliography

Manuscript Sources

Boie, Heinrich. "Erpétologie de Java," manuscript and plates [1822–1823]. Archive Natuurkundige Commissie, 1819–1850. Naturalis Biodiversity Center Archives, Leiden (RMNH).

Curatoren Universiteit Leiden. Correspondence and manuscripts. AC2, Archieven Universiteit Leiden: Curatoren, 1815–1877. Leiden University Libraries, Special Collections (UBL).

Falck, Anton Reinhard. Correspondence. Collectie 048 Falck, 2.21.006.48, Archieven van mr. A. R. Falck [levensjaren 1777–1843] en enkele familieleden, (1737) 1756–1850. National Archives of the Netherlands, The Hague (NL-HaNA).

Falck, Anton Reinhard. Correspondence and decrees. Algemeen Rijksarchief (ARA), 2.14.03, 1819–1820. National Archives of the Netherlands, The Hague (NL-HaNA).

Levaillant, François. Letters. Manuscripts Department, Slg. Darmstaedter Afrika, 1780. Staatsbibliothek zu Berlin (SBB).

Marum, Martinus van. Correspondence. Archive Martinus van Marum 529. Briefwisseling, 5–29, 1776–1837. Nord-Hollands Archief, Haarlem (NHA).

Ministry of Home Affairs, the Netherlands. Correspondence with Coenraad Jacob Temminck. Archief van het Ministerie van Binnenlandse Zaken (BiZa), 2.04.01, 1813–1848. National Archives of the Netherlands, The Hague (NL-HaNA).

Natuurkundige Commissie. Correspondence and manuscripts. Archive Natuurkundige Commissie, 1819–1850. Naturalis Biodiversity Center Archives, Leiden (RMNH).

Pel, Hendrik Severinus. "Register van uitgaande stukken van den agent van het Succursale-depôt te Coomassie." Nederlandse Bezittingen ter Kuste van Guinea, 1658–1872, 1.05.14, inv.nr. 1109. National Archives of the Netherlands, The Hague (NL-HaNA).

Reinwardt, Caspar George Carl. Correspondence. Archief Caspar George Carl Reinwardt, UBL-BPL 2922. Leiden University Libraries, Special Collections (UBL).

Rijksmuseum van Natuurlijke Historie te Leiden (RMNH). Correspondence, decrees, and manuscripts. Archief van het Rijksmuseum voor Natuurlijke Historie, 3.12.17, (1815) 1820–1859. National Archives of the Netherlands, The Hague (NL-HaNA).

Temminck, Coenraad Jacob. "Catalogue du Cabinet de C. J. Temminck." Archive C. J. Temminck, 1820–1858. Naturalis Biodiversity Center Archives, Leiden (RMNH), NL-LdnNAT_Temminck_47.

Temminck, Coenraad Jacob. Correspondence (1820–1850). Temminck Family Archives, A. M. van Lynden-de Bruïne. Private collection.

Temminck, Coenraad Jacob. Correspondence and manuscripts. Archive C. J. Temminck, 1820–1858. Naturalis Biodiversity Center Archives, Leiden (RMNH).

Temminck, Coenraad Jacob. "Galerie zoögraphique du Musée des Pays-Bas: Catalogue des mammifères exposés dans les galeries du Musée des Pays-Bas," [1825], Archive C. J. Temminck, 1820–1858. Naturalis Biodiversity Center Archives, Leiden (RMNH), NL-LdnNAT_Temminck_27.

Temminck, Coenraad Jacob. "Jaarverslagen 1820–1858": Reports to the Ministry of Public Education, National Industry and Colonies, 1820–1823; Reports to the Ministry of Internal Affairs, 1824–1858. Naturalis Biodiversity Center Archives, Leiden (RMNH).

Temminck, Coenraad Jacob. "Uit het kabinet van wijlen professor Brugmans," 1820. Archive C. J. Temminck, 1820–1858. Naturalis Biodiversity Center Archives, Leiden (RMNH), NAT_ARC_TEM_02460-64.

Temminck, Coenraad Jacob. "Verslag van Temminck aan de Minister van Binnenlandse Zaken over zijn wetenschappelijke werkzamheden tijdens zijn verblijf in Parijs," 30 September 1835. Archive C. J. Temminck, 1820–1858. Naturalis Biodiversity Center Archives, Leiden (RMNH), NL-LdnNAT_Temminck_49.

Printed Sources

Acton, John Emerich Edward. *Lectures in Modern History*. London: MacMillan and Co., Limited, 1906.

Ahrens, T. G. "The Ornithological Collections of the Berlin Museum." *Auk* 42 (1925): 241–45.

Aimassi, Giorgio. "The Original Description of Bonelli's Eagle *Aquila fasciata* Vieillot (Aves: Accipitridae)." *Zoological Bibliography* 4, December (2015): 1–14.

Alberti, Samuel J. M. M. "Museum Nature." In *Worlds of Natural History*, edited by Emma C. Spary, Helen Anne Curry, James Andrew Secord, and Nicholas Jardine, 348–62. Cambridge: Cambridge University Press, 2018.

Alberti, Samuel J. M. M. *Nature and Culture: Objects, Disciplines, and the Manchester Museum*. Manchester: Manchester University Press, 2012.

Allen, David. "Amateurs and Professionals." In *The Cambridge History of Science. Volume 6: The Modern Biological and Earth Sciences*, edited by John V. Pickstone and Peter J. Bowler. The Cambridge History of Science, 13–33. Cambridge: Cambridge University Press, 2009. doi:10.1017/CHOL9780521572019.003.

"Analytical Notices of Books: Monographies de mammalogie, ou descriptions de quelques genres de mammifères dont les espèces ont été observées dans les différens musées de l'Europe. Par C. J. Temminck, livraison, 4eme." *Zoological Journal of the Linnean Society* 2 (1826): 526–34.

Anderson, Katharine. "Natural History and the Scientific Voyage." In *Worlds of Natural History*, edited by Emma C. Spary, Helen Anne Curry, James Andrew Secord, and Nicholas Jardine, 304–18. Cambridge: Cambridge University Press, 2018.

Appel, Toby A. *The Cuvier-Geoffroy Debate: French Biology in the Decades before Darwin.* Oxford: Oxford University Press, 1987.

Atran, Scott, Richard W. Burkhardt, Pietro Corsi, et al., eds. *Histoire du concept d'espèce dans les sciences de la vie: Colloque international (mai 1985) organisé par la Fondation Singer-Polignac, Paris.* Paris: Fondation Singer-Polignac, [1987].

Audouin, Jean Victor, Adolphe Brongniart, and Jean-Baptiste Dumas. "Introduction." *Annales des sciences naturelles: Comprenant la physiologie animale et végétale, l'anatomie comparée des deux règnes, la zoologie, la botanique, la minéralogie et la géologie* 1 (1834): v–xvi.

Beaufort, Willem Hendrik de. "Coenraad Jacob Temminck, uit zijne brieven geschetst." *Ardea* 9, no. 2 (1920): 34–43.

Bell, Thomas, John George Children, James de Carle Sowerby, and G. B. Sowerby. "Introduction." *The Zoological Journal* 1, no. 1 (1824): iii–vii.

Berkel, Klaas van. *De stem van de wetenschap: Geschiedenis van de Koninklijke Nederlandse Akademie van Wetenschappen.* Vol. 1. Amsterdam: Bert Bakker, 2011.

Berkel, Klaas van, Albert Van Helden, and Lodewijk C. Palm. *The History of Science in the Netherlands: Survey, Themes and Reference.* Leiden: Brill, 1999.

"Besluit regelende het hooger onderwijs in het koningrijk der Nederlanden: 2 Augustus 1815". In *Staatsblad der Verëenigde Nederlanden*, no. 14. 's-Gravenhage: [n.p.], 1815.

"Bibliographical Notices. *Nouveau recueil de planches coloriées d'oiseaux, pour servir de suite et de complément aux Planches Enluminées de Buffon.* Par M. Temminck et Meiffren Laugier Baron de Chartrouse, Livr. 98–102, 1839." *Annals and Magazine of Natural History; Zoology, Botany, and Geology* 4 (1840): 266–68.

Bicheno, James Ebenezer. "On Systems and Methods in Natural History." *Transactions of the Linnean Society of London* 15, no. 2 (1827): 479–96.

Bierens de Haan, J. A. *De geschiedenis van een verdwenen Haarlemsch museum van natuurlijke historie: Het Kabinet van naturaliën van de Hollandsche Maatschappij der Wetenschappen, 1759–1866.* Haarlem: De Erven F. Bohn N. V., 1941.

Bierens de Haan, J. A. *De Hollandsche Maatschappij der Wetenschappen, 1752–1952.* Haarlem: Tjeenk Willink & Zoon, 1952.

Birkhead, Tim R., and Robert Montgomerie. "A Vile Passion for Altering Names: The Contributions of Charles Thorold Wood jun. and Neville Wood to Ornithology in the 1830s." *Archives of Natural History* 43, no. 2 (2016): 221–36. doi:10.3366/anh.2016.0380.

Birkhead, Tim R., Paul J. Smith, Meghan Doherty, and Isabelle Charmantier. "Willughby's Ornithology." In *Virtuoso by Nature: The Scientific Worlds of Francis Willughby FRS (1635–1672)*. Emergence of Natural History, 268–304. Leiden: Brill, 2016. doi:10.1163/9789004285323_009.

Bleeker, Pieter. “Mémoire sur les poissons de la côte de Guinée.” *Natuurkundige verhandelingen van de Hollandsche Maatschappij der Wetenschappen te Haarlem* 2, no. 18 (1863): 1–136.

Boeseman, Marinus. “Collectors and Fish Collections of the Rijksmuseum van Natuurlijke Historie in Leiden, the Netherlands (1820–1980).” In *Collection Building in Ichthyology and Herpetology*, edited by Theodore W. Pietsch and William D. Anderson Jr., 81–100. Lawrence, KS: American Society of Ichthyologists and Herpetologists, 1997.

Boie, Heinrich. “Manuel d'ornithologie, ou tableau systèmatique des oiseaux qui se trouvent en Europe par C. J. Temminck …” *Heidelbergische Jahrbücher der Literatur* 25–26 (1816): 387–416.

Boitard, Pierre. *Histoire naturelle des oiseaux de proie d'Europe*. Paris: Parmantier, 1824. ark:/12148/bpt6k4416018.

Bonaparte, Charles Lucien. *Observations on the Nomenclature of Wilson's Ornithology*. Philadelphia: Anthony Finley, 1826.

Bonaparte, Charles Lucien. “Observations of the State of Zoology in Europe, as Regards the Vertebrata.” In *Reports on the Progress of Zoology and Botany, 1841, 1842*, 1–43. Edinburgh: Printed for the Ray Society, 1845.

Bonaparte, Charles Lucien. “On the Lorine Genus of Parrots, Eclectus, with the Description of a New Species, Eclectus Cornelia.” *Proceedings of the Zoological Society of London* 17 (1849): 142–46.

Boomgaard, Peter. “The Making and Unmaking of Tropical Science: Dutch Research on Indonesia, 1600–2000.” *Bijdragen tot de taal-, land- en volkenkunde* 162, no. 2/3 (2006): 191–217. doi:10.1163/22134379-90003666.

Bosch, Johannes van den. *Nederlandsche bezittingen in Azia, Amerika en Afrika. In derzelver toestand en aangelegenheid voor dit Rijk, wijsgeerig, staatshuishoudkundig en geographisch beschouwd.* Amsterdam and Den Haag: Gebroeders Van Cleef, 1818.

Botke, IJnte. *Theodorus van Swinderen (1784–1851), hoogleraar en schoolopziener: Gedenktekens door en voor hem gesticht.* Groningen: Universiteitsmuseum Groningen, 1990.

Bowler, Peter J., and John V. Pickstone, eds. *The Modern Biological and Earth Sciences*. Vol. 6. The Cambridge History of Science. Cambridge: Cambridge University Press, 2009. doi:10.1017/CHOL9780521572019.002.

Braun, Alexander. *Betrachtungen über die Erscheinung der Verjüngung in der Natur.* Leipzig: W. Engelmann, 1851.

Breidbach, Olaf, and Michael T. Ghiselin. “Lorenz Oken and ‘Naturphilosophie’ in Jena, Paris and London.” *History and Philosophy of the Life Sciences* 24, no. 2 (2002): 219–47.

Brisson, Mathurin-Jacques. *Ornithologie, ou, Méthode contenant la division des oiseaux en ordres, sections, genres, especes & leurs variétés*. Vol. 4. Paris: J.-B. Bauche, 1760. doi:10.5962/bhl.title.51902.

Browne, Janet. *The Secular Ark: Studies in the History of Biogeography*. New Haven: Yale University Press, 1983.

Brugmans, Sebald Justinus. *De accuratiori plantarum indigenarum notitia maxime commendanda, habita Leidae ... MDCCLXXXVI: Altera valedictoria, de natura soli Frisici exploranda, dicta Franequerae ...* Leiden: S. et J. Luchtmans, 1787.

Buffon, Georges Louis Leclerc. *Histoire naturelle des oiseaux*. 18 vols. Paris: Imprimerie Royale, 1770–1785. doi:10.5962/bhl.title.66490.

Buffon, Georges Louis Leclerc. *Histoire naturelle, générale et particulière, avec la description du Cabinet du Roi*. 44 vols. Paris: Imprimerie Royale, 1749–1804. doi:10.5962/bhl.title.64910.

Buffon, Georges Louis Leclerc, Edme-Louis Daubenton, and François Nicolas Martinet. *Planches enluminées d'histoire naturelle*. 10 vols. Paris: [s.n.], 1765–?[1783]. doi:10.5962/bhl.title.51460.

Cain, Arthur J. "The Development of Systematic Ideas of Variation Illustrated by Malacology." In *History in the Service of Systematics*, edited by A. Wheeler and J. H. Price, 151–156. London: Society for the Bibliography of Natural History, Special Publication, 1981.

Camerini, Jane R. "Evolution, Biogeography, and Maps: An Early History of Wallace's Line." *Isis* 84, no. 4 (1993): 700–27.

Candolle, Augustin Pyramus de. *Elements of the Philosophy of Plants: Containing the Principles of Scientific Botany ... with a History of the Science, and Practical Illustrations*. Translated by Kurt Polycarp Joachim Sprengel. Edinburgh: W. Blackwood, 1821. doi:10.5962/bhl.title.32221.

Candolle, Augustin Pyramus de. "Géographie Botanique." In *Dictionnaire des sciences naturelles, dans lequel on traite méthodiquement des différens êtres de la nature, considérés soit en eux-mêmes, d'après l'état actuel de nos connoissances, soit relativement à l'utilité qu'en peuvent retirer la médecine, l'agriculture, le commerce et les arts*. Vol. 18, edited by Frédéric Cuvier, 359–444. Strasbourg: F. G. Levrault, 1820.

Candolle, Augustin Pyramus de. *Théorie élémentaire de la botanique; ou, Exposition des principes de la classification naturelle et de l'art de décrire et d'étudier les végétaux*. Paris: Déterville, 1813. doi:10.5962/bhl.title.39705.

Capadose, Abraham. *Lofrede op Sebaldus Justinus Brugmans*. Werken der Hollandsche Maatschappij van Fraaie Kunsten en Wetenschappen. Vol. 7. Leiden: Hollandsche Maatschappij van Fraaie Kunsten en Wetenschappen, 1822.

Catalogue de la bibliothèque délaissé par C. J. Temminck: Dont la vente aura lieu le lundi 27 septembre 1858 et deux jours suivants, dans la maison du défunt, (entrée au Aalmarkt, W. IV. no. 97) par les libraires S. J. Brill et van der Hoek frères à Leide. [S.l.]: [s.n.], 1858.

Coleman, William. *Georges Cuvier, Zoologist: A Study in the History of Evolution Theory*. Cambridge, MA: Harvard University Press, 1964.

Colenbrander, H. T., ed. *Gedenkschriften van Anton Reinhardt Falck*. 's Gravenhage: Martinus Nijhoff, 1913.

Crawfurd, John. *History of the Indian Archipelago: Containing an Account of the Manners, Arts, Languages, Religions, Institutions, and Commerce of its Inhabitants.* 3 vols. Edinburgh: Archibald Constable and Co., 1820.

Curry, Helen Anne, and James Andrew Secord. "Natural History and its Histories in the Twenty-First Century." In *Worlds of Natural History*, edited by Emma C. Spary, Helen Anne Curry, James Andrew Secord, and Nicholas Jardine, 535–44. Cambridge: Cambridge University Press, 2018.

Curry, Helen Anne, Nicholas Jardine, James Andrew Secord, and Emma C. Spary, eds. *Worlds of Natural History.* Cambridge: Cambridge University Press, 2018.

Cuvier, Frédéric. "Nouveau recueil de planches coloriées d'oiseaux, pour servir de suite ou de complément aux planches enluminées de Buffon, par M. Temminck, conservateur du cabinet d'histoire naturelle de Leyde, er M. Meiffren Laugier, baron de Chartrouse; 90 livraisons in 40. A Paris, chez Dufour et d'Ocagne." *Journal des savants*, 2, no. 17 (1832): 641–51.

Cuvier, Georges. "Biographical Memoir of M. de Lamarck." *Edinburgh New Philosophical Journal* 20 (1836): 1–21.

Cuvier, Georges. *Discours sur les révolutions de la surface du globe: Et sur les changemens qu'elles ont produits dans le règne animal.* Paris: Chez G. Dufour et Edmond d'Ocagne, 1826. doi:10.5962/bhl.title.96253.

Cuvier, Georges. *Leçons d'anatomie comparée.* 5 vols. Paris: Crochard, 1805. ark:/13960/t1mh1xz3c.

Cuvier, Georges. *Le Règne animal distribué d'après son organisation, pour servir de base à l'histoire naturelle des animaux et d'introduction à l'anatomie comparée.* 4 vols. Paris: Déterville, 1817. doi:10.5962/bhl.title.41460.

Cuvier, Georges. *Tableau élémentaire de l'histoire naturelle des animaux.* Paris: Baudouin, 1798. doi:10.5962/bhl.title.39351.

Cuvier, Georges, and Achille Valenciennes. *Histoire naturelle des poissons.* 22 vols. Paris: Chez F. G. Levrault (vols. 1–12, 1828–1837), Pitois-Levrault et ce. (vols. 13–14, 1839), Ch. Pitois (vol. 15, 1840), P. Bertrand (vols. 16–22, 1842–1849), 1828–1849. doi:10.5962/bhl.title.7339.

Darwin, Charles. *A Monograph on the Sub-Class Cirripedia, with Figures of all the Species.* 2 vols. London: Ray Society, 1851–1854. doi:10.5962/bhl.title.2104.

Darwin, Charles. *On the Origin of Species by Means of Natural Selection.* London: John Murray, 1859. doi:10.5962/bhl.title.68064.

Darwin, Charles, and Alfred Wallace. "On the Tendency of Species to Form Varieties; and on the Perpetuation of Varieties and Species by Natural Means of Selection." *Journal of the Proceedings of the Linnean Society. Zoology* 3, no. 9 (1858): 45–62. doi:10.1111/j.1096-3642.1858.tb02500.x.

Daston, Lorraine, and Peter Louis Galison. *Objectivity.* New York: Zone Books, 2015.

Daudin, Henri. *Cuvier et Lamarck: Les classes zoologiques et l'idée de série animale (1790–1830).* Études d'histoire des sciences naturelles 2. 2 vols. Paris: F. Alcan, 1926.

Daudin, Henri. *De Linné à Jussieu; Méthodes de la classification et idée de série en botanique et en zoologie (1740–1790)*. Études d'histoire des sciences naturelles 1. Paris: F. Alcan, 1926.

Dekker, René W. R. J. "Type Specimens of Birds in the National Museum of Natural History, Leiden. Part 2. Passerines: Eurylaimidae – Eopsaltriidae (Peters's Sequence)." *NNM Technical Bulletin* 6 (2004): 1–142. https://repository.naturalis.nl/pub/214679.

Di Gregorio, Mario A. "Hugh Edwin Strickland (1811–1853) on Affinities and Analogies: or, the Case of the Missing Key." *Ideas and Production* 7 (1987): 35–50.

Dickinson, Edward C. "Systematic Notes on Asian Birds. 9. The 'Nouveau recueil de planches coloriées' of Temminck & Laugier (1820–1839)." *Zoologische verhandelingen* 335 (2001): 7–54. http://www.repository.naturalis.nl/document/46425.

Dickinson, Edward C., Normand David, Leslie K. Overstreet, Frank D. Steinheimer, and Justin J. F. J. Jansen. "Histoire naturelle des pigeons or *Les Pigeons*: Coenraad Jacob Temminck versus Pauline Knip." *Archives of Natural History* 37 (2010): 203–20. doi: 10.3366/anh.2010.0003.

Dickinson, Edward C., and Florence F. J. M. Pieters. "Some Bibliographic Findings on 'Muséum d'histoire naturelle des Pays-Bas. – Revue méthodique et critique des collections déposées dans cet établissement' (1862–1881) edited by Hermann Schlegel." *Zoological Bibliography* 1, no. 3 (2011): 116–35.

Doel, H. W. van den. *Het rijk van Insulinde: Opkomst en ondergang van een Nederlandse kolonie*. Amsterdam: Prometheus, 1996.

Doolittle, W. Ford. "The Practice of Classification and the Theory of Evolution, and What the Demise of Charles Darwin's Tree of Life Hypothesis Means for Both of Them." *Philosophical Transactions of the Royal Society of London. Series B, Biological Sciences* 364, no. 1527 (2009): 2221–28. doi:10.1098/rstb.2009.0032.

Doren, Jan Baptist Jozef van. *Mémoire sur les quadrumanes et les chéiroptères de l'Archipel Indien*. Amsterdam: J. D. Sijbrandi, 1868.

Drapiez, Pierre August Joseph. "Manuel d'ornithologie, ou tableau systématique des oiseaux qui se trouvent en Europe, par C. J. Temminck, 2me édition (1)." *Annales générales des sciences physiques* 6 (1820): 276–85.

Dreier, Rolf P. "Voor museum en vaderland: De oprichtingsjaren van 's Rijksmuseum van Natuurlijke Historie in Leiden (1815–1830)." Master's thesis, Leiden University, 2003.

Dulm, Frits van. *'Zonder eigen gewinne en glorie': Mr. Iman Wilhelm Falck (1736–1785), gouverneur en directeur van Ceylon en Onderhorigheden*. Hilversum: Verloren, 2012.

Dumont d'Urville, Jules-Sébastien-César. *Voyage de la corvette l'Astrolabe exécuté par ordre du roi, pendant les années 1826–1827–1828–1829, sous le commandement de J. Dumont d'Urville*. 13 vols. Paris: J. Tastu, 1830. doi:10.5962/bhl.title.2132.

Duperrey, Louis-Isidore. *Voyage autour du monde: Exécuté par ordre du roi, sur la corvette de sa majesté, la Coquille, pendant les années 1822, 1823, 1824, et 1825*. Paris: Arthus Bertrand, 1826–1830. doi:10.5962/bhl.title.57936.

Ehlers, Ernst Heinrich, Albert von Kölliker, and Karl Theodor Ernst von Siebold. "Prospectus." *Zeitschrift für wissenschaftliche Zoologie* 1, no. 1 (1848): [n.p.].

Elias, Johan Engelebert. *De vroedschap van Amsterdam, 1578–1795*. Amsterdam: N. Israel, 1963.

Elliot, Daniel Giraud. *A Monograph of the "Phasianidae," or family of the pheasants*. New York: [printed for the author], 1872. UUID:ae5f3830-c6d2-012f-1b4e-58d385a7bc34.

Endersby, Jim. "Classifying Sciences: Systematics and Status in mid-Victorian Natural History." In *The Organisation of Knowledge in Victorian Britain*, edited by Martin Daunton, 1–85. Oxford and New York: Oxford University Press, 2005.

Endersby, Jim. "Descriptive and prescriptive taxonomies." In *Worlds of Natural History*, edited by Emma C. Spary, Helen Anne Curry, James Andrew Secord, and Nicholas Jardine, 447–59. Cambridge: Cambridge University Press, 2018.

Endersby, Jim. *Imperial Nature: Joseph Hooker and the Practices of Victorian Science*. Chicago: University of Chicago Press, 2008.

Enenkel, Karl A. E., and Paul J. Smith, eds. *Early Modern Zoology: The Construction of Animals in Science, Literature and the Visual Arts*. 2 vols., Intersections. Leiden: Brill, 2007.

Epen, Didericus Gijsbertus van, ed. *Nederland's Patriciaat 1868–1930*. Vol. 5. 's-Gravenhage: Centraal Bureau voor Genealogie en Heraldiek, 1914. ark:/13960/t3xs8hk56.

Fabricius, Johann Christian. *Philosophia entomologica sistens scientiae fundamenta, adiectis definitionibus, exemplis, observationibus, adumbrationibus*. Hamburg: Impensis Carol. Ernest. Bohnii, 1778. doi:10.5962/bhl.title.10672.

Farber, Paul Lawrence. "Buffon and the Concept of Species." *Journal of the History of Biology* 5, no. 2 (1972): 259–84. doi: 10.1007/BF00346660.

Farber, Paul Lawrence. "The Development of Ornithological Collections in the Late Eighteenth and Early Nineteenth Centuries and Their Relationship to the Emergence of Ornithology as a Scientific Discipline." *Journal of the Society for the Bibliography of Natural History* 9, no. 4 (1980): 391–94. doi:10.3366/jsbnh.1980.9.4.391.

Farber, Paul Lawrence. "The Development of Taxidermy and the History of Ornithology." *Isis* 66, no. 244 (1977): 550–66. doi:10.1086/351874.

Farber, Paul Lawrence. *The Emergence of Ornithology as a Scientific Discipline: 1760–1850*. Dordrecht: D. Reidel, 1982.

Farber, Paul Lawrence. *Finding Order in Nature: The Naturalist Tradition from Linnaeus to E. O. Wilson*. Baltimore, MD: Johns Hopkins University Press, 2000.

Farber, Paul Lawrence. "The Type-Concept in Zoology During the First Half of the Nineteenth Century." *Journal of the History of Biology* 9, no. 1 (1976): 93–119. doi:10.1007/BF00129174.

Finsch, Otto, and Gustav Hartlaub. *Die Vögel Ost-Afrikas. Baron Carl Claus von Deer Decken's Reisen in Ost-Africa*. Vol. 4. Leipzig and Heidelberg: C. F. Winter'sche Verlagshandlung, 1870. doi:10.5962/bhl.title.14155.

Flourens, Pierre. *Résumé analytique des observations de Frédéric Cuvier sur l'instinct et l'intelligence des animaux*. Paris: C. Pitois, 1841.

Forrer, Matthi, and Arlette Kouwenhoven. *Siebold en Japan. Zijn leven en werk*. Leiden: Hotei, 2000.

Forster, Georg. *Ansichten vom Niederrhein, von Brabant, Flandern, Holland, England und Frankreich*. Leipzig: Reclam, 1905.

Fransen, Charles H. J. M., Lipke Bijdeley Holthuis, and J. P. H. M. Adema. "Type-Catalogue of the Decapod Crustacea in the Collections of the Nationaal Natuurhistorisch Museum, with Appendices of pre-1900 Collectors and Material." *Zoologische verhandelingen* 311 (1997): 1–344. https://repository.naturalis.nl/pub/317736.

Fransen, Charles H. J. M., and Martien J. P. van Oijen. "L. B. Holthuis, 'The Institutional Memory' of the Leiden Museum – Obituary." *Contributions to Zoology* 77, no. 3 (2008): 201–04. https://repository.naturalis.nl/pub/286655.

French, John Oliver. "An Inquiry Respecting the True Nature of Instinct, and of the Mental Distinction between Brute Animals and Man." *The Zoological Journal* 1, no. 1 (1824): 1–32.

Freycinet, Louis Claude Desaulses de. *Voyage autour du monde: Entrepris par ordre du roi ... exécuté sur les corvettes de S. M. l'Uranie et la Physicienne, pendant les années 1817, 1818, 1819 et 1820*. 4 vols. Paris: Pillet, 1824?–1844. doi:10.5962/bhl.title.15862.

Gassó Miracle, Maria Eulàlia. "Het beestje bij zijn naam noemen." In *Wonderkamer: Magazine voor wetenschapsgeschiedenis* 1 (2020): 51–53.

Gassó Miracle, Maria Eulàlia. "On Whose Authority? Temminck's Debates on Zoological Classification and Nomenclature: 1820–1850." *Journal of the History of Biology* 44, no. 3 (2011): 445–481. doi:10.1007/s10739-010-9265-1.

Gassó Miracle, Maria Eulàlia. "The Significance of Temminck's Work on Biogeography: Early Nineteenth Century Natural History in Leiden, the Netherlands." *Journal of the History of Biology* 41, no. 4 (2008): 677–716. doi:10.1007/s10739-008-9158-8.

Gärtner, Gottfried, Bernhard Meyer, and Johannes Scherbius. *Oekonomisch-technische Flora der Wetterau*. 3 vols. Frankfurt: P. H. Guilhauman, 1799. doi:10.17192/eb2010.0292.

Gelder, Esther van, ed. *Bloeiende kennis: Groene ontdekkingen in de Gouden Eeuw*. Hilversum: Verloren, 2012.

Geoffroy Saint-Hilaire, Étienne, and Frédéric Cuvier. *Histoire naturelle des mammifères, avec des figures originales, coloriées, dessinées d'après des animaux vivans*. Paris: A. Belin, 1824. doi:10.5962/bhl.title.78766.

Gibson, Susannah. "The Careering Naturalists: Creating Career Paths in Natural History, 1790–1830." *Archives of Natural History* 44, no. 2 (2017): 195–214. doi:10.3366/anh.2017.0444.

Gijzen, Agatha. "'s Rijks Museum van Natuurlijke Historie, 1820–1915." Rotterdam: W. L. & J. Brusse, 1938. https://repository.naturalis.nl/pub/631350.

Glen, Hugh F., and Gerrit Germishuizen. *Botanical Exploration of Southern Africa*, 2nd ed. Pretoria: South African National Biodiversity Institute, 2010.

Global Biodiversity Information Facility (GBIF) Home Page (2020). https://www.gbif.org.

Gloger, Constantin Wilhelm Lambert. *Das Abändern der Vögel durch Einfluss des Klima's*. Breslau: August Schulz & Co., 1833. doi:10.5962/bhl.title.50395.

Gmelin, Johann Friedrich. *Caroli a Linné. Systema naturae per regna tria naturae: Secundum classes, ordines, genera, species, cum characteribus, differentiis, synonymis, locis* … 13th enlarged and revised ed. 2 vols. Leipzig: Georg. Emanuel. Beer, 1788–1793. doi:10.5962/bhl.title.36932.

Gordin, Michael D. *Scientific Babel: The Language of Science from the Fall of Latin to the Rise of English*. London: Profile Books, 2017.

Goss, Andrew. *The Floracrats: State-sponsored Science and the Failure of the Enlightenment in Indonesia*. Madison, WI: University of Wisconsin Press, 2011.

Gould, John. *The Birds of Europe*. 5 vols. London: Richard and John E. Taylor, 1837. doi:10.5962/bhl.title.65989.

Gray, George Robert. *Appendix to a List of the Genera of Birds*. London: Richard and John E. Taylor, 1842. doi:10.5962/bhl.title.119636.

Gray, George Robert. *A List of the Genera of Birds: With their Synonyma and an Indication of the Typical Species of Each Genus*. 2nd ed. London: Richard and John E. Taylor, 1841. doi:10.5962/bhl.title.14202.

Gray, John Edward. "A Revision of the Genera and Species of Viverrine Animals (Viverridae) Founded on the Collection in the British Museum." *Proceedings of the Zoological Society of London* (1864): 502–79.

Greshoff, M. "Kuhl en Van Hasselt: Eene episode in het Nederlandsch-Indisch natuuronderzoek." *Album der natuur* 52, no. 1 (1903): 60–88. https://natuurtijdschriften.nl/pub/562547.

Grondhout, J. A. M. A. "De Ter Meer's." *Jaarboekje voor geschiedenis en oudheidkunde van Leiden en Rijnland* 9 (1912): 153–61.

Haffer, Jürgen. "The History of Species Concepts and Species Limits in Ornithology." *Bulletin of the British Ornithologists' Club* 112A [Centenary volume] (1992): 106–58.

Hallam, Anthony. *Great Geological Controversies*. Oxford: Oxford University Press, 1990.

Hartlaub, Gustav. "Beschreibung einiger neuen, von Herrn H. S. Pel, holländischem Residenten auf der Goldküste, daselbst gesammelten Vögelarten." *Journal für Ornithologie* 3 (1855): 353–60.

Hartlaub, Gustav. *System der Ornithologie Westafrica's*. Bremen: C. Schünemann, 1857. doi:10.5962/bhl.title.105145.

Hartlaub, Gustav. "A Systematic Index to a Series of Descriptions of Birds, Published by C. J. Temminck in his 'Catalogue systématique du cabinet d'ornithologie, etc.,' of the year 1807," in *Contributions to Ornithology for 1848–1853*, edited by William Jardine. Vol. 2, 19–25. Edinburgh: W. H. Lizars, 1848–1853.

Hartlaub, Gustav. "Systematisches Verzeichniss der von Herrn H. S. Pel auf der Goldküste zwischen Cap Tres Puntas und Accrah gesammelten Vögel." *Journal für Ornithologie* 3 (1855): 360–61.

Heiningen, Teunis Willem van. *The Correspondence of C. G. C. Reinwardt*. Tools and Sources for the History of Science in the Netherlands. Vol. 2. The Hague: Huygens Institute for the History of the Netherlands, 2011.

Hempel, Carl G. *Philosophy of Natural Science*. Foundations of Philosophy Series. Englewood Cliffs, N.J.: Prenctice Hall, 1966.

Herklots, Jan Adrianus. "Additamenta ad faunam carcinologicam Africae occidentalis, sive descriptiones specierum novarum e Crustaceorum ordine, quas in Guinea collegit vir strenuus H. S. Pel praefactus residentiis in littore Guineae." PhD dissertation, Leiden University, 1851.

Hes, Ronald. "The History of Animal Preparations in the Collections of Naturalis Biodiversity Center." Student Internship Report. Utrecht University (History and Philosophy of Science). April 2020.

Hoek Ostende, Lars W. van den, René W. R. J. Dekker, and Guido O. Keijl. "Type-Specimens of Birds in the National Museum of Natural History, Leiden. Part 1. Non-Passerines." *NNM Technical Bulletin* 1 (1997): 1–248. https://repository.naturalis.nl/pub/216191.

Hoeven, Jan van der. "Bijdragen tot de natuurlijke geschiedenis van den mensch: 1ste Bijdrage." *Tijdschrift voor natuurlijke geschiedenis* 1, no. 2 (1834): 86–97.

Hoeven, Jan van der. *Handboek der dierkunde, of grondbeginsels der natuurlyke geschiedenis van het dierenrijk*. 2 vols. Delft: J. Allart, 1827–1833.

Hoeven, Jan van der. "Voorberigt." *Tijdschrift voor natuurlijke geschiedenis* 1, no. 1 (1834): i–iv.

Hoeven, Jan van der, and Willem Hendrik de Vriese. "Naschrift van het eerste deel." *Tijdschrift voor natuurlijke geschiedenis* 1, no. 1 (1834): [viii].

Holthuis, Lipke Bijdeley. "Biografische notities betreffende verzamelaars voor het Rijksmuseum van Natuurlijke Historie te Leiden. I. Hendrik Severinus Pel (1818–1876)." *Zoologische bijdragen* 10, no. 1 (1968): 3–32. https://repository.naturalis.nl/pub/317263.

Holthuis, Lipke Bijdeley. *Rijksmuseum van Natuurlijke Historie 1820–1958*. Leiden: Nationaal Natuurhistorisch Museum, 1995. https://repository.naturalis.nl/pub/268714.

Holthuis, Lipke Bijdeley. "The Three Dutch Authors of von Siebold's *Fauna japonica*, with Some Notes on the Artists." In *Ph. F. von Siebold and Natural History of Japan: Crustacea*, edited by Takai Yamaguchi, 689–731. Tokyo: The Carcinological Society of Japan, 1993.

Holthuis, Lipke Bijdeley, and Tsune Sakai. *Ph. F. von Siebold and Fauna japonica; A History of Early Japanese Zoology*. Tokyo: Academic Press of Japan, 1970.

Honig, Pieter, and Frans Verdoorn, eds. *Science and Scientists in the Netherlands Indies*. New York: Board for The Netherlands Indies, Surinam and Curasao, 1945.

Hooker, Joseph Dalton. *Flora Indica: Being a Systematic Account of the Plants of British India, Together With Observations on the Structure and Affinities of their Natural Orders and Genera*. London: W. Pamplin, 1855. doi:10.5962/bhl.title.57706.

Hooker, Joseph Dalton. "Flora Novae-Zelandiae." in *The Botany of the Antarctic Voyage of H. M. Discovery Ships* Erebus *and* Terror *in the Years 1839–1843 under the Command of Captain Sir James Clark Ross*. Vol. 2. London: Reeve Brothers, 1853. doi:10.5962/bhl.title.16029.

Hooker, Joseph Dalton. "Notices of Books: Géographie botanique raisonée, ou exposition des faits principaux et des lois concernant la distribution géographque des plantes de l'époque actuelle; par M. Alph. de Candolle. 2 vols. 8vo. Paris. Concluding Remarks." *Hooker's Journal of Botany and Kew Garden Miscellany* 8 (1856): 248–56.

Hooker, Joseph Dalton. *On the Flora of Australia: Its Origin, Affinities, and Distribution; Being an Introductory Essay to the Flora of Tasmania*. London: Lovell Reeve, 1859. doi:10.5962/bhl.title.60980.

Hoorn, P. F. van. "Levensberigten der in dit jaar afgestorvene medeleden: Levensberigt van Dr. Paulus Godfried van Hoorn." *Jaarboek van de Maatschappij der Nederlandse Letterkunde* 1–9 (1851), 3–10.

Horsburgh, James. *India Directory, or, Directions for Sailing To and From the East Indies, China, New Holland, Cape of Good Hope, Brazil, and the Interjacent Ports, Comp. Chiefly from Original Journals at the East India House, and from Observations and Remarks, Made During Twenty-One Years Experience Navigating in Those Seas*. 2nd ed. 2 vols. London: [printed for the author], 1817.

Horsfield, Thomas. "Description of the *Helarctos euryspilus*; Exhibiting in the Bear from the Island of Borneo, by the Type of a Subgenus of Ursus." *Zoological Journal of the Linnaean Society* 2 (1826): 221–34.

Horsfield, Thomas. *Zoological Researches in Java, and the Neighbouring Islands*. London: Kingsbury, Parbury, & Allen, 1824. doi:10.5962/bhl.title.44848.

Houston, C. Stuart. "The Hudson's Bay Company and its Collectors." In *Naturalists in the Field*, edited by Arthur MacGregor. Emergence of Natural History, 332–64. Leiden: Brill, 2018.

Hubrecht, A. A. W. "Hermann Schlegel." *De Gids* 4 (1884): 536–47.

Huigen, Siegfried. *Knowledge and Colonialism: Eighteenth-Century Travellers in South Africa*. Leiden: Brill, 2009.

Huistra, Hieke. *The Afterlife of the Leiden Anatomical Collections: Hands On, Hands Off*. The History of Medicine in Context Series. London: Routledge, 2019.

Huistra, Hieke. "Preparations on the Move: The Leiden Anatomical Collections in the Nineteenth Century." PhD dissertation, Leiden University, 2013.

Hull, David L. "The Effect of Essentialism on Taxonomy: Two Thousand Years of Stasis (I)." *The British Journal for the Philosophy of Science* 15, no. 60 (1965): 314–26. doi:10.1093/bjps/XV.60.314.

Hull, David L. "The Effect of Essentialism on Taxonomy: Two Thousand Years of Stasis (II)." *The British Journal for the Philosophy of Science* 16, no. 61 (1965): 1–18. doi:10.1093/bjps/XVI.61.1.

Humboldt, Alexander von. "Sur les lois que l'on observe dans la distribution des formes végétales." *Annales de chimie et de physique* 1, no. 2 (1816): 225–39.

Humboldt, Alexander von, and Aimé Jacques Alexandre Bonpland. *Essai sur la géographie des plantes: Accompagné d'un tableau physique des régions équinoxiales.* Paris: Levrault, Schoell et compagnie, 1805. doi:10.5962/bhl.title.9309.

Humboldt, Alexander von, and Aimé Jacques Alexandre Bonpland. *Recueil d'observations de zoologie et d'anatomie comparée, faites dans l'Océan Atlantique, dans l'intérieur du Nouveau Continent et dans la Mer du Sud, pendant les années 1799, 1800, 1801, 1802 et 1803.* 2 vols. Paris: F. Schoell et G. Dufour, 1811–1833. doi:10.5962/bhl.title.43770.

Husson, A. M., and Lipke Bijdeley Holthuis. "The Dates of Publication of 'Verhandelingen over de natuurlijke geschiedenis der Nederlandsche overzeesche bezittingen' edited by C. J. Temminck." *Zoologische mededelingen* 34, no. 2 (1955): 17–25. https://repository.naturalis.nl/pub/318953.

Huxley, Leonard, ed. *Life and Letters of Sir Joseph Dalton Hooker, O.M., G.C.S.I. Based on Materials Collected and Arranged by Lady Hooker.* London: John Murray, 1918.

Illiger, Johann Karl Wilhelm. *Prodromus systematis mammalium et avium; Aditis terminis zoographicis utriusque classis, eorumque versione germanica.* Berlin: C. Salfeld, 1811. ark:/13960/t0sr2hp2d.

Illiger, Johann Karl Wilhelm. *Versuch einer systematischen vollständigen Terminologie für das Thierreich und Pflanzenreich.* Helmstadt: C. G. Fleckeisen, 1800. doi:10.5962/bhl.title.37079.

International Commission on Zoological Nomenclature. *International Code of Zoological Nomenclature.* London: The International Trust for Zoological Nomenclature, 1999. http://www.iczn.org/iczn/index.jsp.

Jansen, Justin J. F. J. "The Ornithology of the Baudin Expedition (1800–1804)." PhD dissertation, Leiden University, 2018.

Jansen, Justin J. F. J. "An Unpublished 'Catalogue du cabinet de C. J. Temminck' (c. 1803–1804)." *Archives of Natural History* 44, no. 2 (2017): 352–55. doi:10.3366/anh.2017.0454.

Jardine, Nicholas, James Andrew Secord, and Emma C. Spary, eds. *Cultures of Natural History.* Cambridge: Cambridge University Press, 1996.

Jardine, William, ed. *Memoirs of Hugh Edward Strickland: With a Selection from his Scientific Writings.* London: John Van Voorst, 1858.

Jardine, William. *The Natural History of Game-Birds.* Edinburgh: W. H. Lizars, 1834. doi:10.5962/bhl.title.101720.

Johnson, Kristin. "Natural History as Stamp Collecting: A Brief History." *Archives of Natural History* 34, no. 2 (2007): 244–58. doi:10.3366/anh.2007.34.2.244.

Jonge, Hans de. "Macht, machinaties en musea: Jan van der Hoeven, Hermann Schlegel en hun strijd om het Rijksmuseum van natuurlijke historie te Leiden." *Bijdragen en mededelingen betreffende de geschiedenis der Nederlanden* 120 (2005): 177–206.

Jonge, Hans de. "Sebald Justinus Brugmans (1763–1819)." Master's thesis, Leiden University, 1999.

Jonge, Hans de. *Het vergeten fenomeen Sebald J. Brugmans 1763–1819, verzamelaar tussen koning, keizer en universiteit.* Leiden: Museum Boerhaave, 2001.

Jussieu, Antoine-Laurent de. *Genera plantarum: Secundum ordines naturales disposita, juxta methodum in horto regio parisiensi exaratam, anno M.DCC.LXXIV.* Paris: Herissant et Theophilum Barrois, 1789. doi:10.5962/bhl.title.284.

Keurs, Pieter ter. "Agency, Prestige and Politics: Dutch Collecting Abroad and Local Responses." In *Unpacking the Collection: Networks of Material and Social Agency in the Museum*, edited by Sarah Byrne, Anne Clarke, Rodney Harrison, and Robin Torrence, 165–82. New York: Springer New York, 2011. doi:10.1007/978-1-4419-8222-3_7.

Kinch, Michael Paul. "Geographical Distribution and the Origin of Life: The Development of Early Nineteenth-Century British Explanations." *Journal of the History of Biology* 13, no. 1 (1980): 91–119. doi:10.1007/BF00125355.

Klaver, Charles John Joseph. *Inseparable Friends in Life and Death: The Life and Work of Heinrich Kuhl (1797–1821) and Johan Conrad van Hasselt (1797–1823), Students of Prof. Theodorus van Swinderen.* Groningen: Barkhuis, 2007.

Knapp, Sandra, Gerardo Lamas, Eimear Nic Lughadha, and Gianfranco Novarino. "Stability or Stasis in the Names of Organisms: The Evolving Codes of Nomenclature." *Philosophical Transactions of the Royal Society of London B: Biological Sciences* 359, no. 1444 (2004): 611–22. doi:10.1098/rstb.2003.1445.

Knip, Pauline and Coenraad Jacob Temminck. *Les pigeons par Madame Knip, née Pauline de Courcelles, le texte par C. J. Themminck.* 2 vols. Paris: Chez Mme Knip, [1808]–1811. doi:10.5962/bhl.title.120559.

Kohler, Robert E. *All Creatures: Naturalists, Collectors, and Biodiversity, 1850–1950.* Princeton and Oxford: Princeton University Press, 2013. doi:10.2307/j.ctt4cgbz8.

Kohler, Robert E. "Reflections on the History of Systematics." In *The Evolution of Phylogenetic Systematics*, edited by Andrew Hamilton. Species and Systematics. Vol. 5., 17–46. Berkeley: University of California Press, 2014. doi:10.1525/california/9780520276581.003.0002.

Kohlstedt, Sally Gregory. "Museums: Revisiting Sites in the History of the Natural Sciences." *Journal of the History of Biology* 28, no. 1 (1995): 151–66.

Koninglijke Maatschappy der Wetenschappen te Haarlem. "Directeuren van de Koninglijke Maatschappy der Wetenschappen: Geschikt naar den tijd der verkiezing." *Natuurkundige verhandelingen van de Koninglijke Maatschappy der Wetenschappen te Haarlem* 5, no. 1 (1809): xxvii–xxxii.

Koninklijk-Nederlandsch Instituut van Wetenschappen Letterkunde en Schoone Kunsten. *Verslagen en mededeelingen, uitgegeven door de vier Klassen van het Koninklijk-Nederlandsch Instituut van Wetenschappen, Letterkunde en Schoone Kunsten.* Vol. 4. Amsterdam: Johannes Müller, 1841.

Koninklijke Orde der Unie. *Gedenkschriften van de Koninklijke Orde der Unie, voor de jaren 1807, 1808 en 1809.* Amsterdam: Johannes Allard, 1810.

Koolmees, J. A. "Caspar Georg Reinwardt (1773–1854)." Master's thesis, Utrecht University, 1979.

Kortenbach, C. J. "Het Hof van Zessen, thans Rijksmuseum van Oudheden." *Jaarboekje van de Historische Vereniging Oud Leiden* 37 (1945): 74–94.

Krikken, Jan, Cees van Achterberg, Piet H. van Doesburg, Rienk de Jong, and K. W. Zwart. "Samuel Constant Snellen van Vollenhoven (1816–1880) and His Entomological Work." *Tijdschrift voor entomologie* 124 (1981): 235–68.

Kuhl, Heinrich, and Johan Conrad van Hasselt. "Pjihorjavor, aan den voet van den Pangerango, 18 Julij 1821." *Algemeene konst- en letterbode* 1, no. 7 (1822): 99–102.

La Fontaine, Jean de. "Le geai paré des plumes du paon." In *Fables choisies mises en vers.* Vol. 1, livre iv, fable 9. Paris: Barbin, 1668.

La Vergata, Antonello. "Au nom de l'espèce: Classification et nomenclature au XIXe siècle." In *Histoire du concept de l'espèce dans les sciences de la vie: Colloque international (mai 1985) organisé par la Fondation Singer-Polignac, Paris,* edited by Scott Atran, 193–225. Paris: Fondation Singer-Polignac, [1987].

Lamarck, Jean-Baptiste de. *Encyclopédie méthodique: Botanique.* 8 vols. Paris and Liège: Panckoucke, Plomteux, 1783–1808. doi:10.5962/bhl.title.824.

Lamarck, Jean-Baptiste de. *Hydrogéologie, ou Recherches sur l'influence qu'ont les eaux sur la surface du globe terrestre; sur les causes de l'existence du bassin des mers, de son déplacement et de son transport successif sur les différens points de la surface du globe: Enfin sur les changemens que les corps vivans exercent sur la nature et l'état de cette surface.* Paris: Agasse et Maillard, 1802.

Lamarck, Jean-Baptiste de. *Philosophie zoologique, ou exposition des considérations relatives à l'histoire naturelle des animaux; à la diversité de leur organisation et des facultés qu'ils en obtiennent; aux causes physiques qui maintiennent en eux la vie et donnent lieu aux mouvemens qu'ils exécutent; enfin, à celles qui produisent les unes le sentiment, et les autres l'intelligence de ceux qui en sont doués.* Nouvelle éd. [1830 re-issue of the 1809 edition]. Paris: Baillière, 1830. doi:10.5962/bhl.title.50194.

Lamarck, Jean-Baptiste de. "Sur l'histoire naturelle en général." *Journal d'histoire naturelle* 1 (1792): 3–6.

Land, Jacob van der, ed. *The History of Natural History in Leiden*. Leiden: Naturalis, 2001.

Larson, James. "Not without a Plan: Geography and Natural History in the Late Eighteenth Century." *Journal of the History of Biology* 19, no. 3 (1986): 447–88.

Latham, John. *A General History of Birds*. 10 vols. Winchester: Jacob and Johnson, 1821–1828. doi:10.5962/bhl.title.62572.

Latham, John. *A General Synopsis of Birds*. Vol. 1. London: Benj. White, 1781. doi:10.5962/bhl.title.49894.

Latham, John. *Index ornithologicus, sive Systema ornithologiae; Complectens avium divisionem in classes, ordines, genera, species, ipsarumque varietates: Adjectis synonymis, locis, descriptionibus, &c.* 2 vols. London: Leigh et Sotheby, 1790.

Lauts, Ulrich Gerard. "Nieuw uitgekomen boeken: Coup-d'oeil général sur les possessions néerlandaises dans l'Inde Archipélagique par C. J. Temminck, 1846." *Algemeene konst- en letterbode* 1, no. 12 (1847): 182–99.

Lauts, Ulrich Gerard. "Nieuw uitgekomen boeken: Coup-d'oeil général sur les possessions néerlandaises dans l'Inde Archipélagique par C. J. Temminck, 1847." *Algemeene konst- en letterbode* 1, no. 9 (1849): 137–44.

Lesson, René-Primevère. "Réclamation de M. Vieillot, relative au *Manuel d'ornithologie*." *Bulletin des sciences naturelles et de géologie* 15 (1828): 142–44.

Levaillant, François. *Histoire naturelle des oiseaux d'Afrique*. 6 vols. Paris: J. J. Fuchs, 1799–1808. doi:10.5962/bhl.title.63636.

Levaillant, François. *Histoire naturelle d'une partie d'oiseaux nouveaux et rares de l'Amérique et des Indes*. Paris: J. E. Gabriel Dufour, 1801. doi:10.5962/bhl.title.102769.

Levaillant, François. *Second voyage de F. Levaillant dans l'intérieur de l'Afrique par le Cap de Bonne-Espérance, pendant les années 1783, 84 et 85*. 2 vols. Paris: H. J. Jansen et compagnie, [1795 or 1796]. doi:10.5962/bhl.title.101758.

Levaillant, François. *Voyage de M. Le Vaillant dans l'intérieur de l'Afrique: Par le Cap de Bonne-Espérance, dans les années 1780, 81, 82, 83, 84 & 85*. Vol. 1. Paris: Leroy, 1790. doi:10.5962/bhl.title.101757.

Levaillant, François, and Jacques Barraband. *Histoire naturelle des perroquets*. 2 vols. Paris: Levrault, 1801–1805. doi:10.5962/bhl.title.60852.

Limoges, Camille. "The Development of the Muséum d'Histoire Naturelle of Paris, c. 1800–1914." In *The Organization of Science and Technology in France 1808–1914*, edited by Robert Fox and George Weisz, 211–40. Cambridge: Cambridge University Press, 2008.

Lindroth, Carl H. "Systematics Specializes between Fabricius and Darwin: 1800–1859." In *History of Entomology*, edited by Ray F. Smith, Thomas E. Mittler, and Carrol N. Smith. Palo Alto, CA: Annual Reviews Inc. and the Entomological Society of America, 1973.

Linnaeus, Carolus. *Philosophia botanica in qua explicantur fundamenta botanica cum definitionibus partium, exemplis terminorum, observationibus rariorum, adjectis figuris æneis*. Stockolm: G. Kiesewetter, 1751. doi:10.5962/bhl.title.37652.

Linnaeus, Carolus. *Species plantarum: Exhibentes plantas rite cognitas ad genera relatas, cum differentiis specificis, nominibus trivialibus, synonymis selectis, locis natalibus, secundum systema sexuale digestas*. 2 vols. Stockholm: Holmiae, 1753. doi:10.5962/bhl.title.37656.

Linnaeus, Carolus. *Systema naturae per regna tria naturae, secundum classes, ordines, genera, species, cum characteribus, differentiis, synonymis, locis*. Leiden: Theodor Haak, 1735.

Lyell, Charles. *Principles of Geology: Being an Attempt to Explain the Former Changes of the Earth's Surface, by Reference to Causes Now in Operation*. Vol. 2. London: John Murray, 1832.

Lynden-de Bruïne, Alberta Maria van. *In vogelvlucht door Europa: De reisjournalen van Dionysia Catharina Temminck-Cau, 1807–1824*. Zutphen: Walburg, 2001.

MacGregor, Arthur, ed. *Naturalists in the Field*. Emergence of Natural History. Vol. 2. Leiden: Brill, 2018.

Macleay, William Sharp. *Horae Entomologicae or, Essays on the Annulose Animals*. Vol. 1. London: S. Bagster, 1819. doi:10.5962/bhl.title.48636.

Macleay, William Sharp. "Remarks on the Identity of Certain General Laws which have been Lately Observed to Regulate the Natural Distribution of Insects and Fungi." *Transactions of the Linnean Society of London* 14 (1825): 46–68.

Manse, Maarten. "Kennis is macht: De veelzijdige expedities van botanicus Pieter Willem Korthals (1807–1892)." *Studium* 6, no. 1 (2013): 42–50. doi:10.18352/studium.8837.

Mayr, Ernst. *The Growth of Biological Thought: Diversity, Evolution, and Inheritance*. Cambridge, MA: The Belknap Press of Harvard University Press, 1982.

Mayr, Ernst. "Illiger and the Biological Species Concept." *Journal of the History of Biology* 1, no. 2 (1968): 163–78.

McOuat, Gordon. "Cataloguing Power: Delineating 'Competent Naturalists' and the Meaning of Species in the British Museum." *The British Journal for the History of Science* 34, no. 1 (2001): 1–28. doi:10.1017/S0007087400004179a.

McOuat, Gordon. "The Origins of 'Natural Kinds': Keeping 'Essentialism' at Bay in the Age of Reform." *Intellectual History Review* 19, no. 2 (2009): 211–30. doi:10.1080/17496970902981694.

McOuat, Gordon. "Species, Rules and Meaning: The Politics of Language and the Ends of Definitions in 19th Century Natural History." *Studies in History and Philosophy of Science Part A* 27, no. 4 (1996): 473–519. doi:10.1016/0039-3681(95)00060-7.

Mees, Gerlof Fokko. "Vogelkundig onderzoek op Nieuw Guinea in 1828: Terugblik op de ornithologische resultaten van de reis van Zr. Ms. Korvet Triton naar de zuid-west kust van Nieuw-Guinea." *Zoologische bijdragen* 40, no. 1 (1994): 3–64. https://repository.naturalis.nl/pub/317250.

Meisner, Friedrich. *Systematisches Verzeichniss der Vögel, welche die Schweiz entweder bewohnen, oder theils zu bestimmten, theils zu unbestimmten Zeiten besuchen, und sich auf der Gallerie der Bürger-Bibliothek in Bern ausgestopft befinden.* Bern: Hallersche Buchhandlung, 1804.

Melville, Richard V. *Towards Stability in the Names of Animals: A History of the International Commission on Zoological Nomenclature, 1895–1995.* London: International Trust for Zoological Nomenclature, 1995.

Milne-Edwards, Henri. *Cours élémentaire d'histoire naturelle: A l'usage des collèges et des maisons d'éducation, rédigé conformément au programme de l'université du 14 septembre 1840. Zoologie.* Paris: Fortin, Masson et Cie, 1841. doi:10.5962/bhl.title.41891.

Minelli, Alessandro, and Giancarlo Contrafatto, eds. *Biological Science Fundamentals and Systematics.* Biological, Physiological and Health Sciences. 4 vols. Oxford, U.K.: EOLSS Publishers, 2009.

Modera, Justin. *Verhaal van eene reize naar en langs de zuid-westkust van Nieuw-Guinea gedaan in 1828: Door Z.M. corvet Triton, en Z.M. coloniale schoener de Iris.* Haarlem: Vincent Loosjes, 1830.

Moes, Ernst Wilhelm, and Eduard van Biema. *De Nationale Konste-Gallery en het Koninklijk Museum. Bijdrage tot de geschiedenis van het Rijksmuseum.* Amsterdam: F. Muller, 1909. ark:/13960/t3126pz73.

Molhuysen, P. C., and P. J. Blok. "Smissaert." In *Nieuw Nederlandsch biografisch woordenboek.* Vol. 2. Leiden: A. W. Sijthoff, 1912.

Molina, Miquel. "More Notes on the Verreaux Brothers." *Pula: Botswana Journal of African Studies* 16, no. 1 (2002): 30–36.

"Monographies de mammalogie, ou descriptions de quelques genres de mammiferes dont les especes ont eté observées dans les differens musées de l'Europe. Par J. C. Temminck. Livraisons 1–3. pp. 72. pl. viii." *The Zoological Journal* 1, no. 4 (1824): 574–76.

Mörzer Bruyns, W. F. J., red. *Met de Triton en Iris naar de zuidwestkust van Nieuw Guinea in 1828: De reisverhalen van Justin Modera en Arnoldus Johannes van Delden.* Werken van de Linschoten-Vereeniging; CXVII. Zutphen: Walburg Pers, 2018.

Müller, Salomon. "Aanteekeningen, over de natuurlijke gesteldheid van een gedeelte der westkust en binnenlanden van Sumatra." *Tijdschrift voor natuurlijke geschiedenis en physiologie* 2, no. 4 (1835): 315–55.

Müller, Salomon. *Reizen en onderzoekingen in den Indischen archipel: Gedaan op last der Nederlandsche Indische regering, tusschen de jaren 1828 en 1836.* Amsterdam: Frederik Muller, 1857.

Müller, Salomon. "Ueber den Character der Thierwelt auf den Inseln des indischen Archipels, ein Beitrag zur zoologischen Geographie." *Archiv für Naturgeschichte* 12, no. 1 (1846): 109–28.

Murray, Andrew. *The Geographical Distribution of Mammals*. London: Day and Son, limited, 1866. doi:10.5962/bhl.title.15762.

Naumann, Johann Andreas, and Johann Friedrich Naumann. *Johann Andreas Naumann's Naturgeschichte der Vögel Deutschlands, nach eigenen Erfahrungen entworfen*. Leipzig: G. Fleischer, 1820. doi:10.5962/bhl.title.49752.

Naumann, Johann Friedrich. *Taxidermie, oder, Die Lehre Thiere aller Klassen: Am einfachsten und zweckmässigsten für Kabinette auszustopfen und aufzubewahren*. Halle: Hemmerde und Schwetschke, 1815. doi:10.5962/bhl.title.43737.

Nelson, Gareth. "From Candolle to Croizat: Comments on the History of Biogeography." *Journal of the History of Biology* 11, no. 2 (1978): 269–305. doi:10.1007/BF00389302.

Nieuwendijk, J. G. *Zoo was Artis, zo is Artis*. Amsterdam: De Bussy, 1970.

"Nouveau recueil de planches coloriéés d'oiseaux. Par MM. Temminck et Meiffren Laugier." *Annales générales des sciences physiques* 8 (1821): 395–96.

"Nouveau recueil de planches coloriéés d'oiseaux pour servir de complement aux planches enluminés de Buffon. Par MM. C. J. Temminck, d'Amsterdam, et Meiffren Laugier, baron de Chartrouse, de Paris." *Annales générales des sciences physiques* 7 (1820): 361–71.

Nozeman, Cornelis, Martinus Houttuyn, Christiaan Sepp, and Jan Christiaan Sepp. *Nederlandsche vogelen volgens hunne huishouding, aert en eigenschappen beschreeven* … 5 vols. Amsterdam: Jan Christiaan Sepp, 1770–1829.

Nyhart, Lynn K. *Modern Nature: The Rise of the Biological Perspective in Germany*. Chicago: University of Chicago Press, 2009.

Nyhart, Lynn K. "Natural History and the 'New' Biology." In *Cultures of Natural History*, edited by Nicholas Jardine, James Andrew Secord, and Emma C. Spary, 426–43. Cambridge: Cambridge University Press, 1996.

O'Hara, Robert J. "Representations of the Natural System in the Nineteenth Century." *Biology and Philosophy* 6, no. 2 (1991): 255–74. doi:10.1007/bf02426840.

Oehser, Paul H. "Louis Jean Pierre Vieillot (1748–1831)." *Auk* 65, no. 4 (1948): 568–76. doi:10.2307/4080607.

Ogilby, William. "Further Observations on 'Rules for Nomenclature'." *Magazine of Natural History* 2, no. 2 (1838): 275–84.

Ogilby, William. "Letter in Reference to Mr. Strickland's Observations on the Application of the Term Simia." *Magazine of Natural History* 2, no. 2 (1838): 492–94.

Ogilby, William. "Observations on 'Rules for Nomenclature'." *Magazine of Natural History* 2, no. 2 (1838): 150–57.

Oken, Lorenz. *Allgemeine Naturgeschichte für alle Stände*. 8 vols. Stuttgart: Hoffmann'sche Verlags-Buchhandlung, 1833–1843. doi:10.5962/bhl.title.36892.

Oken, Lorenz. *Grundriss der Naturphilosophie der Theorie der Sinne und der darauf gegründeten Classification der Thiere*. Frankfurt-am-Main: [printed for the author], 1804.

Outram, Dorinda. *Georges Cuvier: Vocation, Science, and Authority in Post-Revolutionary France*. Manchester: Manchester University Press, 1984.

Outram, Dorinda. *The Letters of Georges Cuvier: A Summary Calendar of Manuscript and Printed Materials Preserved in Europe, the United States of America, and Australasia*. Chalfont St Giles: The British Society for the History of Science, 1980.

Outram, Dorinda. "New Spaces in Natural History." In *Cultures of Natural History*, edited by Nicholas Jardine, James Andrew Secord, and Emma C. Spary, 249–65. Cambridge: Cambridge University Press, 1996.

Pallas, Peter Simon. *Zoographia Rosso-Asiatica: sistens omnium animalium in extenso Imperio Rossico, et adjacentibus maribus observatorum recensionem, domicilia, mores et descriptiones, anatomen atque icones plurimorum*. 3 vols. Petropoli: In officina Caes. Acadamiae Scientiarum Impress, 1831.

Pickstone, John V. "Museological Science? The Place of the Analytical/Comparative in Nineteenth-Century Science, Technology and Medicine." *History of Science* 32, no. 2 (1994): 111–38. doi:10.1177/007327539403200201.

Pieters, Florence F. J. M. "Het schatrijke naturaliënkabinet van Stadhouder Willem V onder directoraat van topverzamelaar Arnout Vosmaer." In *Het verdwenen museum: Natuurhistorische verzamelingen, 1750–1850*, edited by Bert C. Sliggers and Marijke H. Besselink, 19–44. Haarlem: Teylers Museum, 2002.

Poot, Nathalie, Geert Vanpaemel, and Siska Waelkens. *Een walvis in de stad: De collecties van de Leuvense Faculteit Wetenschappen*. Leuven: Leuven University Press, 2014.

Prichard, James Cowles. *Researches into the Physical History of Man*. London: J. and A. Arch, 1813.

Queiroz, Kevin de. "Systematics and the Darwinian Revolution." *Philosophy of Science* 55, no. 2 (1988): 238–59.

R. K. D. Nederland Instituut voor Kunstgeschiedenis. "Portret van Marinus Adriaan Perpetuus Smissaert (1773–1819)." Last modified January, 2015, https://rkd.nl/nl/explore/images/151688.

Raat, Alexis J. P. "Alexander von Humboldt and Coenraad Jacob Temminck." *Zoologische bijdragen* 21, no. 1 (1976): 19–38.

Raat, Alexis J. P. "Coenraad Jacob Temminck (1778–1858): Een biografische schets." *De negentiende eeuw* 2, no. 2 (1978): 88–104.

Raffles, Thomas Stamford. *A History of Java*. 2 vols. London: Black, Parbury, and Allen, 1817.

Ray, John. *Historia plantarum: Species hactenus editas aliasque insuper multas noviter inventas & descriptas complectens*. 3 vols. London: Clark, 1686–1704. doi:10.5962/bhl.title.61178.

Reinwardt, Caspar Georg Carl. *Redevoering van C. G. C. Reinwardt, over hetgeen het onderzoek van Indië tot uitbreiding der natuurlijke historie heeft toegebragt: Gehouden den 3 mei 1823, bij het plegtig aanvaarden van het gewoon hoogleeraarsambt in de scheikunde, kruidkunde en natuurlijke historie aan de Hoogeschool te Leyden.* Amsterdam: J. van der Hey, 1823.

"A Retrospect of the Literature of British Ornithology, from the Close of the 17th Century to the Present Time." *The Analyst: A Quarterly Journal of Science, Literature, and the Fine Arts* 3, no. 3 (1836): 78–99.

Richardson, John. "Report on North American Zoology." *Report of the British Association for the Advancement of Science* 5 (1836): 121–224.

Rijksbureau voor Kunsthistorische Documentatie. "Directeuren van Artis: Reindert Draak, 1838–1840." RKD – Netherlands Institute for Art History, 2021, https://artis.rkdstudies.nl/bijlagen/a-directeuren-van-artis.

Roberts, T. R. "The Freshwater Fishes of Java, as Observed by Kuhl and van Hasselt in 1820–23." *Zoologische verhandelingen* 285, no. 1 (1915): 1–94. https://repository.naturalis.nl/pub/317766.

Roger, Jacques. *Buffon: A Life in Natural History*. Translated by Sarah Lucille Bonnefoi. Ithaca and London: Cornell University Press, 1997.

Rookmaaker, L. C. *Calendar of the Scientific Correspondence of Hugh Edwin Strickland*. Cambridge: University Museum of Zoology, 2010.

Rookmaaker, L. C. "The Early Endeavours by Hugh Edwin Strickland to Establish a Code for Zoological Nomenclature in 1842–1843." *Bulletin of Zoological Nomenclature* 68, no. 1 (2011): 29–40.

Rookmaaker, L. C. *The Zoological Exploration of Southern Africa, 1650–1790*. Rotterdam: A. Balkema, 1989.

Rookmaaker, L. C., P. A. Morris, I. E. Glenn, and P. J. Mundy. "The Ornithological Cabinet of Jean-Baptiste Bécoeur and the Secret of the Arsenical Soap." *Archives of Natural History* 33, no. 1 (2006): 146–58. doi:10.3366/anh.2006.33.1.146.

Rudwick, Martin J. S. *Georges Cuvier, Fossil Bones, and Geological Catastrophes: New Translations & Interpretations of the Primary Texts*. Chicago: University of Chicago Press, 1997.

Saunders, Howards, ed. *Vieillot's Analyse d'une nouvelle ornithologie élémentaire*. London: The Willughby Society, 1883. doi:10.5962/bhl.title.12613.

Schilthuizen, Menno. "Early Entomological Art: The Colour Plates of Snellen van Vollenhoven." *Tijdschrift voor entomologie* 150, no. 2 (2007): 263–69. doi:10.1163/22119434-900000229.

Schlegel, Gustav. "Levensschets van Hermann Schlegel." *Jaarboek van de Koninklijke Akademie van Wetenschappen* (1884): 1–95. doi:10.5962/bhl.title.119796.

Schlegel, Hermann. *Abhandlungen aus dem Gebiete der Zoologie und vergleichenden Anatomie*. Leiden: A. Arnz & comp., 1841–1843. doi:10.5962/bhl.title.16080.

Schlegel, Hermann. *Essai sur la physionomie des serpens*. La Haye: Van Stockum, 1837. doi:10.5962/bhl.title.4273.

Schlegel, Hermann. *Muséum d'histoire naturelle des Pays-Bas, par H. Schlegel, revue méthodique et critique des collections déposées dans cet établissement*. 9 vols. Vol. 1. Leiden: E. J. Brill, 1862.

Schlegel, Hermann. "Notice sur l'Erpétologie de l'ile de Java; par M. Boié (Ouvrage manuscrit)." *Bulletin des sciences naturelles et de géologie* 9 (1826): 233–42.

Schlegel, Hermann. "Over eenige nieuwe soorten van vergiftige slangen van de Goudkust." *Verslagen en mededeelingen der Koninklijke Akademie van Wetenschappen* 3 (1855): 312–17.

Schuh, Randall T., and Andrew V. Z. Brower. *Biological Systematics*. Ithaca, NY: Cornell University Press, 2009.

Secord, James Andrew. "The Crisis of Nature." In *Cultures of Natural History*, edited by Nicholas Jardine, James Andrew Secord, and Emma C. Spary, 447–59. Cambridge: Cambridge University Press, 1996.

Secord, James Andrew. *Visions of Science Books and Readers at the Dawn of the Victorian Age*. Chicago: University of Chicago Press, 2015.

Shufeldt, R. W. "Taxidermical Methods in the Leyden Museum, Holland." *Report of the U.S. National Museum for 1895* (1897): 1031–37.

Siebold, Philipp Franz von, ed. *Fauna japonica, sive, Descriptio animalium, quae in itinere per Japoniam, jussu et auspiciis, superiorum, qui summum in India Batava imperium tenent, suscepto, annis 1823–1830 …* 5 vols. Leiden: J. G. La Lau (vols 1, 3), Arnz et socios (vol. 2, 4, 5), 1833–1850. doi:10.5962/bhl.title.124951.

Simpson, Michael G. *Plant Systematics*. 2nd ed. Amsterdam: Academic Press, 2010.

Sirks, Marius Jacob. "Indisch natuuronderzoek." [Amsterdam]: Amsterdamsche boeken steendrukkerij v/h. Ellerman, Harms & co., 1915. doi:10.5962/bhl.title.4435.

Sliggers, Bert C. "Henry Hope's vermakelijke buitenleven en de Haarlemmerhout." In *Paviljoen Welgelegen 1789–1989: Van buitenplaats van de bankier Hope tot zetel van de provincie Noord-Holland*, edited by F. W. A. Beelaerts van Blokland, 27–39. Haarlem: Schuyt & Co, 1989.

Sliggers, Bert C. "Het naturaliënkabinet van de Hollandsche Maatschappij der Wetenschappen 1759–1866." In *Het verdwenen museum: Natuurhistorische verzamelingen, 1750–1850*, edited by Bert C. Sliggers and Marijke H. Besselink, 46–142. Haarlem: Teylers Museum, 2002.

Sliggers, Bert C. "Van individu tot instituut: De opkomst van institutionele verzamelingen." In *Het verdwenen museum: Natuurhistorische verzamelingen, 1750–1850*, edited by Bert C. Sliggers and Marijke H. Besselink, 8–18. Haarlem: Teylers Museum, 2002.

Sliggers, Bert C., and Marijke H. Besselink. *Het verdwenen museum: Natuurhistorische verzamelingen, 1750–1850*. Haarlem: Teylers Museum, 2002.

Smeenk, Chris. "Hermann Schlegel 1804–1884: Ein reiches Leben in einer reichen Sammlung." *Mauritiana* 15, no. 2 (1995): 73–87.

Smit, Pieter. "The Rijksherbarium and the Scientific and Social Conditions which Influenced its Foundation." *Blumea* 25 (1979): 5–11. https://repository.naturalis.nl/pub/525007.

Smit, Pieter, A. P. M. Sanders, and J. P. F. van der Veer, eds. *Hendrik Engel's Alphabetical List of Dutch Zoological Cabinets and Menageries*. 2nd ed. Amsterdam: Rodopi, 1986.

Snellen van Vollenhoven, Samuel Constantinus. "Coenraad Jacob Temminck." *Algemeene konst- en letterbode* 9 (1858): 67–69.

Sonnerat, Pierre. *Voyage aux Indes orientales et a la Chine, Fait par ordre du Roi, depuis 1774 jusqu'en 1781* … 2 vols. Paris: Chez l'Auteur, 1782.

Stark, John. *Elements of Natural History: Adapted to the Present State of the Science, Containing the Generic Characters of Nearly the Whole Animal Kingdom, and the Descriptions of the Principal Species*. Edinburgh: W. Blackwood, 1828.

Steinheimer, Frank D. "Martin Hinrich Carl Lichtenstein and his Ornithological Purchases at the Auction of William Bullock's Museum in 1819." *Archives of Natural History* 35 (2008): 88–99. doi:10.3366/E0260954108000089.

Stevens, Peter Francis. *The Development of Biological Systematics: Antoine-Laurent de Jussieu, Nature, and the Natural System*. New York: Columbia University Press, 1994.

Stresemann, Erwin. "Analyse von C. J. Temmincks 'Catalogue systématique' (1807)." *Zoologische mededelingen* 31, no. 29 (1953): 319–31. https://repository.naturalis.nl/pub/319192.

Stresemann, Erwin. "Aus C. J. Temmincks Briefen an H. Lichtenstein." *Ardea* 39, no. 4 (1951): 253–59.

Stresemann, Erwin. "Aus dem Briefwechsel von C. J. Temminck mit dem Grafen von Hoffmannsegg und C. Illiger 1810–1814." *Ardea* 44, no. 4 (1956): 253–63.

Stresemann, Erwin. "Die Entwicklung der Vogelsammlung des Berliner Museums unter Illiger und Lichtenstein." *Journal für Ornithologie* 70 (1922): 498–509.

Stresemann, Erwin. "Histoire des origines des 'Planches coloriées' de Temminck et Laugier." *L'oiseau et la revue française d'ornithologie* 21 (1951): 33–47.

Stresemann, Erwin. *Ornithology from Aristotle to the Present*. Translated by Hans J. Epstein and Cathleen Epstein. Cambridge, MA: Harvard University Press, 1975.

Stresemann, Erwin, and Peter Thomsen. "Ornithologen-Briefe aus den Jahren 1816 bis 1820 gewechselt zwischen J. F. Naumann und C. J. Temminck." *Centaurus* 2 (1952): 97–139.

Strickland, Hugh Edwin. "A Few Words of Explanation in Reference to Mr. Ogilby's Letter." *Magazine of Natural History* 2, no. 2 (1838): 555–56.

Strickland, Hugh Edwin. "Observations upon the Affinities and Analogies of Organized Beings." *Magazine of Natural History* 4, no. 5 (1840): 219–26.

Strickland, Hugh Edwin. "On the Inexpedience of Altering Established Terms." *Magazine of Natural History and Journal of Zoology, Botany, Mineralogy, Geology, and Meteorology* 1 (1837): 127–31.

Strickland, Hugh Edwin. "On the True Method of Discovering the Natural System in Zoology and Botany." *Annals and Magazine of Natural History* 6, no. 36 (1840): 184–94.

Strickland, Hugh Edwin. "Remarks on Mr. Ogilby's 'Further Observations on Rules for Nomenclature'." *Magazine of Natural History* 2, no. 2 (1838): 326–33.

Strickland, Hugh Edwin. "Reply to Mr. Ogilby's 'Observations on Rules for Nomenclature'." *Magazine of Natural History* 2, no. 4 (1838): 198–204.

Strickland, Hugh Edwin. "Report on the Progress and the Present State of Ornithology." *Report of the British Association for the Advancement of Science* 14 (1845): 170–221.

Strickland, Hugh Edwin. "Rules for Zoological Nomenclature." *Magazine of Natural History and Journal of Zoology, Botany, Mineralogy, Geology, and Meteorology* 1 (1837): 173–76.

Strickland, Hugh Edwin, Charles Darwin, John Stevens Henslow, Leonard Jenyns, William Ogilby, John Phillips, John Richardson, and John Obadiah Westwood. "Report of a Committee Appointed to Consider of the Rules by which the Nomenclature of Zoology may be Established on a Uniform and Permanent Basis." *Report of the British Association for the Advancement of Science* 12 (1843): 105–21.

Striekwold, Robbert J. "Naam maken: Wat de dood van twee negentiende-eeuwse wetenschappers ons leert over de natuurhistorie." *Ex Tempore* 37, no. 3 (2018): 213–26.

Sundevall, Carl Jakob. "Kritisk framställning af fogelarterna uti äldre ornithologiska arbete." *Kongliga Svenska Vetenskaps-Akademiens handlingar* 2, no. 3 (1857): 23–60.

Susanna, Joannes Andreas. "Levensschets van Coenraad Jacob Temminck." *Jaarboek van de Maatschappij der Nederlandse Letterkunde* (1858): 47–78.

Susanna, Joannes Andreas. *Levensschets van Hendrik Boie, en hulde aan zijne deugden en verdiensten, benevens eenige door hem geschrevene brieven, gedurende zijne reis en verblijf in Oost-Indië*. Amsterdam: P. Meijer Warnars, 1834.

Swainson, William. *On the Natural History and Classification of Birds*. The Cabinet Cyclopædia: Natural History. 2 vols. London: John Taylor, 1836–1837. doi:10.5962/bhl.title.64988.

Swainson, William. *A Preliminary Discourse on the Study of Natural History*. London: Longman, Rees, Orme, Brown, Green & Longman, 1834. doi:10.5962/bhl.title.10710.

Swainson, William. *Taxidermy, with the Biography of Zoologists and Notices of Their Works*. The Cabinet Cyclopædia: Natural History. London: Longman, Orme, Brown, Green & Longmans, 1840. doi:10.5962/bhl.title.61651.

Swainson, William. *A Treatise on the Geography and Classification of Animals*. The Cabinet Cyclopædia: Natural History. London: Longman, Rees, Orme, Brown, Green & Longman, Taylor, 1835.

Swainson, William. *Zoological Illustrations, or, Original Figures and Descriptions of New, Rare, or Interesting Animals: Selected Chiefly from the Classes of Ornithology, Entomology, and Conchology, and Arranged on the Principles of Cuvier and other Modern Zoologists*. London: Baldwin, Cradock, and Joy, 1820. doi:10.5962/bhl.title.42279.

Swinderen, Theodorus van. *Index rerum naturalium quae conservantur in Museo Academico Groningano: Naamlijst der voorwerpen van natuurlijke historie welke bewaard worden in het Akademisch Museum te Groningen*. Groningen: J. Oomkens, 1822. doi:10.5962/bhl.title.99949.

Temminck, Coenraad Jacob. "Algemeen overzigt van de orde der Cheiroptera, en monographie van de geslachten Harpyia en Cephalotes." *Bijdragen tot de natuurkundige wetenschappen* 6 (1831): 301–34.

Temminck, Coenraad Jacob. "Aperçu général et spécifique sur les mammifères qui habitent le Japon et les îles qui en dépendent." In *Fauna japonica, sive, Descriptio animalium, quae in itinere per Japoniam, jussu et auspiciis, superiorum, qui summum in India Batava imperium tenent, suscepto, annis 1823–1830* … edited by Philipp Franz von Siebold, Vol. 5, 1–9. Leiden: A. Arnz et socios, 1842.

Temminck, Coenraad Jacob. "Binnenlandsche berigten: 's Gravenhage, den 3den November." *Nederlandsche Staatscourant*, November 4, 1835, p. 3.

Temminck, Coenraad Jacob. "Blik op de dierlijke bewoners van de Sunda-eilanden en van de overige Nederlandsche bezittingen in Indië." *Bijdragen tot de natuurkundige wetenschappen* 3 (1828): 62–78.

Temminck, Coenraad Jacob. *Catalogue systématique du cabinet d'ornithologie et de la collection de quadrumanes avec une courte description des oiseaux non-décrits suivi: d'une nôte d'oiseaux doubles et de quelques autres objets d'histoire naturelle offerts en échange*. Amsterdam: C. Sepp Jansz., 1807.

Temminck, Coenraad Jacob. *Coup-d'oeil général sur les possessions néerlandaises dans l'Inde archipélagique*. 3 vols. Leiden: A. Arnz & comp, 1846–1849.

Temminck, Coenraad Jacob. "Coup-d'oeil sur la faune des Iles de la Sonde et de l'empire du Japon: Discours préliminaire destiné a servir d'introduction a la faune du Japon." In *Fauna japonica, sive, Descriptio animalium, quae in itinere per Japoniam, jussu et auspiciis, superiorum, qui summum in India Batava imperium tenent, suscepto, annis 1823–1830* … edited by Philipp Franz von Siebold. Vol. 1, i–xxx. Leiden: J. G. La Lau, 1836.

Temminck, Coenraad Jacob. "Description d'un nouveau genre de mammifères (Urotrichus)." *Magasin de zoologie, d'anatomie comparée et de palaeontologie* 4 (1842): 1–5.

Temminck, Coenraad Jacob. *Esquisses zoologiques sur la côte de Guiné. Pt. 1: Les mammifères*. Leiden: E. J. Brill, 1853. doi:10.5962/bhl.title.14828.

Temminck, Coenraad Jacob. *Histoire naturelle générale des pigeons et des gallinacés*. 3 vols. Amsterdam: J. C. Sepp & fils; Paris: G. Dufour, 1813–1815. doi:10.5962/bhl.title.64844.

Temminck, Coenraad Jacob. *Manuel d'ornithologie ou Tableau systématique des oiseaux qui se trouvent en Europe*. Amsterdam: J. C. Sepp & fils; Paris: G. Dufour, 1815.

Temminck, Coenraad Jacob. *Manuel d'ornithologie, ou Tableau systématique des oiseaux qui se trouvent en Europe; précédé d'une analyse du système général d'ornithologie, et suivi d'une table alphabétique des espèces*. 2nd ed. 4 vols. Paris: Gabriel Dufour (vols. 1 and 2, 1820), Edmond d'Ocagne (vol. 3, 1835), H. Cousin (vol. 4, 1840), 1820–1840. doi:10.5962/bhl.title.130481.

Temminck, Coenraad Jacob. "Monographie over een nieuw geslacht van knaagdier, onder den naam van Nyctocleptes." *Bijdragen tot de natuurkundige wetenschappen* 7 (1832): 1–8.

Temminck, Coenraad Jacob. *Monographies de mammalogie, ou description de quelques genres de mammifères, dont les espèces ont été observées dans les différens musées de l'Europe*. 2 vols. Paris and Amsterdam: G. Dufour et Edmond d'Ocagne (vol. 1, 1827); Paris: G. Dufour et Edmond d'Ocagne; Leiden: C. C. van der Hoek (vol. 2, 1835–1841), 1827–1841.

Temminck, Coenraad Jacob. "Nieuwstijdingen: Nederland." *Middelburgsche Courant*, November 7, 1835, p. 1.

Temminck, Coenraad Jacob. *Observations sur la classification méthodique des oiseaux, et remarques sur l'Analyse d'une nouvelle ornithologie élémentaire par L. P. Vieillot, auteur de divers ouvrages d'ornithologie, et un des collaborateurs du Nouveau dictionnaire d'histoire naturelle*. Amsterdam and Paris: Gabriel Dufour, 1817.

Temminck, Coenraad Jacob. "Over de kennis en de verbreiding der zoogdieren van Japan." *Tijdschrift voor natuurlijke geschiedenis en physiologie* 5, no. 4 (1838): 273–93.

Temminck, Coenraad Jacob. "Over eenige geslachten van zoogdieren, een deel der fauna der Japan uitmakende." *Verslagen en mededeelingen uitgegeven door de vier Klassen van het Koninklijk Nederlandsche Instituut van Wetenschappen, Letterkunde en Schoone Kunsten* 1, no. 1 (1841): 208–16.

Temminck, Coenraad Jacob. "Over eenige nieuwe soorten van eekhoorns van den Indischen Archipel." *Natuurkundig tijdschrift voor Nederlandsch Indië* 4 (1854): 343–48.

Temminck, Coenraad Jacob. "Van de Directeur van 's Rijks Museum." *'s Gravenhaagsche Courant*, May 8, 1822, p. 2.

Temminck, Coenraad Jacob. ed. *Verhandelingen over de natuurlijke geschiedenis der Nederlandsche overzeesche bezittingen*. 3 vols. Leiden: S. en J. Luchtmans en C. C. van der Hoek, 1839–1842. doi:10.5962/bhl.title.114730.

Temminck, Coenraad Jacob. *Voorschrift, hoedanig te handelen met voorwerpen van natuurlijke historie, ten einde dezelve behoorlijk te verzenden en voor bederf te bewaren: Ten gebruike van het 's Rijks Museum van Natuurlijke Historie te Leyden*. Leiden: [printed for the author], 1825.

Temminck, Coenraad Jacob. "Zoologische schetsen van eenige soorten van hoenderachtige vogelen." *Bijdragen tot de dierkunde* 6, no. 1 (1854): 49–50.

Temminck, Coenraad Jacob, and Pauline Knip-de Courcelles. *Histoire naturelle générale des pigeons*. Paris: Garnery, 1808.

Temminck, Coenraad Jacob, and Guillaume Michel Jerôme Meiffren Laugier de Chartrouse. *Nouveau recueil de planches coloriées d'oiseaux: Pour servir de suite et de complément aux planches enluminées de Buffon*. 5 vols., 102 livr., 600 pls. Paris: F. G. Levrault, 1820–1838. doi:10.5962/bhl.title.51468.

Temminck, Coenraad Jacob, and Hermann Schlegel. "Mammalia." In *Fauna japonica, sive, Descriptio animalium, quae in itinere per Japoniam, jussu et auspiciis, superiorum, qui summum in India Batava imperium tenent, suscepto, annis 1823–1830* ... edited by Philipp Franz von Siebold. Vol. 5. Leiden: A. Arnz et socios, 1842.

Temminck, Coenraad Jacob, and Jean-Charles Werner. *Les oiseaux d'Europe décrits par C. J. Temminck; Atlas de 530 planches dessinées par J.-C. Werner*. Paris: Baillière, 1848. doi:10.5962/bhl.title.104704.

Theunissen, Bert. "Biologie en Bildung: Jan van der Hoeven (1802–1868)." *Gewina* 21 (1998): 108–26.

Theunissen, Bert. *'Nut en nog eens nut': Wetenschapsbeelden van Nederlandse natuuronderzoekers, 1800–1900*. Hilversum: Verloren, 2000.

Thomson, William, and Hugh Blackburn, eds. *Sir Isaac Newton's Principia: Reprinted for Sir William Thomson ... and Hugh Blackburn*. 3rd ed. Glasgow: J. Maclehose, 1871.

Thunberg, Carl Peter. *Reise durch einen Theil von Europa, Afrika und Asien, hauptsächlich in Japan, in den Jahren 1770 bis 1779*. Berlin: Haude und Spener, 1792.

Tjon Sie Fat, L. A., and G. J. C. M. van Vliet. *Philipp von Siebold: Zijn Japanse flora en fauna*. Haarlem: Becht, 1990.

Trustees of the British Museum. "Annual Report of the Trustees of the British Museum, 1822." *The Edinburgh Review or Critical Journal for Feb. 1823–May 1823* 38, no. 76, v (1823): 379–98.

Ueno, Masuzō. "The Western Influence on Natural History in Japan." *Monumenta Nipponica* 19, no. 3/4 (1964): 315–39.

Veth, Huibert Johannes. "Overzicht van hetgeen, in het bijzonder door Nederland, gedaan is voor de kennis der fauna van Nederlandsch Indië." Leiden: S. C. van Doesburgh, 1879.

Vieillot, Louis Pierre. "L'Aigle à queue barrée. 9. A. Fasciata." In *Tableau encyclopédique et méthodique des trois règnes de la nature*, edited by Pierre Joseph Bonnaterre. Vol. 3, 1192. Paris: Mme veuve Agasse, 1823.

Vieillot, Louis Pierre. *Analyse d'une nouvelle ornithologie élémentaire*. Paris: Deterville, 1816.

Vieillot, Louis Pierre. "Le Coucou a ventre noir." In *Nouveau dictionnaire d'histoire naturelle, appliquée aux arts, à l'agriculture, à l'économie rurale et domestique, à la médecine, etc.*, edited by Charles Sigisbert Sonini. Vol. 8, 236–237. Paris: Deterville, 1817.

Vieillot, Louis Pierre. *Histoire naturelle des oiseaux de l'Amérique Septentrionale, contenant un grand nombre d'espèces décrites ou figurées pour la première fois*. 2 vols. Paris: Desray, 1807). doi:10.5962/bhl.title.100831.

Vieillot, Louis Pierre. *Histoire naturelle des plus beaux oiseaux chanteurs de la zone torride*. Paris: J. E. Gabriel Dufour, 1805. doi:10.5962/bhl.title.148906.

Vieillot, Louis Pierre. "Ornithologie." In *Nouveau dictionnaire d'histoire naturelle, appliquée aux arts, à l'agriculture, à l'économie rurale et domestique, à la médecine, etc.*, edited by Charles Sigisbert Sonini. Vol. 24, 68–144. Paris: Deterville, 1818.

Vigors, Nicholas Aylward. "Some Observations on the Nomenclature of Ornithology; Particularly with Reference to the Admission of New Genera." *The Zoological Journal* 1, no. 2 (1824): 180–200.

Vigors, Nicholas Aylward. "Observations on the Natural Affinities that Connect the Orders and Families of Birds." *Transactions of the Linnean Society of London* 14, no. 3 (1825): 395–517.

Visser, Rob P. W. "Het Rijksmuseum van Natuurlijke Historie in de 19de eeuw." In *Het verdwenen museum: Natuurhistorische verzamelingen, 1750–1850*, edited by Bert C. Sliggers and Marijke H. Besselink, 175–186. Haarlem: Teylers Museum, 2002.

Völkel, Hans. *Herman H. ter Meer: Ein Leben als Dermoplastiker und Künstler*. Leipzig: Leipziger Universitätsverlag, 2004.

Vriese, Willem Hendrik de. *Ons streven naar waren roem: Eenige woorden van dankbare herinnering bij het afsterven van den Hoogleeraar C. G. C. Reinwardt, gericht tot de kweekelingen van Leydens Hoogeschool, op 13 Maart 1854*. Leiden: C. C. van der Hoek, 1854.

Vriese, Willem Hendrik de. *Reis naar het oostelijk gedeelte van den Indischen Archipel in het jaar 1821 door Caspar Georg Carl Reinwardt*. Amsterdam: Frederik Muller, 1858.

Vrolik, Willem. "Levensbericht van Coenraad Jacob Temminck." *Jaarboek van de Maatschappij der Nederlandse Letterkunde* (1857): 65–80.

Vrolik, Willem. "Levensberigt van Wilhem de Haan." *Verslagen en mededeelingen der Koninklijke Akademie van Wetenschappen (afdeeling Natuurkunde)* 3 (1855): 399–408.

Wallace, Alfred Russel. "Attempts at a Natural Classification of Birds." *Annals and Magazine of Natural History Including Zoology, Botany, and Geology* 18 (1856): 193–216.

Wallace, Alfred Russel. *The Malay Archipelago: The Land of the Orang-Utan, and the Bird of Paradise*. New York: Harper & Brothers, 1869. doi:10.5962/bhl.title.11170.

Wallace, Alfred Russel. "On the Habits of the Orang-Utan of Borneo." *Annals and Magazine of Natural History Including Zoology, Botany, and Geology* 18 (1856): 26–32.

Wallace, Alfred Russel. "On the Orang-Utan or Mias of Borneo." *Annals and Magazine of Natural History Including Zoology, Botany, and Geology* 17 (1856): 471–76.

Wallace, Alfred Russel. "On the Tendency of Varieties to Depart Indefinitely from the Original Type." *Journal of the Proceedings of the Linnean Society. Zoology* 3, no. 9 (1858): 53–62. doi:10.1111/j.1096-3642.1858.tb02500.x.

Wallace, David Rains. *Mountains and Marshes: Exploring the Bay Area's Natural History*. Berkeley: Counterpoint, 2015.

Walls, Laura Dassow. *The Passage to Cosmos: Alexander von Humboldt and the Shaping of America*. Chicago: University of Chicago Press, 2009.

Weber, Andreas. "Collecting Colonial Nature: European Naturalists and the Netherlands Indies in the Early Nineteenth Century." *Low Countries Historical Review* 134, no. 3 (2019): 1–25. doi:10.18352/bmgn-lchr.10741.

Weber, Andreas. "Hybrid Ambitions: Science, Governance, and Empire in the Career of Caspar C. G. Reinwardt (1773–1854)." PhD dissertation, Leiden University, 2012.

Weiss, Martin Paul Michael. "The Masses and the Muses: A History of Teylers Museum in the Nineteenth Century." PhD dissertation, Leiden University, 2013.

Westwood, John Obadiah. *The Entomologist's Text Book: An Introduction to the Natural History, Structure, Physiology, and Classification of Insects, Including the Crustacea and Arachnida*. London: W. S. Orr and Company, 1838.

Wilkins, John S. *Species: A History of the Idea*. Species and Systematics. Vol. 1. Berkeley: University of California Press, 2009.

Willdenow, Karl Ludwig. *Grundriss der Kräuterkunde zu Vorlesungen*. Berlin: Haude und Spener, 1792. doi:10.5962/bhl.title.8054.

Wingerden, Pieter van. "Science on the Edge of Empire: E. A. Forsten (1811–1843) and the Natural History Committee (1820–1850) in the Netherlands Indies." *Centaurus* 62, no. 4 (2020): 797–821. doi:10.1111/1600-0498.12346.

Winsor, Mary P. "The Creation of the Essentialism Story." *History and Philosophy of the Life Sciences* 28 (2006): 149–74.

Winsor, Mary P. "The Development of Linnaean Insect Classification." *Taxon* 25, no. 1 (1976): 57–67.

Winsor, Mary P. *Reading the Shape of Nature: Comparative Zoology at the Agassiz Museum*. Science and Its Conceptual Foundations. Chicago: The University of Chicago Press, 1991.

Winsor, Mary P. *Starfish, Jellyfish, and the Order of Life: Issues in Nineteenth-Century Science*. New Haven: Yale University Press, 1976.

Witkamp, Pieter H. "Vier-en-twintig voorgangers." *Jaarboekje Natura Artis Magistra* 18, (1869): 137–202.

Witteveen, Joeri. "Naming and Contingency: The Type Method of Biological Taxonomy." *Biology and Philosophy* 30, no. 4 (2015): 569–86. doi:10.1007/s10539-014-9459-6.

Witteveen, Joeri. "Suppressing Synonymy with a Homonym: The Emergence of the Nomenclatural Type Concept in Nineteenth Century Natural History." *Journal of the History of Biology* 49, no. 1 (2016): 135–189. doi:10.1007/s10739-015-9410-y.

Wolf, Johann, and Bernhard Meyer. *Naturgeschichte der Vögel Deutschlands in getreuen Abbildungen und Beschreibungen*. Nürnberg: J. F. Frauenholz, 1805–[1821]. doi:10.5962/bhl.title.62380.

Wood, Neville. *The Ornithologist's Text-Book: Being Reviews of Ornithological Works, with an Appendix, Containing Discussions on Various Topics of Interest*. London: John W. Parker, 1836. doi:10.5962/bhl.title.62717.

Yamaguchi, Takai, ed. *Von Siebold and Natural History of Japan: Studies on the Unpublished Specimens of Japanese Animals Collected by Von Siebold and Still Preserved in the Museums of the Netherlands*. Kumamoto: Kumamoto University, 1987.

Zachos, Frank E. *Species Concepts in Biology: Historical Development, Theoretical Foundations and Practical Relevance*. Cham (Switzerland): Springer, 2016. Adobe Digital Editions EPUB. doi:10.1007/978-3-319-44966-1.

Zimmermann, Eberhard August Wilhelm von. *Specimen zoologiae geographicae, quadrupedum domicilia et migrationes sistens*. Leiden: Apud Theodorum Haak, et Socios, 1777. doi:10.5962/bhl.title.155837.

Index of Subjects